15th Workshop on Biomedical Natural Language Processing (BioNLP 2016)

Held at ACL 2016

Berlin, Germany
12 August 2016

ISBN: 978-1-5108-2766-0

ACL 2016

BioNLP 2016

Proceedings of the 15th Workshop on Biomedical Natural Language Processing

August 12, 2016
Berlin, Germany

Introduction

The past year has been an exciting and productive time for biomedical natural language processing. A search for natural language processing or text mining in PubMed®/MEDLINE® limited to 2015 and 2016 returns over 800 results. The number of corpora available in the domain continues to increase, and the past year has seen two hackathons devoted to biomedical corpora, with another two planned for this coming year. A variety of shared tasks have led to increases in the shared knowledge of the community, with more to come.

The high level of activity in biomedical natural language processing includes a number of good conferences. Among those, the BioNLP meeting has now been ongoing for 15 years, and the quality of submissions continues to impress the program committee and the organizers—and to increase. BioNLP 2016 received 38 exceptional submissions, of which 13 were accepted for oral presentation and 15 as poster presentations; increasing the rejection rate to 30% this year.

The themes in this year's papers and posters continue showing equal interest in clinical text and in biological language processing. The morning sessions focus on extraction of entities, relations and events. The afternoon sessions present disambiguation, classification, vocabulary development and syntactic analysis. The invited talks present overviews of two community-wide evaluations: BioNLP-ST 2016 and BioASQ 2016.

As always, we are profoundly grateful to the authors who chose BioNLP as venue for presenting their innovative research. The authors' willingness to continue sharing their work through BioNLP consistently makes the workshop noteworthy and stimulating. We are equally indebted to the program committee members (listed elsewhere in this volume) who produced three thorough reviews per paper on a tight review schedule and with an admirable level of insight.

Invited Talks

The BioNLP-ST challenges on information extraction and knowledge acquisition in biology

Speakers: Robert Bossy and Jin-Dong Kim

Robert Bossy is a research engineer at INRA, the French national institute for agronomy, agriculture and food science. His main interests are the design of Natural Language Processing, Information Extraction, Information Retrieval, and Knowledge Acquisition methods and services in the domains of biology and food science. His domains of expertise are NLP workflows and software development for knowledge engineering. He also has a wide experience in the dialogue between biology experts and NLP method providers. He has organized the Bacteria Biotope and Bacteria Genic Interaction tasks in the BioNLP-ST challenges 2011 and 2013. He has a MSc in Populations Biology and Taxonomy and a PhD in bioinformatics from Pierre et Marie Curie (Paris, France).

Jin-Dong Kim is a project associate professor of DBCLS (Database Center for Life Science). He received his Ph.D from Korea University in 2001. He is the main author of Genia resources and a regular organizer of BioNLP Shared Task series. He is also the chief organizer of the BLAH (annual Biomedical Linked Annotation Hackathon) series. His recent projects include PubAnnotation, TextAE and LODQA.

BioASQ: A challenge on large-scale biomedical semantic indexing and question answering

Speaker: Anastasia Krithara

Dr. Anastasia Krithara has been a post-doctoral researcher in the Institute of Informatics and Telecommunications at National Center for Scientific Research (NCSR) "Demokritos" since 2008, where she is involved in national and international projects. Before, she was a research engineer in Xerox Research Centre Europe, in Grenoble, France, where she carried out research in the area of machine learning. She holds a BSc in Informatics from Athens University of Economics and Business, an MSc in Machine Learning and Data Mining from University of Bristol and a PhD in Machine Learning from Pierre and Marie Curie University (Paris VI). Her research interests include Machine Learning, Information Retrieval, Bioinformatics and Natural Language Processing. She is a program committee member of several international conferences and workshops and her work has been published in international journals, conferences and books. She is co-organizing the BioASQ challenges.

Organizers:

Kevin Bretonnel Cohen, University of Colorado School of Medicine, USA
Dina Demner-Fushman, US National Library of Medicine
Sophia Ananiadou, National Centre for Text Mining and University of Manchester, UK
Jun-ichi Tsujii, National Institute of Advanced Industrial Science and Technology, Japan

Program Committee:

Sophia Ananiadou, National Centre for Text Mining and University of Manchester, UK
Eiji Aramaki, University of Tokyo, Japan
Alan Aronson, US National Library of Medicine
Asma Ben Abacha, US National Library of Medicine
Olivier Bodenreider, US National Library of Medicine
Kevin Bretonnel Cohen, University of Colorado School of Medicine, USA
Aaron Cohen, Oregon Health and Science University
Dina Demner-Fushman, US National Library of Medicine
Filip Ginter, University of Turku, Finland
Cyril Grouin, LIMSI - CNRS, France
Antonio Jimeno Yepes, IBM, Melbourne Area, Australia
Halil Kilicoglu, US National Library of Medicine
Robert Leaman, US National Library of Medicine
Ulf Leser, Humboldt-Universität zu Berlin, Germany
Zhiyong Lu, US National Library of Medicine
Timothy Miller, Children's Hospital Boston, USA
Makoto Miwa, Toyota Technological Institute, Japan
Danielle L Mowery, VA Salt Lake City Health Care System, USA
Yassine M'Rabet, US National Library of Medicine
Aurelie Neveol, LIMSI - CNRS, France
Nhung Nguyen, The University of Manchester, Manchester
Naoaki Okazaki, Tohoku University, Japan
Sampo Pyysalo, University of Cambridge, UK
Bastien Rance, Hopital Europeen Georges Pompidou, France
Fabio Rinaldi, University of Zurich, Switzerland
Thomas Rindflescht, US National Library of Medicine
Kirk Roberts, The University of Texas Health Science Center at Houston, USA
Angus Roberts, The University of Sheffield, UK
Yoshimasa Tsuruoka, University of Tokyo, Japan
Karin Verspoor, The University of Melbourne, Australia
Byron C. Wallace, University of Texas at Austin, USA
W John Wilbur, US National Library of Medicine
Pierre Zweigenbaum, LIMSI - CNRS, France

Table of Contents

Conference Program

Friday August 12, 2016

8:30–8:40   **Opening remarks**

8:40–10:30   **Session 1: Entity extraction and representation**

8:40–9:00   *A Machine Learning Approach to Clinical Terms Normalization*
Jose Castano, María Laura Gambarte, Hee Joon Park, Maria del Pilar Avila
Williams, David Perez, Fernando Campos, Daniel Luna, Sonia Benitez, Hernan
Berinsky and Sofía Zanetti

9:00–9:20   *Improved Semantic Representation for Domain-Specific Entities*
Mohammad Taher Pilehvar and Nigel Collier

9:20–9:40   *Identification, characterization, and grounding of gradable terms in clinical text*
Chaitanya Shivade, Marie-Catherine de Marneffe, Eric Fosler-Lussier and Albert
M. Lai

9:40–10:00   *Graph-based Semi-supervised Gene Mention Tagging*
Golnar Sheikhshab, Elizabeth Starks, Aly Karsan, Anoop Sarkar and Inanc Birol

10:00–10:30   **Invited Talk: "The BioNLP-ST challenges on information extraction and
knowledge acquisition in biology" – Robert Bossy and Jin-Dong Kim**

10:30–11:00   *Coffee Break*

Friday August 12, 2016 (continued)

11:00–12:30   Session 2: Event and Relation Extraction

11:00–11:20   *Feature Derivation for Exploitation of Distant Annotation via Pattern Induction against Dependency Parses*
Dayne Freitag and John Niekrasz

11:40–12:00   *Inferring Implicit Causal Relationships in Biomedical Literature*
Halil Kilicoglu

12:00–12:20   *SnapToGrid: From Statistical to Interpretable Models for Biomedical Information Extraction*
Marco A. Valenzuela-Escárcega, Gus Hahn-Powell, Dane Bell and Mihai Surdeanu

12:20–12:40   *Character based String Kernels for Bio-Entity Relation Detection*
Ritambhara Singh and Yanjun Qi

12:40–14:00   *Lunch break*

14:00–15:40   Session 3: Disambiguation, Classification, and more

14:00–14:20   *Disambiguation of entities in MEDLINE abstracts by combining MeSH terms with knowledge*
Amy Siu, Patrick Ernst and Gerhard Weikum

14:20–14:40   *Using Distributed Representations to Disambiguate Biomedical and Clinical Concepts*
Stephan Tulkens, Simon Suster and Walter Daelemans

14:40–15:00   *Unsupervised Document Classification with Informed Topic Models*
Timothy Miller, Dmitriy Dligach and Guergana Savova

15:00–15:20   *Vocabulary Development To Support Information Extraction of Substance Abuse from Psychiatry Notes*
Sumithra Velupillai, Danielle L Mowery, Mike Conway, John Hurdle and Brent Kious

15:20–15:40   *Syntactic analyses and named entity recognition for PubMed and PubMed Central — up-to-the-minute*
Kai Hakala, Suwisa Kaewphan, Tapio Salakoski and Filip Ginter

15:40–16:00   *Coffee Break*

16:00–16:30 **Invited Talk: "BioASQ: A challenge on large-scale biomedical semantic indexing and question answering" – Anastasia Krithara**

16:30–17:30 **Poster Session**

Improving Temporal Relation Extraction with Training Instance Augmentation
Chen Lin, Timothy Miller, Dmitriy Dligach, Steven Bethard and Guergana Savova

*Using Centroids of Word Embeddings and Word Mover's Distance for Biomedical
Document Retrieval in Question Answering*
Georgios-Ioannis Brokos, Prodromos Malakasiotis and Ion Androutsopoulos

*Measuring the State of the Art of Automated Pathway Curation Using Graph Algo-
rithms - A Case Study of the mTOR Pathway*
Michael Spranger, Sucheendra Palaniappan and Samik Gosh

Construction of a Personal Experience Tweet Corpus for Health Surveillance
Keyuan Jiang, Ricardo Calix and Matrika Gupta

*Modelling the Combination of Generic and Target Domain Embeddings in a Con-
volutional Neural Network for Sentence Classification*
Nut Limsopatham and Nigel Collier

PubTermVariants: biomedical term variants and their use for PubMed search
Lana Yeganova, Won Kim, Sun Kim, Rezarta Islamaj Doğan, Wanli Liu, Donald C
Comeau, Zhiyong Lu and W John Wilbur

This before That: Causal Precedence in the Biomedical Domain
Gus Hahn-Powell, Dane Bell, Marco A. Valenzuela-Escárcega and Mihai Surdeanu

Syntactic methods for negation detection in radiology reports in Spanish
Viviana Cotik, Vanesa Stricker, Jorge Vivaldi and Horacio Rodriguez

How to Train good Word Embeddings for Biomedical NLP
Billy Chiu, Gamal Crichton, Anna Korhonen and Sampo Pyysalo

*An Information Foraging Approach to Determining the Number of Relevant Fea-
tures*
Brian Connolly, Benjamin Glass and John Pestian

*Assessing the Feasibility of an Automated Suggestion System for Communicating
Critical Findings from Chest Radiology Reports to Referring Physicians*
Brian E. Chapman, Danielle L Mowery, Evan Narasimhan, Neel Patel, Wendy
Chapman and Marta Heilbrun

Building a dictionary of lexical variants for phenotype descriptors
Simon Kocbek and Tudor Groza

Applying deep learning on electronic health records in Swedish to predict healthcare-associated infections
Olof Jacobson and Hercules Dalianis

Identifying First Episodes of Psychosis in Psychiatric Patient Records using Machine Learning
Genevieve Gorrell, Sherifat Oduola, Angus Roberts, Tom Craig, Craig Morgan and Rob Stewart

Relation extraction from clinical texts using domain invariant convolutional neural network
Sunil Sahu, Ashish Anand, Krishnadev Oruganty and Mahanandeeshwar Gattu

A Machine Learning Approach to Clinical Terms Normalization

José Castaño[1], Hernán Berinsky[2], Hee Park[2], David Pérez[2], Pilar Ávila[2],
Laura Gambarte[2], Sonia Benítez[2], Daniel Luna[2], Fernando Campos[2] and Sofía Zanetti[2]

[1]Depto. Computación, FCEyN, Universidad de Buenos Aires
`jcastano@dc.uba.ar`

[2]Departamento de Informática en Salud, Hospital Italiano de Buenos Aires
`{firstname.lastname}@hospitalitaliano.org.ar`

Abstract

We propose a machine learning approach for semantic recognition and normalization of clinical term descriptions. Clinical terms considered here are noisy descriptions in Spanish language written by health care professionals in our electronic health record system. These description terms contain clinical findings, family history, suspected disease, among other categories of concepts. Descriptions are usually very short texts presenting high lexical variability containing synonymy, acronyms, abbreviations and typographical errors. Mapping description terms to normalized descriptions requires medical expertise which makes it difficult to develop a rule-based knowledge engineering approach. In order to build a training dataset we use those descriptions that have been previously matched by terminologists to the hospital thesaurus database. We generate a set of feature vectors based on pairs of descriptions involving their individual and joint characteristics. We propose an unsupervised learning approach to discover term equivalence classes including synonyms, abbreviations, acronyms and frequent typographical errors. We evaluate different combinations of features to train MaxEnt and XGBoost models. Our system achieves an F_1 score of 89% on the Hospital Italiano de Buenos Aires (HIBA) problem list.

1 Introduction

Some electronic health records (EHR) implementations allow users to introduce free text descriptions to capture clinical problems information enabling higher level of expressiveness and flexibility to physicians. Those descriptions must be encoded according to their meaning in order to allow the information to be consumed by other systems. Descriptions are grouped into concepts according to the meaning. The following descriptions correspond to the same concept[1]:

(1)   neoplasia maligna de pulmón
      neoplasia malign of lung
      'Malignant tumor of lung'

(2)   cáncer pulmonar
      cancer lung-of
      'lung cancer'

(3)   ca pulmonar
      ca lung-of
      'lung cancer'

(4)   cáncer de pulmón desde 2009
      cancer of lung   since 2009
      'cancer of the lung since 2009'

Free text descriptions written by health care professionals contain typos as in *cancer plumnoar*: a variation of description (2). It should be noted also that description (4) does not represent a synonym in a strict terminological sense. However it represents the same concept because the string *desde 2009* (since 2009) does not add relevant information from a *problem list* perspective (in the sense of EHR and terminology tradition (Van Vleck et al., 2008)).

Mapping strings to concepts has been a long standing problem in BioNLP, string similarity techniques as well as machine learning approaches have been applied. Automatic mapping of key concepts from text in clinical notes to a reference terminology is an important task to achieve, in order to extract clinical information present in notes and patient reports. One of the problems of bio-

---

[1]In these examples the Spanish description is followed by the word-for-word English gloss and then the English translation.

1

Proceedings of the 15th Workshop on Biomedical Natural Language Processing, pages 1–11,
Berlin, Germany, August 12, 2016. ©2016 Association for Computational Linguistics

medical data integration is variation of terms usage. Exact string matching often fails to associate a string with its bio-medical concept (represented by an ID or accession number in the database) due to differences of string occurrences. Soft string matching algorithms are able to find the relevant concept by considering the string similarity between candidate strings. However, the accuracy of soft matching highly depends on the similarity measure employed. String similarity techniques have been applied to a variety of problems in BioNLP, such as UMLS concepts normalization (Aronson and Lang, 2010; Wellner et al., 2005; Rudniy et al., 2014), UMLS clinical terms (Kate, 2016), disease normalization (Leaman et al., 2013; Kang et al., 2013), gene and protein names (Kim and Park, 2004; Tsuruoka et al., 2007; Tsuruoka et al., 2008; Fang et al., 2006; Wermter et al., 2009), to interface terminologies (Rosenbloom et al., 2006) and different databases (Sun, 2004).

String similarity can be used for named entity recognition (SNOMED-CT taggers) and reference resolution (Castaño et al., 2002; Lin and Liang, 2004; C. et al., 2003) alias extraction (Yu and Agichtein, 2003), acronym-expansion extraction, e.g. (Pustejovsky et al., 2001).

On a similar view there are a number of works on automated clinical coding (Friedman et al., 2004; Pakhomov et al., 2006; Patrick et al., 2006; Suominen et al., 2008; Stanfill et al., 2010; Perotte et al., 2014).

This work explores traditional soft string matching methods along with n-gram character and word features in a machine learning approach using MaxEnt and XGBoost classifiers. An unsupervised learning approach to generate new features by detecting synonyms, abbreviations and typos is presented to improve classification performance. The models are compared to a baseline obtained by a vector space model configuration based on character n-grams and a TF-IDF weighting scheme, implemented in Apache Lucene.

The remainder of this paper is structured as follows: in Section 2, we describe the data set we used. In Section 3, we discuss the similarity metrics and similarity features for machine learning algorithms. In Section 4 we discuss the experimentation and results. Finally in Section 5 we report our conclusions and expected future work.

2 Description Terms Data-set

We build a data-set based on the problem list of Hospital Italiano de Buenos Aires (HIBA) interface terminology (Lopez Osornio et al., 2007; Gambarte et al., 2007) which includes adequate synonym coverage and it is linked to the HIBA thesaurus. This thesaurus is built upon the Spanish version of SNOMED-CT, while extending it with new concepts and additional synonym terms.

Following SNOMED-CT and other thesauri, terms in the thesaurus are grouped by concepts. The following terms are associated to the same concept.

(5)   tabaquismo
        smoking

(6)   abuso   de tabaco
        tobacco   consumption

We selected those clinical concepts that had at least 10 terms and no more than 100 for a given concept.[2] The set is composed of 151,513 terms and 5,222 concepts. The set of descriptions (D) was split in a training set (T) 70%, and an evaluation test set (E) 30%.

Descriptions in T were used to build a new data-set T_1 consisting of pairs of descriptions samples of the form $(d_1, d_2, value)$. Positive and negative samples were constructed in the following way:

- For each pair of descriptions $d_i, d_j \in T$ with $i \neq j$ such that d_i and d_j are associated to the same concept, we create a sample $(d_i, d_j, 1)$

- We split the set of descriptions T in corpus and query sets. We indexed with Apache Lucene the corpus set of descriptions using TF-IDF weights on n-gram characters. Using a description d as a query, a set of relevant and non-relevant results are retrieved. Relevant results are those descriptions d_i already stored as samples of pair of terms describing the same concept: $(d, d_i, 1)$. Non-relevant results are those results d_j for which there is not a sample $(d, d_j, 1)$ and therefore a sample $(d, d_j, 0)$ is created.

The training data-set (T_1) has 1,173,617 instances with 777,585 negative and 396,032 positive samples.

---

[2]Those concepts that had more than 100 terms were noisy, and were not considered relevant.

3 Methods for Computing Term Normalization

String similarity methods can be either character-based or token-based. Character-based approaches typically consist of variations of the edit-distance metric, like Levenshtein distance or longest common subsequence. Token-based approaches include the Jaccard similarity metric and the cosine similarity based on TF-IDF weighting schema. There are also hybrid token and character-based approaches. Soft-TFIDF (Cohen et al., 2003) includes not only exact matches but also close matches, using a threshold. Another approach uses n-grams of the target strings instead of the tokens (Cohen et al., 2003; Moreau et al., 2008; Köpcke and Rahm, 2010).

Many works have also focused on automatic methods for combining these string similarity measures using machine learning (Cohen and Richman, 2001; Belenko and Mooney, 2003; Wellner et al., 2005; Moreau et al., 2008).

In this section we explore a hybrid soft-TFIDF approach based on an n-gram character vector space model as well as other character-based and token-based similarity metrics. Next, we mention some limitations of combining the previous metrics due to information redundancy and lack of semantic information which produces false positive and false negative instances. We propose an usupervised machine learning approach which allows to capture semantic information.

3.1 Information retrieval and TF-IDF

We use an information retrieval (*IR*) Soft-TFIDF approach (Cohen et al., 2003) to match a new description to those terms already existing in the hospital thesaurus database. First, the set of known terms in the thesaurus are indexed with Lucene, where the collection of terms is represented in a Vector Space Model (VSM) using TF-IDF weights based on character n-grams. A new description is used as a query and the set of ranked descriptions terms with the corresponding scores is retrieved, being the highest ranked description the candidate term to associate the query with. The cosine similarity measure is used to obtain similarity scores.

However this approach will outcome both false positive and false negative results such as:

(7)   sospecha de laringitis alérgica (query)
       suspected (of) alergic   laringytis

(8)   sospecha de faringitis alérgica (false positive)
       suspected (of) alergic   pharyngitis

Due to the high string similarity score between *sospecha de laringitis alérgica* and *sospecha de faringitis alérgica* if either of them is not indexed as a concept, then the returned result is considered a match and therefore a false positive instance is obtained.

(9)   neoplasia maligna de pulmón (query)
       malignant tumor of lung

(10)   cáncer pulmonar (false negative, not retrieved)
       lung   cancer

A low similarity score between *neoplasia maligna de pulmón* and *cáncer pulmonar* implies that the target string is not retrieved (i.e. it is not ranked above the threshold). Since the concept is just represented by *cáncer pulmonar*, the string *neoplasia maligna de pulmón* is a false negative instance.

Figure 1 shows overlapping distribution of scores. The positive match curve represents the score (cosine similarity) distribution of query and retrieved string pairs that represent the same concept. It shows higher average score than negative match. As threshold score increases, false negative cases increase and false positive cases decrease.

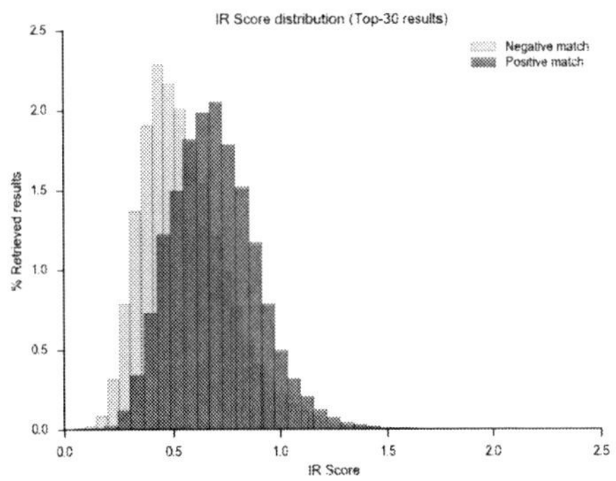

Figure 1: IR Score distribution (normalized histograms)

Given a query, it is not known whether relevant information exists or not in the indexed dataset, and the term with the highest score is not necessarily a desired result. The performance of matching the query with the highest ranked term can be measured using *precision*, *recall* and F_1 metrics.

In a Soft-TFIDF approach is possible to control precision/recall trade-off considering a threshold t as shown in Figure 2. The algorithm returns the highest ranked term if $score \geq t$. Higher values of t increase precision but recall is decreased.

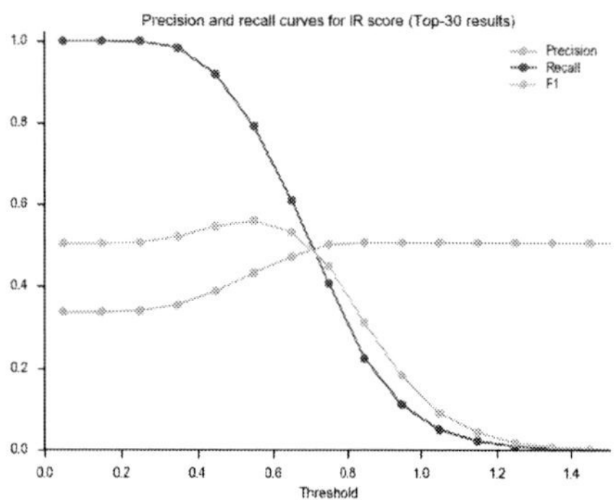

Figure 2: Precision, recall and F1 measure
Precision(t), recall(t) and $F_1(t)$ show measures for the results returned by the IR system where $score \geq t$ (when increasing t, precision(t) increases and recall(t) decreases)

3.2 String similarity metrics

String similarity metrics have been used combined with *IR* TF-IDF approaches. Using traditional string similarity and distance metrics like *Damerau-Levenshtein* or *Longest common subsequence* allow to increase to some extent precision. It is possible to score results using a string metric or combine together with *IR* scores using some rules or formulas and thresholds. Using a set of scores it is possible to use machine learning models to classify relevant/non-relevant result.

Even though there are other string metric measures that can be combined, some of them are very much related. For example Damerau-Levenshtein ($DamLev_{dist}$) distance allows an additional edit operation respect to Levenshtein (Lev_{dist}) distance, then $DamLev_{dist} \leq Lev_{dist}$. Also Jaccard and Sorensen-Dice similarity metrics present a high correlation. In Table 1 we show pairwise correlations between Damerau-Levenshtein ratio, Longest common sub-sequence, Sorensen-Dice, Jaro-Winkler and Jaccard coefficient metrics.

Due to high correlation between Jaccard and Sorensen-Dice, we can choose one of them, and in the same way with Damerau-Levenshtein and Longest common subsequence to fit a classification model using machine learning approach.

Metric	DamLev	LCS	SorDic	JarWin	Jac
DamLev	1.00	0.96	0.78	0.73	0.77
LCS		1.00	0.84	0.73	0.82
SorDic			1.00	0.65	0.99
JarWin				1.00	0.64
Jac					1.00

Table 1: Correlation (Pearson) between string metrics
Damerau-Levenshtein ratio (DamLev), Longest common subsequence (LCS), Sorensen-Dice (SorDic), Jaro-Winkler (JarWin) and Jaccard (Jac). Damerau-Levenshtein ratio is a transformation of Damerau-Levenshtein distance d using the formula M is the maximum lenght of s_1 and s_2.

By computing the principal components, the eigenvalues show that using the first k components the cumulative variance explained is 76% ($k = 1$), 93% ($k = 2$), 98% ($k = 3$), 99% ($k = 4$). This means that k new variables (linear combination of original metrics) explain those proportions of the total variance and we can also reduce redundant information.

Limitations to this approach are present both in false positive and false negative cases. It is a quantitative improvement but cases like those presented in examples (7-10) above, require a more sophisticated approach. Such approach must consider which modifications in a clinical term changes its meaning.

3.3 A machine learning approach to string matching

As it has been already observed in many other works, abbreviations, acronyms, synonyms and typos are sources of variation that generate terms with the same meaning. Table shows some examples from the Spanish dataset:[3]

Many pairs of description terms are very similar but they have different meaning as described by the following non-synonym pairs where only a character difference in a long string entails a different meaning:

(11)   a.   *sospecha de laringitis alérgica*
       b.   *sospecha de faringitis alérgica*

(12)   a.   *duelo por fallecimiento de madre*
       b.   *duelo por fallecimiento de padre*

(13)   a.   *sospecha de hi potiroidismo*
       b.   *sospecha de hi pertiroidismo*

(14)   a.   *artr itis de tobillo*
       b.   *artr osis de tobillo*

---

[3]We do not include the translations from now on because the relevant information is the string similarity.

Alternative forms	'meaning'
sd, sme, sind, sindr	*sindrome*
izq, izqdo, iz	*izquierdo*
mmii	*miembros inferiores*
grado 2, 2do grado,	*segundo grado*
2 grado, gr II, G2, GII	
hta	*hipertensión arterial*
ira	*insuficiencia renal aguda*
oma	*otitis media aguda*
AF, antec fliar, atc familiar	*antecedente familiar*
fratura	*fractura*
trauatismo	*traumatismo*
dematitis	*dermatitis*
litisis	*litiasis*
Reynaud/Raynoud	*Raynaud*
Hodkin, Hodking	*Hodgkin*
dolores de cabeza	*dolor de cabeza*
fallecimiento , muerte	*deceso*
de hígado	*hepático*
biológico	*natural*
cáncer	*neoplasia maligna*

Table 2: Examples of equivalent strings.

We use a machine learning approach to learn whether a pair of descriptions is a match or not. We create a family of features to train classification algorithms. Hyper-parameters were adjusted using 5-fold cross-validation. Models are based on different combinations of feature sets explained in next subsection.

3.4 Features

Features are organized in sets $S1, ..., S10$ and then different set combinations are used to generate the corresponding models. In Table 3 d_1 and d_2 are the description strings that are compared, where d_1 vectors represent queries and d_2 a retrieved string. The training corpus was used to adjust the corresponding d_2 vectors, using both word unigrams and character bi-grams.[4]

Feature set $S1$ represents string metrics to obtain differences in string characteristics between a pair of description terms. Features in $S2, S3, S4, S5$ are traditional representations in vector space model of d_1 and d_2 based on unigram word and bi-gram character representation with TF-IDF, binary occurrence and term frequency weights. In S_6 and S_7 we consider differences in descriptions (d_{12} and d_{21}), and S_9, S_{10} considers context (c) also.

We define $w(d)$ as the set of words in d, and d_{12}, d_{21} and c as follows

$$d_{12} = w(d_1) \setminus w(d_2)$$

$$d_{21} = w(d_2) \setminus w(d_1)$$

---

[4]Features d_{12} and d_{21} are explained below.

Set	Feature
S1	$L_1 = length(d_1)$ $L_2 = length(d_2)$ $m = min(L_1, L_2)$ $M = max(L_1, L_2)$ $ratio_{length} = \frac{m}{M}$ $difference_{length} = |M - m|$ $Levenshtein_{ratio}(L_1, L_2)$ $Jaccard(L_1, L_2)$
S2	Vector of unigram word occurrence in d_1 Vector of unigram word occurrence in d_2
S3	Vector of unigram word TF-IDF in d_1 Vector of unigram word TF-IDF in d_2
S4	Vector of bigram character frequency in d_1 Vector of bigram character frequency in d_2
S5	Vector of bigram character TF-IDF in d_1 Vector of bigram character TF-IDF in d_2
S6	Vector of unigram word occurrence in d_{12} Vector of unigram word occurrence in d_{21}
S7	Vector of bigram character frequency in d_{12} Vector of bigram character frequency in d_{21}
S8	Vector of unigram word occurrence in d_{12} Vector of unigram word occurrence in d_{21} Vector of unigram word occurrence in c
S9	Vector of bigram character frequency in d_{12} Vector of bigram character frequency in d_{21} Vector of bigram character frequency in c
S10	Vector of group of words in d_{12} Vector of group of words in d_{21} Vector of group of words in c

Table 3: Feature-sets.

$$c = w(d_1) \cap w(d_2)$$

For example:

d_1 = fractura de rodilla izquierda,
d_2 = fractura de rodilla izq then
$w(d_1) = \{fractura, de, rodilla, izquierda\}$,
$w(d_2) = \{fractura, de, rodilla, izq\}$,
$d_{12} = \{izquierda\}$
$d_{21} = \{izq\}$ and $c = \{fractura, de, rodilla\}$

3.5 Unsupervised Learning of Synonyms, Abbreviations and Typos

In this section we present an approach to detect word synonyms, abbreviations, acronyms and frequent typographical errors. We explain how the set of features $S10$ was generated.

Unsupervised algorithms were studied widely in the literature to detect relationships between words in order to improve results of NLP tasks such us chunking or named entity recognition. Clustering to detect word equivalence classes from unlabeled corpus were studied in (Kneser and Ney, 1993) and (Turian et al., 2010).

We introduce a procedure to generate sets of semantically equivalent strings from term descriptions using a graph algorithm.

Given a set of positive description pair matchings such as

(15) d_1: sospecha de infección urinaria
        d_2: probable infección urinaria

(16) d_1: urticaria en cara
        d_2: urticaria en rostro

(17) d_1: duelo por fallecimiento de padre
        d_2: duelo por muerte de padre

(18) d_1: duelo por deceso de padre
        d_2: duelo por muerte de padre

The following semantically equivalent pairs can be inferred using word differences between pairs of descriptions:

$\{sospecha, probable\}$, $\{cara, rostro\}$
$\{fallecimiento, muerte\}$ and $\{deceso, muerte\}$

Therefore it is possible to replace, for example, the terms *sospecha* and *probable* by a concept representing this class with some label. Using the concept class label instead of a term as a feature in a vector space model we can deal with synonymy problems.

Since this approach only infers direct associations, we cannot detect the pair $\{deceso, muerte\}$ using this approach.

Semantically equivalent pairs can be extended to larger sets (semantically equivalence classes), building an undirected weighted graph, considering terms as vertices and equivalent pairs as edges. Connected components in the graph can be detected and terms can be clustered in some cases.

An undirected weighted graph $G = (V, E, W)$ is generated creating an edge $(d_{12}, d_{21}, w) \in E$ for each pair of descriptions d_1, d_2 such that $\mid d_{12} \mid = \mid d_{21} \mid = 1$. For example, the pair of descriptions *duelo por fallecimiento de padre* and *duelo por muerte de padre* generates the *fallecimiento* and *muerte* connection. In the same way, *deceso* and *muerte* are connected. The weight associated with each edge is the frequency of the corresponding pair in T_1.

The graph constructed under this approach is composed of different connected components. Figure 3 shows some connected components in the final graph once all edges are generated us-

ing T_1 and considering only edges with minimum frequency of 20 (lower frequency thresholds are very sensible to noisy data while higher values results in loss of information). Vertices in the same connected component are potentially equivalent. The connected components of G can be computed in linear time using either depth-first search or breadth-first search approach.

Since some terms can be ambiguous, they can be connected to some non-equivalent terms, like *od* which can be connected to *ojo derecho* (*right eye*) and *oido derecho* (*right ear*). In those cases, the connected component containing an ambiguous term, includes more than one concept. In a vector space model, in some cases disambiguation can be obtained from the context. For example in *otitis od* the *od* term refers to *oido*, while in *conjuntivitis od* refers to *ojo*. It would still be desirable to partition the connected component breaking edges like *ojo derecho* and *oido derecho*.

We used the label propagation algorithm described in (Raghavan et al., 2007). It is a clustering algorithm intended to be applied in social communities detection in large-scale networks and biochemical networks among other domains. This

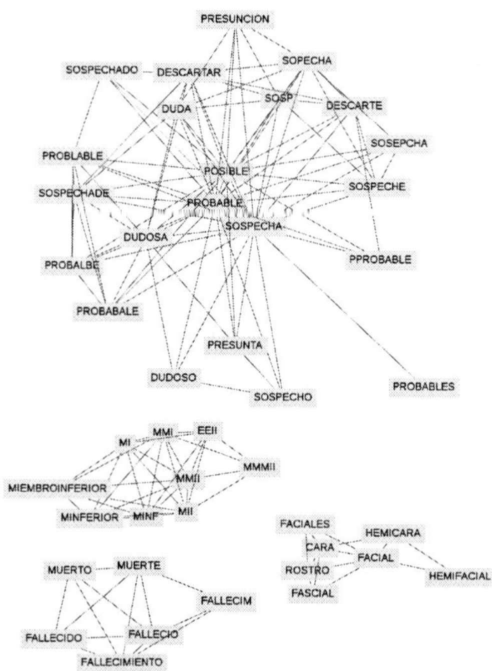

Figure 3: Word graph connected components example

clustering algorithm computes clusters based on the network structure and -unlike other approaches like k-means or DBSCAN- there is no requirement to specify the number of clusters or the neighbourhood size as parameters. The algorithm initializes each node with a unique identifier, and iteratively assigns to each node the label that most of its neighbors currently have. We run this label propagation algorithm to obtain a clustering analysis on large connected components that contained different word meanings.

As a final example, combining different terms from the final set of equivalence classes, we obtain a unique representation of the following 72 possible ways to express *duelo por fallecimiento de padre biológico debido a cáncer renal*:

$$\text{Duelo por} \left\{ \begin{array}{c} \text{fallecimiento} \\ \text{muerte} \\ \text{deceso} \end{array} \right\} \text{de padre} \left\{ \begin{array}{c} \text{biológico} \\ \text{natural} \end{array} \right\}$$

$$\left\{ \begin{array}{c} \text{debido a} \\ \text{a causa de} \end{array} \right\} \left\{ \begin{array}{c} \text{cáncer} \\ \text{ca} \\ \text{neoplasia maligna} \end{array} \right\} \left\{ \begin{array}{c} \text{de riñón} \\ \text{renal} \end{array} \right\}$$

4 Experiments and Results

Our experiments were conducted by using *scikit-learn* machine learning library (Pedregosa et al., 2011) with *liblinear* (Fan et al., 2008) solver for MaxEnt, considering L2 regularization. Hyperparameter C was determined by 5-fold cross-validation considering F1 measure. We trained XGBoost model, with binary logistic objective and F1 score as evaluation metric, by using XGBoost library described in (Chen and He, 2015). Connected components in graph and label propagation algorithm for graph clustering were conducted by using *igraph* library.

In order to generate word equivalence classes for $S10$ we found 278,555 concepts in the thesaurus with at least two associated descriptions which generate 5,956,368 potential pairs of descriptions connected to the same concept. Filtering pairs of descriptions such that both shares the same words except one, we obtain 505,447 word associations. By taking the connected components of G we get $505,447$ edges and 805 groups. Finally, clustering connected components for which more than one meaning are represented, we obtain 4,711 words in 957 group of words.

We compare the predictive power to classify a pair of descriptions as a positive match by calculating the *F1* measure on different models. Also,

we compare the ability to rank the retrieved results using the classification model probability as scoring by calculating P@1, R@1 and the mean reciprocal rank (*MRR*).

By using IR score with some fixed threshold we define a classifier algorithm with its respective precision and recall (as threshold increases, recall decreases and precision increase). Figure 4 shows IR score precision-recall curve against string metrics features based fitted models. Table 4 shows MaxEnt and XGBoost F1 score for string features based models.

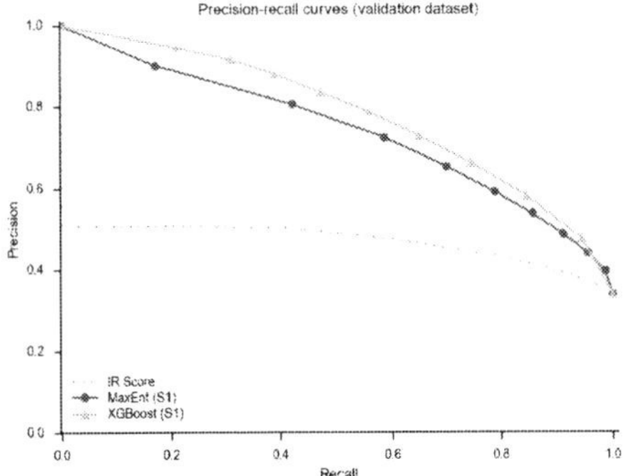

Figure 4: Precision-recall curves (String metrics features)
IR comparison vs MaxEnt and XGBoost models based on string metrics features.

Featureset		Source	MaxEnt	XGBoost
String metrics	(S1)	d_1, d_2	0.67	0.70

Table 4: MaxEnt and XGBoost F1-score over string metrics

By considering F1 measure on string metrics ($S1$) and vector space model representation of descriptions ($S2, S3, S4, S5$), XGBoost showed a considerable improvement on bi-gram character features based (see Table 5) either on frequency ($S4$) or TF-IDF ($S5$) weight schemas, outperforming MaxEnt.

	Source	Weight	MaxEnt	XGBoost
(S4)	d_1, d_2	freq.	0.57	0.76
(S5)	d_1, d_2	tf-idf	0.56	0.74
(S7)	d_{12}, d_{21}	freq.	0.58	0.76
(S9)	d_{12}, d_{21}, c	freq.	0.72	0.77

Table 5: MaxEnt and XGBoost F1-score over bigram character features $S4, S5, S7, S9$

Each XGBoost bi-gram character features based model (dashed lines with markers) outperforms the word features based models (solid lines). Precision values are given across all recall levels (Figure 5).

Figure 5: Precision-recall curves (XGBoost)

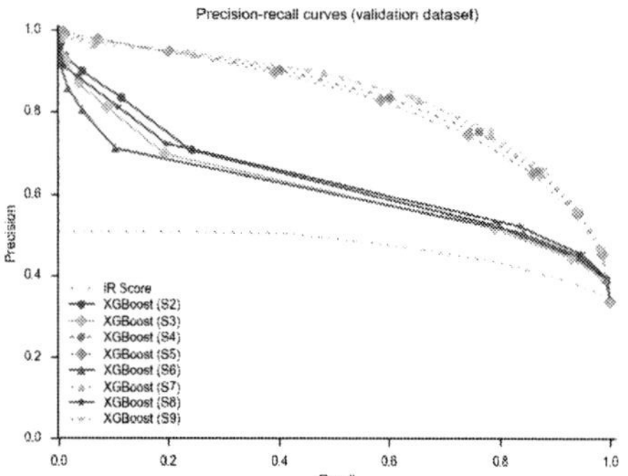

Markers are present on word and bi-gram features curves. IR curve has no markers. Word features are represented by solid lines while bi-gram character features are represented in dashed lines. Marker type indicates a specific source, for example $S6$ and $S7$ triangle correspond to d_{12}, d_{21} source.

s

Figure 6: Precision-recall curves (MaxEnt)

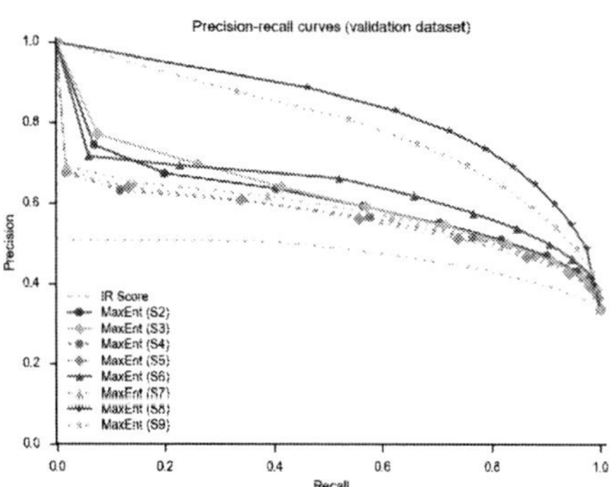

Markers are present on word and bi-gram features curves. IR curve has no markers. Word features are represented by solid lines while bi-gram character features are represented in dashed lines. Marker type indicates a specific source, for example $S6$ and $S7$ triangle correspond to d_{12}, d_{21} source.

We can see in Figure 6 that word features based models improves performance over bi-gram character feature based models using MaxEnt (each source is represented by a marker type, e.g. a triangle for the source d_{12}, d_{21}). $S8$ and $S9$ features outperform the others features on MaxEnt. Results on word features are detailed in Table 6.

With respect to the set of features considering word difference between pairs of descriptions ($S6, S7$), XGBoost also performs the task better when consider bi-gram character features ($S7$)

	Source	Weight	MaxEnt	XGBoost
(S2)	d_1, d_2	binary	0.59	0.59
(S3)	d_1, d_2	tf-idf	0.59	0.58
(S6)	d_{12}, d_{21}	binary	0.63	0.62
(S8)	d_{12}, d_{21}, c	binary	0.76	0.62

Table 6: MaxEnt and XGBoost F1-score over unigram word features $S2, S3, S4, S6, S8$

as shown in Table 5, while MaxEnt works better on word features ($S6$) as shown in Table 6. When context vector is present along with word difference representation ($S6$ vs $S8$ and $S7$ vs $S9$), MaxEnt showed a considerable improvement in $S8$ respect to $S6$ (see Table 6) but XGBoost achieves a slightly improvement in $S9$ compared to $S7$ (see Table 5, $S6$ vs $S8$ and $S7$ vs $S9$). Word difference vector representation worked better in MaxEnt, than combining string metric and traditional word or n-gram based representation of descriptions, while XGBoost achieves similar performance when consider that combination.

When word equivalence classes features based models are considered ($S10$), MaxEnt and XGBoost achieves similar performance (see Table 7).

Featureset	Source	MaxEnt	XGBoost
(S10)	d_1, d_2	0.69	0.69

Table 7: MaxEnt and XGBoost F1-score over word equivalence class features

By combining ($S1$, $S8$, $S10$) features MaxEnt achieves an F1 score of 0.87, while XGBoost achieves an F1 score of 0.86 by combining ($S1$, $S9$, $S10$) as showed in Table 8 and Figure 7 improving the previous models.

Figure 7: Precision-recall curves

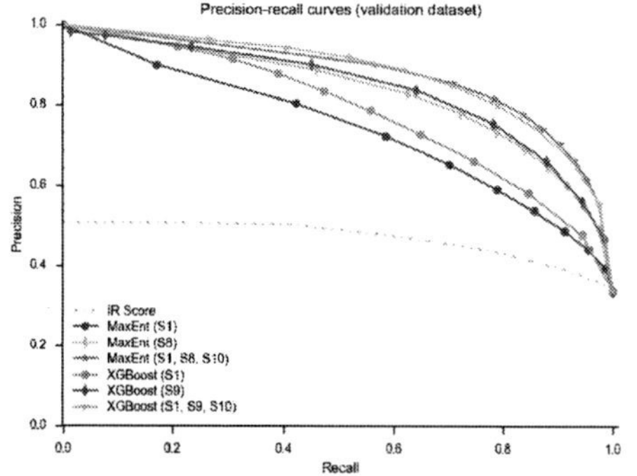

MaxEnt vs XGBoost comparison. Circle markers represent string based features ($S1$) models, diamond d_{12}, d_{21}, c based models. Models combining string, d_{12}, d_{21}, c and $S10$ features are represented by curves with star markers.

8

Model	Prec	Rec	F1
IR	0.73	0.76	0.74
MaxEnt (S1, S8, S10)	0.80	0.95	0.87
XGBoost (S1, S9, S10)	0.77	0.96	0.86

Table 8: MaxEnt and XGBoost F1-score over feature sets combination

To evaluate these models performance on ranked results, we compute the P@1, R@1 and mean reciprocal rank (MRR) metrics showed in Table 9.

Model	P@1	R@1	F1	MRR
IR	0.73	0.76	0.74	0.84
MaxEnt (S1, S8, S10)	0.87	0.91	0.89	0.94
XGBoost (S1, S9, S10)	0.87	0.91	0.89	0.94

Table 9: MaxEnt and XGBoost F1-score over feature sets combination

5 Conclusions and future work

We presented a hybrid Soft-TFIDF and machine learning approach to bio-medical terms normalization. This technique can be used in different problems such as automatic coding of descriptions and reference resolution in general. Our approach neither requires any additional resource like acronyms/abbreviations, alias and synonyms lists nor a spell checker because that ability is acquired from examples by defining a scoring function learned from data. As a result, our approach shows very good F1 score and mean reciprocal rank results. Even though the data set was in Spanish, we did not use any specific resource for that language, therefore our approach can be replicated in any language.

Creation of new features based on differences between descriptions and its context, in addition to the more traditional features, allow machine learning models to improve detection of pairs of semantically equivalent descriptions with low syntactic similarity and discard non semantically equivalent ones with high syntactic similarity by learning semantic equivalence from pairs of descriptions examples. As result, the false negative and false positive rates were reduced.

By generating a clustering of words to find synonyms, specially from indirect associations between words from descriptions across different concepts from direct associations, the semantic feature space generated improved the performance of machine learning models increasing F1 measure.

Finally, MaxEnt and XGBoost models showed to be effective for the task with some minor differences in the set of features returning best results.

Our work was based on the performance of the text search engine results. Then, this approach can not consider results that were not retrieved by the search engine. To overcome this limitation it is possible to use a query expansion approach. Alternatively, words in the terms can be transformed to a canonical form, both at index and query time. We also plan to expand this work to other biomedical domains such as procedures or drugs.

References

Alan R Aronson and François-Michel Lang. 2010. An overview of metamap: historical perspective and recent advances. *Journal of the American Medical Informatics Association*, 17(3):229–236.

Mikhail Belenko and Raymod J. Mooney. 2003. Adaptive duplicate detection using learnable string similarity measures. In *Proceedings of the 9th ACM SIGKDD International Conference on Knowledge Discovery and Datamining*, pages 39–48, Washington D.C.

Blaschke C., Hirschman L., Yeh A., and Valencia A. 2003. Critical assessment of information extraction systems in biology. *Comparative and Functional Genomics*, pages 674–677.

J. Castaño, J. Zhang, and J. Pustejovsky. 2002. Anaphora resolution in biomedical literature. In *International Symposium on Reference Resolution*, Alicante, Spain.

Tianqi Chen and Tong He. 2015. Higgs boson discovery with boosted trees. In *Cowan et al., editor, JMLR: Workshop and Conference Proceedings*, number 42.

William Cohen and Jacob Richman. 2001. Learning to match and cluster entity names. In *ACM SIGIR-2001 Workshop on Mathematical/Formal Methods in Information Retrieval*, New Orleans, LA, September.

William W. Cohen, Pradeep Ravikumar, and Stephen Fienburg. 2003. A comparison of string metrics for matching names and records. In *KDD Workshop on Data Cleaning and Object Consolidation*.

Rong-En Fan, Kai-Wei Chang, Cho-Jui Hsieh, Xiang-Rui Wang, and Chih-Jen Lin. 2008. LIBLINEAR: A library for large linear classification. *Journal of Machine Learning Research*, 9:1871–1874.

Haw-ren Fang, Kevin Murphy, Yang Jin, Jessica S. Kim, and Peter S. White. 2006. Human gene name normalization using text matching with automatically extracted synonym dictionaries. In *Proceedings of the Workshop on Linking Natural Language Processing and Biology: Towards Deeper Biological Literature Analysis*, BioNLP '06, pages 41–48, Stroudsburg, PA, USA. Association for Computational Linguistics.

Carol Friedman, Lyudmila Shagina, Yves Lussier, and George Hripcsak. 2004. Automated encoding of clinical documents based on natural language processing. *Journal of the American Medical Informatics Association*, 11(5):392–402.

Maria Laura Gambarte, Alejandro Lopez Osornio, Marcela Martinez, Guillermo Reynoso, Daniel Luna, and Fernan Gonzalez Bernaldo de Quiros. 2007. A practical approach to advanced terminology services in health information systems. *Medinfo 2007: Proceedings of the 12th World Congress on Health (Medical) Informatics; Building Sustainable Health Systems*, page 621.

Ning Kang, Bharat Singh, Zubair Afzal, Erik M van Mulligen, and Jan A Kors. 2013. Using rule-based natural language processing to improve disease normalization in biomedical text. *Journal of the American Medical Informatics Association*, 20(5):876–881.

Rohit J Kate. 2016. Normalizing clinical terms using learned edit distance patterns. *Journal of the American Medical Informatics Association*, 23(2):380–386.

Jung-Jae Kim and Jong C. Park. 2004. Bioar: Anaphora resolution for relating protein names to proteome database entries. In Sanda Harabagiu and David Farwell, editors, *ACL 2004: Workshop on Reference Resolution and its Applications*, pages 79–86, Barcelona, Spain, July. Association for Computational Linguistics.

Reinhard Kneser and Hermann Ney. 1993. Forming word classes by statistical clustering for statistical language modelling. In *Contributions to Quantitative Linguistics*, pages 221–226. Springer.

Hanna Köpcke and Erhard Rahm. 2010. Frameworks for entity matching: A comparison. *Data & Knowledge Engineering*, 69(2):197 – 210.

Robert Leaman, Rezarta Islamaj Doan, and Zhiyong Lu. 2013. Dnorm: disease name normalization with pairwise learning to rank. *Bioinformatics*, 29(22):2909–2917.

Y. Lin and T. Liang. 2004. Pronominal and sortal anaphora resolution for biomedical literature. In *Proceedings of ROCLING XVI: Conference on Computational Linguistics and Speech Processing*, Taipei, Taiwan.

Alejandro Lopez Osornio, Daniel Luna, Maria Laura Gambarte, Adrian Gomez, Guillermo Reynoso, and Fernan Gonzalez Bernaldo de Quiros. 2007. Creation of a local interface terminology to snomed ct. *Medinfo 2007: Proceedings of the 12th World Congress on Health (Medical) Informatics; Building Sustainable Health Systems*, page 765.

Erwan Moreau, François Yvon, and Olivier Cappé. 2008. Robust similarity measures for named entities matching. In *Proceedings of the 22Nd International Conference on Computational Linguistics - Volume 1*, COLING '08, pages 593–600, Stroudsburg, PA, USA. Association for Computational Linguistics.

Serguei VS Pakhomov, James D Buntrock, and Christopher G Chute. 2006. Automating the assignment of diagnosis codes to patient encounters using example-based and machine learning techniques. *Journal of the American Medical Informatics Association*, 13(5):516–525.

Jon Patrick, Yefeng Wang, and Peter Budd. 2006. Automatic mapping clinical notes to medical terminologies. In *Proceedings of the 2006 Australasian Language Technology Workshop (ALTW 2006)*, pages 75–82.

F. Pedregosa, G. Varoquaux, A. Gramfort, V. Michel, B. Thirion, O. Grisel, M. Blondel, P. Prettenhofer, R. Weiss, V. Dubourg, J. Vanderplas, A. Passos, D. Cournapeau, M. Brucher, M. Perrot, and E. Duchesnay. 2011. Scikit-learn: Machine learning in Python. *Journal of Machine Learning Research*, 12:2825–2830.

Adler Perotte, Rimma Pivovarov, Karthik Natarajan, Nicole Weiskopf, Frank Wood, and Noémie Elhadad. 2014. Diagnosis code assignment: models and evaluation metrics. *Journal of the American Medical Informatics Association*, 21(2):231–237.

J. Pustejovsky, J. Castaño, B. Cochran, M. Kotecki, and M. Morrell. 2001. Automatic extraction of acronym-meaning pairs from medline databases. In *Proceedings of Medinfo, London*.

Usha Nandini Raghavan, Réka Albert, and Soundar Kumara. 2007. Near linear time algorithm to detect community structures in large-scale networks. *Physical Review E*, 76(3):036106.

S Trent Rosenbloom, Randolph A Miller, Kevin B Johnson, Peter L Elkin, and Steven H Brown. 2006. Interface terminologies: facilitating direct entry of clinical data into electronic health record systems. *Journal of the American Medical Informatics Association*, 13(3):277–288.

Ale Rudniy, Min Song, and James Geller. 2014. Mapping biological entities using the longest approximately common prefix method. *Bioinformatics*, 15.

Mary H Stanfill, Margaret Williams, Susan H Fenton, Robert A Jenders, and William R Hersh. 2010. A systematic literature review of automated clinical coding and classification systems. *Journal of the American Medical Informatics Association*, 17(6):646–651.

Yao Sun. 2004. Methods for automated concept mapping between medical databases. *Journal of Biomedical Informatics*, 37(3):162 – 178.

Hanna Suominen, Filip Ginter, Sampo Pyysalo, Antti Airola, Tapio Pahikkala, S Salanter, and Tapio Salakoski. 2008. Machine learning to automate the assignment of diagnosis codes to free-text radiology reports: a method description. In *Proceedings of the ICML/UAI/COLT Workshop on Machine Learning for Health-Care Applications*.

Yoshimasa Tsuruoka, John McNaught, Jun'i;chi Tsujii, and Sophia Ananiadou. 2007. Learning string similarity measures for gene/protein name dictionary look-up using logistic regression. *Bioinformatics*, 23(20):2768–2774.

Yoshimasa Tsuruoka, John McNaught, and Sophia Ananiadou. 2008. Normalizing biomedical terms by minimizing ambiguity and variability. *BMC Bioinformatics*, 9(3):1–10.

Joseph Turian, Lev Ratinov, and Yoshua Bengio. 2010. Word representations: a simple and general method for semi-supervised learning. In *Proceedings of the 48th annual meeting of the association for computational linguistics*, pages 384–394. Association for Computational Linguistics.

Tielman T Van Vleck, Adam Wilcox, Peter D Stetson, Stephen B Johnson, and Noémie Elhadad. 2008. Content and structure of clinical problem lists: A corpus analysis.

Ben Wellner, José Castaño, and James Pustejovsky. 2005. Adaptive string similarity metrics for biomedical reference resolution. In *Proceedings of the ACL-ISMB Workshop on Linking Biological Literature, Ontologies and Databases: Mining Biological Semantics*, ISMB '05, pages 9–16, Stroudsburg, PA, USA. Association for Computational Linguistics.

Joachim Wermter, Katrin Tomanek, and Udo Hahn. 2009. High-performance gene name normalization with geno. *Bioinformatics*, 25(6):815–821.

H. Yu and E. Agichtein. 2003. Extracting synonymous gene and protein terms from biological literatur e. *Bioinformatics*, pages 340–349.

Improved Semantic Representation for Domain-Specific Entities

Mohammad Taher Pilehvar and **Nigel Collier**
Language Technology Lab
Department of Theoretical and Applied Linguistics
University of Cambridge
Cambridge, UK
{mp792,nhc30}@cam.ac.uk

Abstract

Most existing corpus-based approaches to semantic representation suffer from inaccurate modeling of domain-specific lexical items which either have low frequencies or are non-existent in open-domain corpora. We put forward a technique that improves word embeddings in specific domains by first transforming a given lexical item to a sorted list of representative words and then modeling the item by combining the embeddings of these words. Our experiments show that the proposed technique can significantly improve some of the recent word embedding techniques while modeling a set of lexical items in the biomedical domain, i.e., phenotypes.

1 Introduction

Semantic representation is one of the oldest, yet most active, research areas in Natural Language Processing (NLP) owing to the central role it plays in many applications (Pilehvar and Navigli, 2015). The field has experienced a resurgence of interest in recent years with the introduction of low-dimensional continuous space models that leverage neural networks for learning semantic representations. Word2vec (Mikolov et al., 2013) is a good example which despite its recent invention has found its way prominently into literature, mainly thanks to its ability to be quickly and effectively trained on large amounts of text.

However, since most of these corpus-based techniques base their representation only on the co-occurrence statistics derived from text corpora, they fall short of effectively modeling lexical items for which not many statistical clues can be obtained from the underlying corpus. Several attempts have been made to improve word embed-dings with the help of knowledge derived from other resources (Yu and Dredze, 2014; Bian et al., 2014; Faruqui et al., 2015) or by including arbitrary contexts in the training process (Levy and Goldberg, 2014). However, most of these techniques still suffer from another deficiency of word embeddings that they inherit from their count-based ancestors: they conflate the different meanings of a word into a single vector representation. Attempts have been made to tackle the meaning conflation issue of word-level representations. A series of approaches cluster the context of a word prior to representation (Reisinger and Mooney, 2010; Huang et al., 2012; Neelakantan et al., 2014) whereas others exploit lexical knowledge bases for sense-specific information (Rothe and Schütze, 2015; Chen et al., 2014; Iacobacci et al., 2015; Camacho-Collados et al., 2015).

We propose a model that addresses both these issues through a mapping of a lexical item to a sorted list of representative words that brings about two advantages. Firstly, it pinpoints with an inherent disambiguation the meaning of the given lexical item at a deeper semantic level. Secondly, by casting the representation of the item as that of a set of potentially more frequent words, our approach can provide a more reliable representation of domain-specific items based on significantly more statistical knowledge. Our experiments show that the proposed model can provide a considerable improvement over some of the state-of-the-art word embedding approaches in a semantic similarity-based task.

Data. The final goal of this paper is to improve the semantic representation of domain-specific terms and phrases which usually have low frequencies (or are non-existent) in open-domain corpora and hence have a lower chance of being effectively modeled by existing word representation

Proceedings of the 15th Workshop on Biomedical Natural Language Processing, pages 12–16,
Berlin, Germany, August 12, 2016. ©2016 Association for Computational Linguistics

techniques. Therefore, for our experiments we retrieved terms and phrases from a domain-specific ontology in the biomedical domain. Specifically, as our dataset in the experiments we opted for Human Phenotype Ontology (Sebastian Khler, 2014, HPO) which is a standardized vocabulary of phenotypic abnormalities encountered in human disease. Semantic modeling of phenotypes has several applications in the biomedical domain such as profiling heritable diseases or understanding the genetic origins of diseases (Collier et al., 2013).

2 Improved Semantic Representation

In this section we explain how our technique builds on top of pre-trained word embeddings to provide a more accurate semantic representation.

2.1 Disambiguation

As mentioned in the Introduction, one of the drawbacks of word-level representations is that they conflate different meanings of a word into a single vector. Our technique constructs a more accurate semantic representation of a lexical item by constraining its semantics through a set of relevant words. Interestingly, we achieve this on the basis of the same set of word-level representations. To this end, we first disambiguate the content word(s) in a given lexical item. In our experiments, we used Babelfy (Moro et al., 2014) which is a state-of-the-art WSD system based on the BabelNet sense inventory. BabelNet is a merger of Wikipedia and WordNet, among other resources (Navigli and Ponzetto, 2012). Let $t = $ *flexion contracture of digit* be the phrase we are interested in modeling. The disambiguation phase transforms the phrase to three BabelNet concepts corresponding to the intended meanings of the content words {*flexion, contracture,* and *digit*}. Disambiguating with respect to BabelNet provides us with an additional benefit: it links a content word to the corresponding Wikipedia page of its intended meaning, giving us the chance to draw additional context for improving its representation.

2.2 Representative list

Let the set of disambiguated concepts for a lexical item t be $\mathcal{C}_t$. We further enrich this set by adding all the BabelNet concepts that have a semantic link (in the semantic network of BabelNet) to any of the concepts in $\mathcal{C}_t$. Let the enriched set of concepts be $\mathcal{C}_t^*$. Our goal here is to map $\mathcal{C}_t^*$ to a set of most relevant words that can represent its semantics. We achieve this by exploiting the fact that these concepts are linked to relevant Wikipedia articles. Let D_t be the set of Wikipedia articles retrieved for t (i.e., the set of articles that are associated with the concepts in $\mathcal{C}_t^*$). We analyze the textual content of these articles by leveraging the method proposed by Camacho-Collados et al. (2015) and retrieve a sorted list of salient words. Specifically, we use lexical specificity and contrast word frequency statistics between D_t and all articles in Wikipedia. Lexical specificity (Lafon, 1980) is a statistical measure based on the hypergeometric distribution which can be used to compute the semantic importance of an arbitrary vocabulary word w for D_t as:

$$Spec(H; h; G; g) = -log_{10}P(X \geq g) \qquad (1)$$

where H and h are the respective aggregate frequencies of all words in all Wikipedia articles and D_t, and G and g are the respective frequencies of w in all Wikipedia articles and D_t. For a given lexical item t, we construct the set of semantically representative words $\mathcal{R}_t$ by keeping the words that are relevant to D_t with a minimum confidence of 99% according to the hypergeometric distribution, i.e., $P(X \geq 0.01)$.

For our example phenotype *flexion contracture of digit*, the representative list $\mathcal{R}_t$ comprises of around 1300 weighted words, with the top ones being *muscle, finger, spasticity, toe, hand, patient,* and *spastic*. Please note that our technique mapped an ambiguous term *digit* to a set of more semantically constrained keywords such as *finger, toe,* and *hand*. This enables us to construct a sense-specific representation of the word by leveraging word-level representations.

2.3 Vector construction

So far, we mapped a given lexical item t to a set of relevant concepts $\mathcal{C}_t^*$ and obtained for this set the sorted list $\mathcal{R}_t = \{r_1, ..., r_m\}$ of the most semantically representative words. The final step is to construct a vector representation V_t for t. We do this by combining the vectors for the words in $\mathcal{R}_t$. Let $\mathcal{V}(x)$ be the vector representation given by a model such as Word2vec for the word x. We compute the weight for the ith dimension of the vector V_t, i.e., v_i, as:

$$v_i = \sum_{j=1}^{m} e^{-\lambda j} \mathcal{V}(r_j)_i \quad i = 1, ..., n \qquad (2)$$

sim.	**Flexion contracture of digit**	sim.	**Bipolar affective disorder**	sim.	**Chaotic rapid conjugate ocular movements**
0.94	Flexion contracture of finger	0.80	Personality disorder	0.85	Abnormal conjugate eye movement
0.92	Flexion contracture of thumb	0.85	Schizophrenia	0.80	Jerky ocular pursuit movements
0.91	Congenital finger flexion contractures	0.85	Psychosis	0.76	Slow saccadic eye movements

sim.	**Hydranencephaly** (A defect of development of the brain characterized by replacement of greater portions of the cerebral hemispheres [...].)
0.81	Porencephaly (A disorder of the brain in which a cyst or cavity filled with cerebrospinal fluid develops in the cerebral hemisphere.)
0.79	Dandy-walker malformation (A congenital brain malformation typically characterized by incomplete formation of the cerebellar vermis, dilation of [...].)
0.77	Ventriculomegaly (An increase in size of the ventricular system of the brain.)

Table 1: The most similar phenotypes (among 11,591) to four phenotypes in the HPO database together with their similarity scores. We also show the definitions for more technical terms in parentheses.

where $\mathcal{V}(r_j)_i$ is the weight of the ith dimension of the base vector for the jth word in $\mathcal{R}_t$ and $e^{-\lambda j}$ is a decay function (with the decay constant λ) that gives more importance to the higher ranking terms in $\mathcal{R}_t$. In our experiments, we did not perform a tuning on the value of λ which was set to $\frac{1}{5}$. Please note that the dimensionality of V_t is identical to that of the base word representations, i.e., n. Table 1 shows the top-3 most similar phenotypes for four phenotypes in the HPO ontology when Word2vec was used as the base representation.

3 Experiments

We evaluate our model in the semantic representation of phenotypes in the HPO ontology.

3.1 Dataset

As of February 2016, the HPO ontology comprises of 11,591 phenotypic abnormalities. Each of these concepts is provided with a title (with an average length of four words) and about 35% of all these concepts are associated with synonymous titles (by average, each of these concepts has 1.94 synonyms). For example, *Keratoconjunctivitis sicca* is a phenotype for which three synonymous titles are provided by the ontology: *Dry eye syndrome*, *Keratitis sicca*, and *Xerophthalmia*.

3.2 Tasks

Based on the ontological structure of HPO, we propose two tasks in the framework of semantic similarity measurement.

Synonym identification. Let $\mathcal{P}$ be the set of all phenotypes in the HPO ontology. Let $\mathcal{P}^* = \{p_1, ..., p_k\} \ (\subset \mathcal{P})$ be the subset of k phenotypes for which at least one synonymous phenotype is provided in HPO and $\mathcal{S}_{p_i} = \{s_{p_i}^1, ..., s_{p_i}^l\}$ be the set of l synonymous phenotypes for phenotype p_i. Given a s_{p_i}, the task here is simply to identify the corresponding phenotype (i.e., p_i). In other words, the system has to identify the set of synonymous phenotypes to a given phenotype. Specifically, we compare the representation of s_{p_i} with those of all the phenotypes in $\mathcal{P}$, obtaining a sorted list of most similar phenotypes. Ideally, the concept containing the synonymous title should appear at the top of this list. The higher the rank of p_i for a given s_{p_i}, the better has the system captured the semantics of the phenotypes. For this task we have 7193 synonymous titles ($\sum_{i=1}^k |\mathcal{S}_{p_i}|$) that are to be matched with their corresponding phenotypes (among a total of 11,591 phenotypes).

Hypernym identification. Similarly to the previous experiment, a system's task here is to identify the hypernym of a given phenotype. The aim of this experiment is to have a broader evaluation that can also cover all those concepts that do not provide synonymous titles (the dataset comprises of 11,590 phenotypes that have a hypernym).

3.3 Baselines

As baseline, we benchmark our improved representations against Word2vec. We use the 300-dimensional vectors trained on the Google News corpus (about 100B tokens). We also report results for the Word2vec vectors when retrofitted using the approach of Faruqui et al. (2015) to the Paraphrase Database (Ganitkevitch et al., 2013, PPDB) and SNOMED-CT[1]. The latter is a comprehensive clinical terminology from which we extracted 108K synonymous sets, each comprising an average of 2.7 synonyms. We also compare our representations against the 300-dimensional GloVe vectors (Pennington et al., 2014) trained on the Wikipedia 2014 + Gigaword 5 corpus (6B tokens).

We were also interested in verifying how Word2vec and GloVe would perform if trained on

---

[1] https://www.nlm.nih.gov/snomed/

System	Description	Mean rank	Median rank	First match
Word2vec	Trained on open-domain data (Google News)	1343.6	11	22%
Word2vec (2nd order)		**664.1**	**6**	**28%**
Word2vec	Trained on in-domain data (PubMed)	224.1	4	32%
Word2vec (2nd order)		**198.2**	**3**	**36%**
GloVe	Trained on open-domain data (Wikipedia + Gigaword)	1326.4	9	24%
GloVe (2nd order)		**673.5**	**6**	**28%**
GloVe	Trained on in-domain data (PubMed)	701.4	4	34%
GloVe (2nd order)		**493.5**	**3**	**36%**
Word2vec	Trained on Google News, retrofitted to PPDB	1357.4	8	26%
Word2vec	Trained on Google News, retrofitted to SNOMED-CT	1346.2	9	25%
Random baseline	Random selection of the synonymous phenotype	5473.0	5473.0	0%

Table 2: Evaluation results for the synonym identification task. We report mean and median rank (lower better) and the percentage of phenotypes for which the rank was equal to one (first match; higher better).

an in-domain corpus. Thankfully, the biomedical domain is a rich domain for which large amounts of textual data are available. We retrieved a corpus of 4B tokens from article abstracts indexed in PubMed[2]. We then trained Word2vec and GloVe with window size of 5 words and the same dimensionality as the open-domain vectors (i.e., 300). For Word2vec we opted for the skip-gram model.

3.4 Results and discussion

Table 2 shows the evaluation results. We report mean and median rank of the target phenotype in the sorted list of most semantically similar phenotypes as well as the percentage of target phenotypes for which this rank was equal to one, i.e., the synonymous title was computed as the most similar item (*first match* in the table). As a reference, we also report the performance of a baseline which randomly picks the target phenotype.

We can see that a considerable performance improvement was gained when our technique was used for improving Word2vec and GloVe representations trained on open-domain corpora. Interestingly, even when the vectors were trained on an in-domain corpus (PubMed) that covers a large portion of the phenotypes with high frequencies, our model was still able to provide statistically significant improvements according to mean rank over the vanilla Word2vec and GloVe.[3] The retrofitting of the vanilla vectors improved median rank and first match irrespective of the resource but did not match the performance of our model.

The substantial improvement of our approach in the open-domain setting should be attributed to

its mapping of domain-specific phenotypes with lower frequencies to a set of more frequent representative terms. In fact, only around 60% of the unique tokens of the phenotypes in the HPO ontology were covered by the vanilla Word2vec and GloVe models, which left around 5% of all the phenotypes with no representation. The token coverage raised to 91% when the two models were trained on PubMed, resulting in the generation of representations for 99.7% of all phenotypes. In this setting, the respective relative mean rank improvements of 11.4% and 29.7% of our approach with respect to Word2vec and GloVe should be attributed to the additional semantic information that our model introduces to the vectors as well as the more accurate representation of concepts, thanks to the disambiguation phase and the semantically constraining keywords.

For the hypernym identification task we observed a very similar trend where our model improved Word2vec and GloVe from the respective mean ranks of 1034.1 and 1021.5 to 606.1 and 556.7 on the open-domain corpus and from 317.2 and 424.9 to 277.6 and 309.5 on PubMed.

4 Conclusions and future work

We proposed an approach for enhancing the representation capability of existing word modeling techniques in specific domains and showed that consistent improvement can be gained over Word2vec and GloVe even when they are trained on domain-specific corpora. We plan to enhance our technique by making it sensitive to syntax and different parts of speech, such as in the manner of Baroni and Zamparelli (2010). We also plan to carry out a deeper analysis to better understand

---

[2] http://www.ncbi.nlm.nih.gov/pubmed/
[3] According to *t*-test with 95% confidence interval.

the potential of our model and to identify places in which it can be improved.

Acknowledgments

The authors gratefully acknowledge the support of the MRC grant No. MR/M025160/1 for PheneBank.

References

Marco Baroni and Roberto Zamparelli. 2010. Nouns are vectors, adjectives are matrices: Representing adjective-noun constructions in semantic space. In *Proceedings of EMNLP*, pages 1183–1193, Cambridge, Massachusetts.

Jiang Bian, Bin Gao, and Tie-Yan Liu. 2014. Knowledge-powered deep learning for word embedding. In *Proceedings of the European Conference on Machine Learning and Knowledge Discovery in Databases*, pages 132–148, Nancy, France.

José Camacho-Collados, Mohammad Taher Pilehvar, and Roberto Navigli. 2015. A Unified Multilingual Semantic Representation of Concepts. In *Proceedings of ACL-IJCNLP*, pages 741–751, Beijing, China.

Xinxiong Chen, Zhiyuan Liu, and Maosong Sun. 2014. A unified model for word sense representation and disambiguation. In *Proceedings of EMNLP*, pages 1025–1035, Doha, Qatar.

Nigel Collier, Anika Oellrich, and Tudor Groza. 2013. Toward knowledge support for analysis and interpretation of complex traits. *Genome Biology*, 14(9):1–11.

Manaal Faruqui, Jesse Dodge, Sujay Kumar Jauhar, Chris Dyer, Eduard Hovy, and Noah A. Smith. 2015. Retrofitting word vectors to semantic lexicons. In *Proceedings of NAACL-HLT*, pages 1606–1615, Denver, Colorado.

Juri Ganitkevitch, Benjamin Van Durme, and Chris Callison-Burch. 2013. PPDB: The paraphrase database. In *Proceedings of NAACL-HLT*, pages 758–764, Atlanta, Georgia.

Eric H. Huang, Richard Socher, Christopher D. Manning, and Andrew Y. Ng. 2012. Improving word representations via global context and multiple word prototypes. In *Proceedings of ACL*, pages 873–882, Jeju Island, Korea.

Ignacio Iacobacci, Mohammad Taher Pilehvar, and Roberto Navigli. 2015. SensEmbed: Learning sense embeddings for word and relational similarity. In *Proceedings of ACL-IJCNLP*, pages 95–105, Beijing, China.

Pierre Lafon. 1980. Sur la variabilité de la fréquence des formes dans un corpus. *Mots*, 1:127–165.

Omer Levy and Yoav Goldberg. 2014. Dependency-based word embeddings. In *Proceedings of ACL (Volume 2: Short Papers)*, pages 302–308, Baltimore, Maryland.

Tomas Mikolov, Kai Chen, Greg Corrado, and Jeffrey Dean. 2013. Efficient estimation of word representations in vector space. *CoRR*, abs/1301.3781.

Andrea Moro, Alessandro Raganato, and Roberto Navigli. 2014. Entity Linking meets Word Sense Disambiguation: a Unified Approach. *Transactions of the Association for Computational Linguistics (TACL)*, 2:231–244.

Roberto Navigli and Simone Paolo Ponzetto. 2012. BabelNet: The automatic construction, evaluation and application of a wide-coverage multilingual semantic network. *Artificial Intelligence*, 193:217–250.

Arvind Neelakantan, Jeevan Shankar, Alexandre Passos, and Andrew McCallum. 2014. Efficient non-parametric estimation of multiple embeddings per word in vector space. In *Proceedings of EMNLP*, pages 1059–1069, Doha, Qatar.

Jeffrey Pennington, Richard Socher, and Christopher Manning. 2014. Glove: Global vectors for word representation. In *Proceedings of EMNLP*, pages 1532–1543, Doha, Qatar.

Mohammad Taher Pilehvar and Roberto Navigli. 2015. From senses to texts: An all-in-one graph-based approach for measuring semantic similarity. *Artificial Intelligence*, 228:95–128.

Joseph Reisinger and Raymond J. Mooney. 2010. Multi-prototype vector-space models of word meaning. In *Proceedings of NAACL-HLT*, pages 109–117, Los Angeles, California.

Sascha Rothe and Hinrich Schütze. 2015. AutoExtend: Extending word embeddings to embeddings for synsets and lexemes. In *Proceedings of ACL-IJCNLP*, pages 1793–1803, Beijing, China.

Christopher J. Mungall Sebastian Bauer Helen V. Firth et al. Sebastian Khler, Sandra C Doelken. 2014. The Human Phenotype Ontology project: linking molecular biology and disease through phenotype data. *Nucleic Acids Research*, 42:966–974.

Mo Yu and Mark Dredze. 2014. Improving lexical embeddings with semantic knowledge. In *Proceedings of ACL (Volume 2: Short Papers)*, pages 545–550, Baltimore, Maryland.

Identification, characterization, and grounding of gradable terms in clinical text

Chaitanya Shivade[†], Marie-Catherine de Marneffe[§], Eric Fosler-Lussier[†], Albert M. Lai[*]
[†]Department of Computer Science and Engineering,
[§]Department of Linguistics,
[*]Department of Biomedical Informatics,
The Ohio State University, Columbus OH 43210, USA.
shivade@cse.ohio-state.edu, mcdm@ling.ohio-state.edu
fosler@cse.ohio-state.edu, albert.lai@osumc.edu

Abstract

Gradable adjectives are inherently vague and are used by clinicians to document medical interpretations (e.g., *severe* reaction, *mild* symptoms). We present a comprehensive study of gradable adjectives used in the clinical domain. We automatically identify gradable adjectives and demonstrate that they have a substantial presence in clinical text. Further, we show that there is a specific pattern associated with their usage, where certain medical concepts are more likely to be described using these adjectives than others. Interpretation of statements using such adjectives is a barrier in medical decision making. Therefore, we use a simple probabilistic model to ground their meaning based on their usage in context.

1 Introduction

Expressions used in a language are said to be vague if they do not convey a precise meaning. Sentences using vague expressions do not give rise to precise truth conditions (Kennedy, 2007). Consider the following sentence: "The patient was maintained on a *high* dose of insulin." Interpreting such statements is a problem since it is unclear what was the exact amount of insulin used. Gradability (Sapir, 1944; Lyons, 1977) is a semantic property that allows a word to describe the intensity of a measure in context, and thus enables comparative constructs. In the above example, the word *high* is said to be gradable since it conveys the meaning associated with the measure - amount.

Gradable adjectives inherently possess a degree of vagueness and are used in a language to express epistemic uncertainties (Kennedy, 2007; Frazier et al., 2008). While judgments are strong in extreme cases, there exist borderline cases, where it is difficult to ascribe an adjective. In the above example, some amounts of insulin would be considered as a high dose by all, other amounts would never be considered a high dose, but there is a middle range where it can be difficult for even experts to judge, if it is a high dose. This is because, different experts may have differing thresholds for what constitutes a high dose.

Broadly, gradable adjectives can be classified into two categories based on their interpretation as measure functions (Bartsch, 1975; Kennedy, 1999). Adjectives such as *tall, heavy, expensive* can be viewed as measurements that are clearly associated with a numerical quantity (height, weight, cost). In contrast, adjectives like *clever, beautiful, naive* are more complex and underspecified for the exact feature being measured. Gradable adjectives have been the focus of several recent studies (de Melo and Bansal, 2013; Ruppenhofer et al., 2014) in the NLP community. Gradablity is property not limited to adjectives and also extends to other parts of speech such as adverbs (Shivade et al., 2015; Ruppenhofer et al., 2015) (e.g., slightly, marginally), nouns (e.g., joy, euphoria), and also verbs (e.g., drizzling, pouring).

In this paper, we conduct a comprehensive study of gradable adjectives used in clinical text. Using a method proposed by Hatzivassiloglou and Wiebe (2000), we identify the gradable adjectives in our dataset of clinical notes. We found that these adjectives have a substantial presence (30%) in our data. Further, we show that there is a specific pattern in which gradable adjectives are used: some medical concepts are more likely to be modified by these adjectives than others. Finally, we focus on a specific subset of gradable adjectives associated with measurements of numerical quantities and demonstrate the use of a simple computational

Proceedings of the 15th Workshop on Biomedical Natural Language Processing, pages 17–26,
Berlin, Germany, August 12, 2016. ©2016 Association for Computational Linguistics

model to ground their meaning.

2 Dataset preparation

We used 58,880 clinical notes on Chronic Lymphocytic Leukemia (CLL), 2,652 notes on prostate cancer (PC) and 14,378 notes on Methicillin-resistant Staphylococcus Aureus (MRSA) representing three different cohorts from our institution as a corpus for our study. Thus we had a total of 75,910 notes with an average word count of 1,476 words per note. In addition, we also had access to 8,192 echocardiograms, which are cardiology reports mostly containing semi-structured data with few lines of free text (avg. word count = 64). All clinical notes were from adult patients collected for a period from 2005 to 2010 with necessary approval of the institutional review board at our institution.

These notes are written by healthcare professionals communicating different aspects of patient care and therefore correspond to different note types. For instance, "Progress Notes" are written by physicians documenting periodic developments in the condition of patients, their diagnosis, and treatment. "Operative Notes" are written by surgeons documenting the pre-operative diagnosis, description of the procedure, and the post-operative condition. Our corpus consists of notes belonging to 98 different note types. The name of each note type is mentioned in the first few lines of a templated document header and often has multiple lexical variations. For instance, a "Progress Note" can be an "Inpatient Progress Note" or an "Outpatient Progress Note." These names were manually normalized to 18 note types, and confirmed by a physician for correctness. Each note from our dataset was thus mapped to one of these normalized types.

Clinical notes have a typical structure: the content is often organized in sections (e.g., "History of Present Illness" followed by "Physical Examination" and ending with "Assessment and Plan"). The beginning of a section is formatted as distinct text with the section name in capital letters followed by a newline characted. We used a simple rule-based system to identify section headers and map the contents of a note to these sections. As with note types, section names also had multiple lexical variations (e.g., "Physical Examination" can be "Physical Exam" or "Physical Assessment" or simply "Exam"). Our corpus had 587 section names which were normalized to 17 note sections with a physician's approval.

3 Identification of gradable adjectives

First, we want to automatically identifiy gradable adjectives in our corpus. We reimplemented the method described in (Hatzivassiloglou and Wiebe, 2000), a log linear regression model that learns the weights associated with two features: 1) Number of times an adjective is used in comparative and superlative constructs, and 2) Number of times an adjective is modified by terms that intensify or diminish the semantic meaning of adjectives (mostly adverbs such as *very, little, somewhat,* etc. and a few nouns such as *bit,* etc.). Hatzivassiloglou and Wiebe (2000) manually created a list of 73 such terms. Their model was generated using the 1987 Wall Street Journal Corpus (Marcus et al., 1993) and tested on a hand curated gold standard dataset of 453 adjectives (235 gradable and 218 non-gradable) created using the Collins Birmingham University International Language Database dictionary, which is annotated for gradable and non-gradable adjectives.

We developed a logistic regression model with the two features described above. For the first feature, a morphology analysis component was developed to identify inflections of adjectives from their base form. This consisted of identifying adjectives in their comparative form using simple parts-of-speech tagging (Toutanova et al., 2003) and regular expression based rules. Although the test set used in (Hatzivassiloglou and Wiebe, 2000) is available, the list of 73 noun phrases and adverbial modifications is not. We therefore compiled this list using ten fold cross validation to capture the second feature. In each fold of training, we found all the adverbs and nouns modifying the gradable adjectives using the Stanford Dependency Parser (version 2.0.4) (de Marneffe et al., 2006). We determined the best subset by choosing an optimal threshold for the ($k = 81$) most frequent modifiers through cross validation. This gave us the second feature for gradability.

Although the method was developed on newswire text, we found that it worked surprisingly well for our clinical corpus. We trained the model on clinical notes and evaluated it on the test set published by Hatzivassiloglou and Wiebe (2000). Of the 453 adjectives in that gold standard test set, we found that 61 adjectives (e.g. *wealthy,*

Study	Corpus	Gradable	Non-gradable	Precision	Recall	F-Score
H & W(2000)	1987 WSJ	235	218	94.15	82.13	87.73
Our study	Clinical notes	217	175	99.51	84.32	91.34

Table 1: Performance of gradable adjective identification on the test set from Hatzivassiloglou and Wiebe (2000).

zesty) were not present in our corpus, resulting in a total of 392 adjectives (217 gradable and 175 non-gradable). Table 1 outlines (does not compare) the performance of classification in the two studies. Since the F-score of our model is reasonably high, we use it to identify the gradable adjectives in our corpus. In addition to the 392 adjectives present in the test set, the model identifies 1,709 gradable adjectives in our data. These were domain-specific words such as *therapeutic, retroperitoneal, edematous*, common adjectives such as *acute, febrile, gentle, pale*, and also some interesting compositions such as *well-nourished, low-normal*, and *near-complete*.

4 Usage characterization

Vagueness induced by gradable adjectives has been studied by researchers in the past. We want to investigate how frequently such language appears in clinical notes, and if there are certain situations where these terms are more likely to be used. In the following sections, we show that not only do gradable adjectives have a substantial presence in clinical text, but there is also a definite pattern in their usage.

4.1 Presence of gradable adjectives

Using the model described in the previous section, we found all gradable adjectives present in our corpus. The percentage of adjectives identified as gradable in the notes across the 18 normalized note types was calculated. This percentage is fairly consistent across different note types, $\mu = 30.85\%, \sigma = 4.9\%$.

In addition to examining the distribution of gradable adjectives across notes types, we performed a finer analysis by calculating their percentage across different sections in a note. The percentage of adjectives identified as gradable across the 17 normalized sections was calculated. Again, it if fairly consistent ($\mu = 31.45\%, \sigma = 6.2\%$) across different sections.

4.2 Usage pattern

In this section, we present statistics that characterize the usage of gradable adjectives in describing medical concepts of different semantic types in clinical notes. The Unified Medical Language System (UMLS) (Lindberg et al., 1993) is a repository of multiple biomedical vocabularies and standards, developed by the US National Library of Medicine. A major component of the UMLS is the Semantic Network which assigns a semantic type to every concept. A semantic type is a high-level category (e.g., "Sign or Symptom," "Pharmacological Substance," "Plant," "Enzyme") analogous to named-entity types and there are 133 such semantic types in the 2013AA version of the UMLS.

MetaMap (Aronson, 2001) is a program that can map words from free text documents to concepts from the UMLS. Using the Stanford Dependency Parser, we identified medical concepts that were modified by a gradable adjective in our corpus and looked up their semantic types. For example: in *extreme fatigue*, the gradable adjective *extreme* modifies the term *fatigue* which has the semantic type "Sign or Symptom," while in *severe stenosis*, the adjective *severe* modifies the term *stenosis* which has the semantic type "Disease or Syndrome."

We hypothesized that gradable adjectives modify certain nouns more often than others. In order to test this hypothesis, we calculated how often nouns of a particular semantic type are modified by gradable adjectives. These frequencies were calculated for the three sets of clinical notes corresponding to three different diagnoses (CLL, PC, and MRSA) in our corpus. Nouns from a certain semantic types were very frequently described using gradable adjectives (e.g., "Finding," "Therapeutic or Preventive Procedure," "Disease or Syndrome"), and hence had high frequency values in all three datasets. Similarly, nouns from a few semantic types were never described by gradable adjectives (e.g., "Reptiles," "Professional Society").

Dataset	CLL	PC	MRSA
CLL	1.00	0.93	0.90
PC	0.93	1.00	0.91
MRSA	0.90	0.91	1.00

Table 2: Spearman's Correlation between clinical notes for semantic type modification by gradable adjectives.

We confirmed this by sampling each dataset into five equal folds and repeating the frequency calculations. The observations for frequency variations were consistent for every fold across each dataset. We performed a simple add-one Laplace smoothing to account for low frequency semantic types across datasets. Since the size of the three datasets were significantly different, we normalized the frequencies by the sum of frequencies across all semantic types within each dataset. The normalized frequency values represent the probability of a semantic type being modified by gradable adjectives in a dataset. We computed the Spearman's correlation for these 133 probabilities across each pair of datasets and found that there was a high correlation between them (Table 2). This high correlation across all three diagnoses suggests a definite pattern for the usage of gradable adjectives in clinical text.

5 Probabilistic Modeling

Gradable adjectives are widely studied as implicit or explicit measurements of certain quantities (Bartsch, 1975; Kennedy, 1999). Moreover, they also participate in a scale. For example, the adjectives (*warm* < *hot* < *scorching*) represent a scalar relationship and implicitly measure temperature. While judgments to associate an adjective with extreme values are very strong, those for borderline cases are difficult. In the above example, certain values of temperature are definitely *warm* and others are definitely considered *hot* (and yet not scorching). But there is always a set of values in between which can be either *warm* or *hot*. In order to capture this intuition, we created a probabilistic model using Bayes rule:

$$P\left(grad|num\right) = \frac{P\left(num|grad\right) \cdot P\left(grad\right)}{P\left(num\right)} \quad (1)$$

where *grad* represents the gradable term and *num* the numerical value.

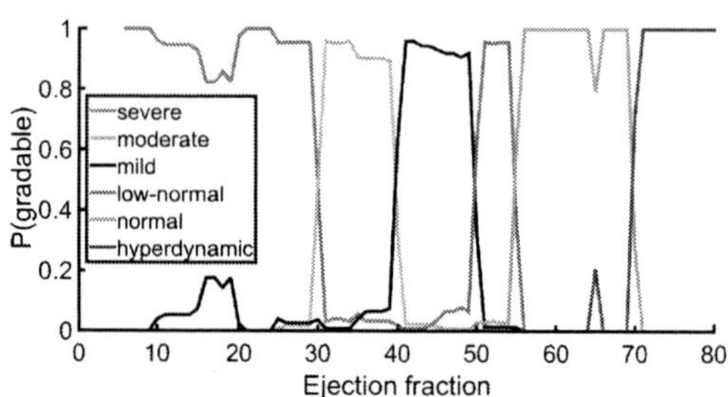

Figure 1: Probabilistic modeling of adjectives describing systolic function.

Clinicians frequently document their assessments for a patient along with evidence to support their claim, e.g., "Mild anemia, Hgb 8.2." This sentence has a medical concept "anemia" being described by a gradable adjective *mild* on the basis of the measurement of a numerical value - hemoglobin. For several medical concepts, we extracted using regular expressions, instances where an assessment for a medical concept was made using a gradable term, along with a numerical evidence to support the claim. Specifically, we indexed all sentences using Lucene and searched for ones containing the medical term (e.g. anemia) and the quantity of interest (e.g. hemoglobin). Finally, numerical values and adjectives were extracted using regular expressions. In the following subsections, we demonstrate that we can ground the meaning of gradable terms using the above model.

5.1 Systolic Function

Systolic function is a measure of how well the lower left pumping chamber of the heart sends blood to the rest of the body. It is measured using a numerical quantity called left ventricular ejection fraction (LVEF) which is documented in an echocardiogram. There is variation among physicians defining the precise threshold for a normal ejection fraction (Sanderson, 2007). While normal values range from 55 to 65, values less than 30 imply that the systolic function is severely compromised. We extracted LVEF values from the echocardiogram reports and their corresponding descriptions of systolic function. Posterior probabilities $P(gradable|LVEF)$ were calculated using equation (1) which resulted in a plot as shown in Figure 1.

From the 8,192 echocardiogram reports, we found six gradable adjectives in association with

LVEF values. While the adjectives *severe, mild* and *moderate* are associated with systolic dysfunction, the adjectives *low-normal, normal* and *hyperdynamic* are associated with systolic function. Although there is discussion in the clinical community regarding qualitative descriptions for ejection fraction (Radford, 2005), there is variation in these recommendations. Moreover, certain terms though used frequently (e.g. *low-normal*) are never a part of such guidelines.

An interesting observation can be made regarding Figure 1, drawing an analogy from the concept of WordNet *dumbbells* (Sheinman et al., 2012). A WordNet dumbbell is a representation involving an antonym pair (e.g. *small* and *large*) as two ends of a semantic scale with semantically similar adjectives arranged in a radial fashion around each adjective. The antonym acting as a centroid and its synonyms as members of a cluster represent words that most likely participate in the same scale. For example, the antonym pair (*small, large*) results in the dumbbell with clusters (*small, tiny, pocket-size, smallish*) and (*large, gigantic, monstrous, huge*) at the two ends. WordNet dumbbells have been used in the past (Sheinman et al., 2013; de Melo and Bansal, 2013) to group gradable adjectives belonging to the same scale. It can be seen that the analogous dumbbell consisting of (*severe, mild, moderate*) and (*low-normal, normal, hyperdynamic*) can be constructed using the modified terms systolic dysfunction and systolic function respectively.

The model captures essential aspects of gradability very well. The scalar relationships (*severe < moderate < mild*) and (*low-normal < normal < hyperdynamic*) can be inferred by imposing an order on the mean values for the posterior distributions of these adjectives. Strong judgments for extreme cases and uncertainty for borderline cases can be observed in the form of flat peaks for specific intervals and overlapping distributions for mid-range values.

5.2 Anemia

Hemoglobin is a protein in the red blood cells (RBCs) that contains iron and carries oxygen from the lungs to the rest of the body. Anemia is a blood disorder, operationally defined as a reduction in the hemoglobin content of blood caused by a decrease in the RBCs below a reference interval of healthy individuals. The range of normal

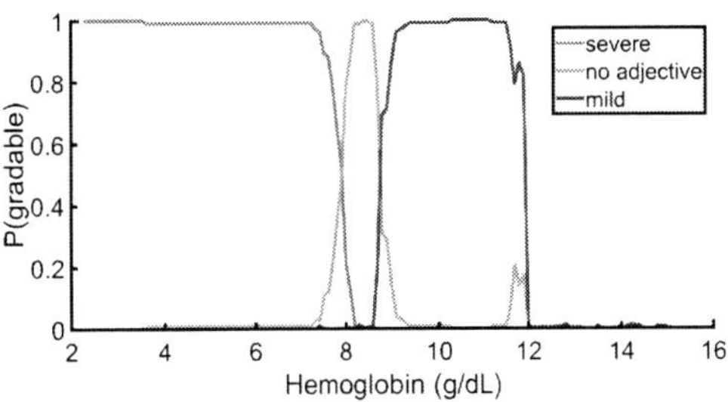

Figure 2: Probabilistic modeling of descriptions for anemia.

hemoglobin values for the laboratories at our institution is from 11.7 to 15.5. We found the two adjectives *severe* and *mild* to be most commonly used for describing anemia. A number of notes also mentioned anemia with no modifier at all. Figure 2 shows the posterior probabilities calculated for the three modifications of anemia: *mild, no adjective*, and *severe* using the model outlined in equation 1.

It is interesting to note that when physicians refer to anemia without an adjective, it is neither severe nor mild, and has a value in between. As with systolic function, we can infer the ordinal relationship (*severe anemia < anemia < mild anemia*), considering the mean values for the posterior distributions of these adjectives. Also, strong judgments for extreme values and uncertainty for borderline cases are evident through flat peaks and overlapping distributions respectively. We also found the adjective *moderate* being used in our data for describing anemia for hemoglobin values between *mild* and *severe*. However, it had few occurrences and hence we did not include *moderate* in our model. Other adjectives such as *significant, marked, slight* and *pernicious* were also found in the data but with low frequency counts.

5.3 Platelet count

Platelets (also known as thrombocytes) are colorless blood cells that help the process of blood clotting. There are about 150,000 to 450,000 platelet per microliter of blood in the human body (Erkurt et al., 2012). While the condition resulting from a lower than normal platelet count is known as *thrombocytopenia*, the condition resulting from a higher than normal platelet count is referred to as *thrombocytosis*. Since the notion of *low* and *high* counts is gradable, we treat equivalent descriptions of thrombocytopenia and thrombocyto-

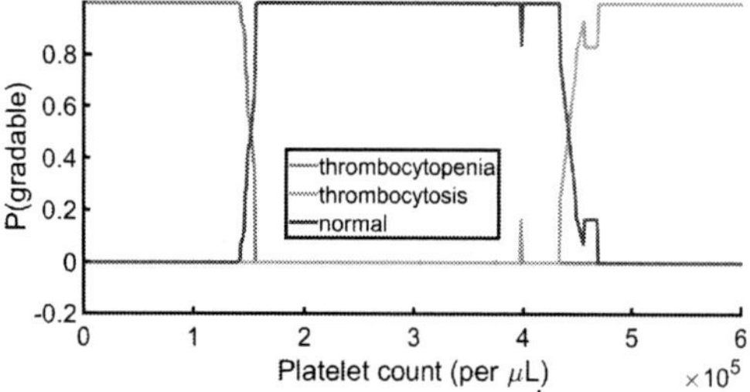

Figure 3: Probabilistic modeling of descriptions for variations in platelet count.

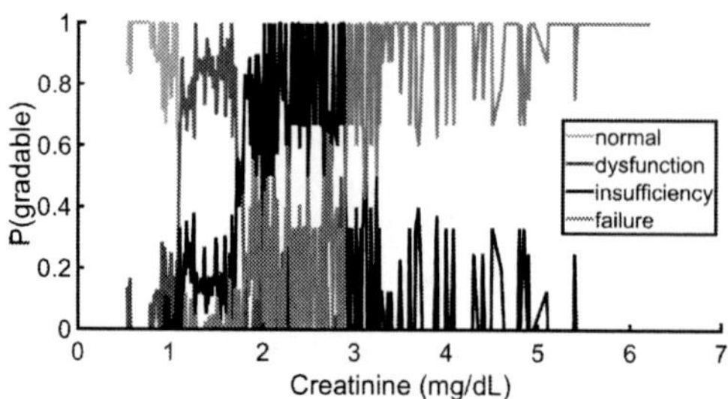

Figure 4: Probabilistic modeling of descriptions for variations in creatinine.

sis as gradable. In addition we also extracted instances of clinical notes where the platelet count was referred to as *normal*. Using these three descriptions, we applied the Bayes rule explained in Equation 1.

Figure 3 shows posterior probabilities calculated for these three descriptions of platelet count. As with previous examples, we can infer the ordinal relationship (*thrombocytopenia < normal < thrombocytosis*) by considering the mean values of their posterior distributions.

5.4　Renal Function

Creatinine is a chemical made by the body and is used to supply energy to the muscles. Creatinine is removed from the body by the kidneys and released through urine. If kidney function (or renal function) is not normal, creatinine level in the body increases (Israni and Kasiske, 2011). Abnormal renal function is referred to through different terminologies such as renal insufficiency, renal failure, and renal dysfunction. The vagueness introduced by the use of these gradable terms is also evident in clinical literature. Hsu and Chertow (2000) in their paper titled "Chronic renal confusion: insufficiency, failure, dysfunction, or disease" propose a set of laboratory values to classify patients as *mild*, *moderate* and *advanced* degrees of chronic renal insufficiency to "facilitate communication among nephrologists and other physicians and provide a framework for comparison of populations." It should be noted that linguistic ambiguity is not the only reason for this confusion and also has medical explanations which are beyond the scope of discussion of our work.

This problem was acknowledged by the medical community. More than 30 new definitions were proposed (Bellomo et al., 2004) and a new standard is now in place (Khwaja, 2012). How-

ever, our data is older (from 2005 to 2010) and has frequent occurrences of these terms. We extracted instances for the gradable terms "normal renal function," "renal failure," "renal insufficiency," "renal failure" and the corresponding creatinine values mentioned by physicians in the text. Further, we computed posterior probabilities for $P(gradable|creatinine)$ using our model (Figure 4). The range of normal creatinine values is between 0.60 to 1.10 for the laboratories at our institution. In comparison with other examples discussed so far, it can be seen from the plot that there is a greater confusion in the use of these terms. This is especially evident in the interval [2,3]. Again, this confirms with the property of uncertainty for borderline cases. However, an ordering (*normal < dysnfunction < insufficiency < failure*) can still be inferred.

5.5　Evaluation

We evaluated the model to determine if it fits the data well. Using leave one out cross validation, we tested if the model was able to predict the adjective for a given numerical value. The gradable mentioned in each text extract was regarded as the gold standard prediction label. While creating a model, we ensured that there were at least three data points for each measurement value of the numerical quantity present in the data. This allowed us to compute priors for all values in the data. In practice, one would either need large amounts of data or employ smoothing (Kneser and Ney, 1995) to ensure prior calculations for all numerical values are possible. Accuracy is calculated across all gradable terms for each medical concept as described in previous sections (Table 3). The models achieve fairly high accuracies which demonstrates that our model fits the data well.

Medical concept	Number of data points	Accuracy (%)
Systolic function	10,201	90.4
Anemia	12,711	88.3
Platelet count	14,234	94.6
Renal function	16,309	74.8

Table 3: Evaluation of probabilistic models to predict gradable terms for numerical values in the data.

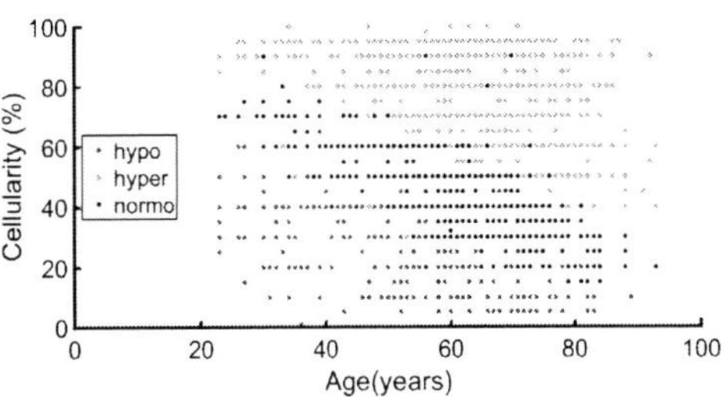

Figure 5: Dependency of gradable terms for BMC on age.

6   Limitations and future work

We illustrated through examples that gradable terms in clinical text can be effectively analyzed through data using a simple probabilistic model. The model is developed for cases where the use of gradable terms is dependent on a single numerical quantity. We included analysis of descriptions for heart function and kidney function. Similar analysis can be conducted for liver function which measures the amount of bilirubin in the body. Common tests such as body mass index, blood pressure and heart rate can also be analyzed in this way. Such a data-driven approach can help in creation of a standard terminology and avoid confusions (Hsu and Chertow, 2000).

However, context sensitivity is an important characteristic of gradable adjectives (Kennedy, 2007). Thus, "John is a *tall* boy" and "John is a *tall* basketball player" convey different meanings despite using the same gradable adjective for the same person (van Rooij, 2011). Similarly, the gradable description of a medical concept may not always be dependent on a single numerical quantity. For example, there is a slight variation in the upper limit of normal (ULN) values for creatinine with gender. The ULN for males is 1.3 while that for females is 1.2 at our institution. Similarly, the lower limit of normal for hemoglobin in males is 11.7 while that for females is 13.2. These variations are small in magnitude. However, this is a problem in cases where the dependency on other variables is much more pronounced. We illustrate this through an example.

Bone Marrow Cellularity (BMC) is the volume ratio of hematopoietic cells (blood cells that give rise to other blood cells) and fat. Pathologists perform a bone marrow analysis and use the three adjectives *hypocellular, normocellular*, and *hypercellular* to describe the sample. However, BMC is largely dependent on age of the patient. It is 100% for newborn infants and reduces with age in adults (Muschler et al., 2001). Therefore, the notion of hypocellular, normocellular, and hypercellular also varies with age. We extracted BMC values and associated adjectives from our data. Figure 5 shows the likelihood plot of BMC values against associated age of patients with three different colors for the adjectives *hypocellular, hypercellular*, and *normocellular*. Although the three gradable descriptions are linearly separable, $P(gradable|BMC)$ cannot be modeled using Equation 1, which ignores the age of the patient.

Time is a very common variable that often plays an important role in clinical assessments. This is most evident in blood sugar values for diabetic patients that vary with every hour depending on times of food consumption. Temporal adjectives are frequently found as descriptions of medical concepts. Some of the commonly found temporal adjectives in our data include *acute, chronic, recent, progressive, worsening, stable, persistent*, and *continued*.

Clinical notes are created and read by different individuals associated with the hospital. Vital decisions such as clinical trial recruitment, adherence to treatment guidelines, etc. are made by healthcare professionals based on their interpretation of these clinical narratives. Introducing automation in these processes is an active area of NLP research (Demner-Fushman et al., 2009). This decision making becomes challenging if language used in the clinical notes is vague and does not deliver a precise meaning. Our work is a small step to illustrate that gradability and its associated vagueness is an important aspect of clinical text which can be modeled through data. Creating a single model that can flexibly incorporate multiple variables and yet capture the properties of gradable adjectives can be an interesting line of research for the future.

7 Related Work

The phenomenon of adjectival modification in biomedical discourse has also been a subject of interest. Through empirical observations, Chute and Elkin (1997) classified frequent modifiers for medical concepts into two types: clinical modifiers (*e.g., chronic, severe, acute*) and administrative qualifiers (*e.g., history of, no evidence of, status post*). Bodenreider and Pakhomov (2003) extended this idea and compared adjectival modifications in biomedical literature and patient records. They found that while patient records contain markers for uncertainty (e.g., *possible, probable*) and non-specific symptoms (e.g., *low back pain, discomfort*), scientific articles are precise about attributes of organisms or age-groups (e.g., human, canine, neonatal).

Adjectives have been studied extensively in computational linguistics. WordNet (Fellbaum, 1998) classifies adjectives into two broad categories: descriptive and relational. Descriptive adjectives (e.g., *big* house, *heavy* bag) ascribe the value of an attribute to a noun, while relational adjectives (e.g., *atomic* bomb, *dental* hygiene) do not. Among the various distinctions between descriptive and relational adjectives, relational adjectives are typically not gradable (Fellbaum, 1998).

Although association between adjectives and numerical quantities has been a topic of research in some studies (Aramaki et al., 2007; Davidov and Rappoport, 2010; Iftene and Moruz, 2010), very few studies have investigated grounding the meaning of adjectives to numerical quantities. de Marneffe et al. (2010) investigated the problem of interpreting implied answers to yes/no questions when the response is not explicit. Specifically, they investigated question-answer pairs in which the question contains an adjective and the answer contains a numerical measure. For example, predicting the correct yes/no answer in (1) involves interpreting a numerical quantity (age) with respect to the gradable adjective *little*.

1.   Q. Are your kids little?

       A. I have a 7 year-old and a 10 year-old.

The authors created logistic regression models for each adjective by querying the web with appropriate keywords ("little kids") and its antonyms ("not little kids"), so that both positive and negative instances can be learned.

Narisawa et al. (2013) explore a closely related problem of learning *numerical common sense* for the task of RTE in Japanese text. They study a broad set of cases that require semantic inference over numerical expressions. They query the web to gather instances of pairs of numerical quantities and corresponding contexts and propose two approaches. The distribution based approach concludes the numerical quantity to be *large* or *small* if it appears in the top or bottom five percent of the distribution generated for the numerical quantity and *normal* if it is in between. The cue-based approach relies on explicit textual cues (e.g., *as large as, only*) for associating a judgment about a numerical expression.

8 Conclusion

We empirically evaluated use of gradable adjectives in clinical documents. We reimplemented a previously published model for identifying gradable adjectives in newswire text and found that it performs surprisingly well with our clinical data. These adjectives have a substantial presence in clinical notes across multiple types of documents, written by different healthcare professionals. Analysis of the frequencies of these adjectives and their association with clinical concepts from UMLS revealed that there is a specific pattern for their usage. Finally, we showed that a simple Bayesian model can be used effectively to ground the meaning of gradable terms when they are used to describe medical concepts involving measurement of numerical quantities. Our data-driven approach can help in development of clinical standards in situations where there is a need to establish a precise relationship between adjectives and measurements.

Acknowledgements

We would like to thank Courtney Hebert and Kelly Regan for their help in this work. Research reported in this publication was supported by the National Library of Medicine of the National Institutes of Health under award number R01LM011116. The content is solely the responsibility of the authors and does not necessarily represent the official views of the National Institutes of Health.

References

Eiji Aramaki, Takeshi Imai, Kengo Miyo, and Kazuhiko Ohe. 2007. UTH: SVM-based Semantic Relation Classification using Physical Sizes. In *Proceedings of the Fourth International Workshop on Semantic Evaluations (SemEval-2007)*, pages 464–467.

Alan R. Aronson. 2001. Effective mapping of biomedical text to the UMLS Metathesaurus: the MetaMap program. In *Proceedings of the Annual AMIA Symposium*, pages 17–21.

Renate Bartsch. 1975. The grammar of relative adjectives and comparison. In *Formal Aspects of Cognitive Processes*, pages 168–185. Springer.

Rinaldo Bellomo, Claudio Ronco, John A Kellum, Ravindra L Mehta, Paul Palevsky, and ADQI workgroup. 2004. Acute renal failure–definition, outcome measures, animal models, fluid therapy and information technology needs: the Second International Consensus Conference of the Acute Dialysis Quality Initiative (ADQI) Group. *Critical care*, 8(4):R204–R212.

Olivier Bodenreider and Serguei V. Pakhomov. 2003. Exploring adjectival modification in biomedical discourse across two genres. In *Proceedings of the ACL 2003 Workshop on Natural Language Processing in Biomedicine*.

Christopher G. Chute and Peter L. Elkin. 1997. A clinically derived terminology: qualification to reduction. In *Proceedings of the AMIA Annual Fall Symposium*.

Dmitry Davidov and Ari Rappoport. 2010. Extraction and approximation of numerical attributes from the web. In *Proceedings of ACL*, pages 1308–1317.

Marie-Catherine de Marneffe, Bill MacCartney, and Christopher D. Manning. 2006. Generating Typed Dependency Parses from Phrase Structure Parses. In *Proceedings of LREC*, pages 449–454.

Marie-Catherine de Marneffe, Christopher D. Manning, and Christopher Potts. 2010. "Was it good? It was provocative". Learning the meaning of scalar adjectives. In *Proceedings of ACL*, pages 167–176.

Gerard de Melo and Mohit Bansal. 2013. Good, Great, Excellent: Global Inferences of Semantic Intensities. *Transactions of the Association of Computational Linguistics*, 1(July):279–290.

Dina Demner-Fushman, Wendy W. Chapman, and Clement J. McDonald. 2009. What can natural language processing do for clinical decision support? *Journal of Biomedical Informatics*, 42(5):760–72, Oct.

Mehmet Ali Erkurt, Emin Kaya, Ilhami Berber, Mustafa Koroglu, and Irfan Kuku. 2012. Thrombocytopenia in adults: review article. *Journal of Hematology*, 1(2-3):44–53.

Christiane Fellbaum. 1998. *WordNet: An Electronic Lexical Database*. Bradford Books.

Lyn Frazier, Charles Clifton, and Britta Stolterfoht. 2008. Scale structure: Processing minimum standard and maximum standard scalar adjectives. *Cognition*, 106(1):299–324.

Vasileios Hatzivassiloglou and Janyce M. Wiebe. 2000. Effects of adjective orientation and gradability on sentence subjectivity. In *Proceedings of the 18th Conference on Computational Linguistics*, pages 299–305.

Chi-yuan Hsu and Glenn M. Chertow. 2000. Chronic renal confusion: insufficiency, failure, dysfunction, or disease. *American Journal of Kidney Diseases*, 36(2):415–418.

Adrian Iftene and Mihai-Alex Moruz. 2010. UAIC participation at RTE-6. In *Proceedings of the Text Analysis Conference (TAC 10)*.

Ajay K. Israni and Bertram L. Kasiske. 2011. Laboratory assessment of kidney disease: glomerular filtration rate, urinalysis, and proteinuria. In *Brenner and Rector's The Kidney*, volume 9, pages 1585–619. Elsevier.

Christopher Kennedy. 1999. *Projecting the adjective: the syntax and semantics of gradability and comparison*. Routledge.

Christopher Kennedy. 2007. Vagueness and grammar: the semantics of relative and absolute gradable adjectives. *Linguistics and Philosophy*, 30(1):1–45, March.

Arif Khwaja. 2012. KDIGO clinical practice guidelines for acute kidney injury. *Nephron Clinical Practice*, 120(4):c179–c184.

Reinhard Kneser and Hermann Ney. 1995. Improved backing-off for m-gram language modeling. In *Proceedings of ICASSP*, pages 181–184.

Donald A Lindberg, Betsy L Humphreys, and Alexa T McCray. 1993. The unified medical language system. *Methods of Information in Medicine*, 32(4):281–291.

John Lyons. 1977. Semantics (Volumes I & II). *Cambridge CUP*.

Mitchell P. Marcus, Mary Ann Marcinkiewicz, and Beatrice Santorini. 1993. Building a Large Annotated Corpus of English: The Penn Treebank. *Computational Linguistics*, 19(2):313–330.

George F. Muschler, Hironori Nitto, Cynthia A. Boehm, and Kirk A. Easley. 2001. Age-and gender-related changes in the cellularity of human bone marrow and the prevalence of osteoblastic progenitors. *Journal of Orthopaedic Research*, 19(1):117–125.

Katsuma Narisawa, Yotaro Watanabe, Junta Mizuno, Naoaki Okazaki, and Kentaro Inui. 2013. Is a 204 cm Man Tall or Small ? Acquisition of Numerical Common Sense from the Web. In *Proceedings of ACL*, pages 382–391.

Martha J. Radford. 2005. ACC/AHA Key Data Elements and Definitions for Measuring the Clinical Management and Outcomes of Patients With Chronic Heart Failure. *Journal of the American College of Cardiology*, 46(6):1179–1207, sep.

Josef Ruppenhofer, Michael Wiegand, and Jasper Brandes. 2014. Comparing methods for deriving intensity scores for adjectives. In *14th Conference of the European Chapter of the Association for Computational Linguistics*, pages 117–122.

Josef Ruppenhofer, Jasper Brandes, Petra Steiner, and Michael Wiegand. 2015. Ordering adverbs by their scaling effect on adjective intensity. In *Proceedings of RANLP*, pages 545–554.

John E Sanderson. 2007. Heart failure with a normal ejection fraction. *Heart*, 93(2):155–158.

Edward Sapir. 1944. Grading, A Study in Semantics. *Philosophy of Science*, 11(2):93–116.

Vera Sheinman, Takenobu Tokunaga, Isaac Julien, Peter Schulam, and Christiane Fellbaum. 2012. Refining WordNet adjective dumbbells using intensity relations. In *Sixth International Global Wordnet Conference*, pages 330–337.

Vera Sheinman, Christiane Fellbaum, Isaac Julien, Peter Schulam, and Takenobu Tokunaga. 2013. Large, huge or gigantic? Identifying and encoding intensity relations among adjectives in WordNet. *Language Resources and Evaluation*, 47(3):797–816, January.

Chaitanya Shivade, Marie-Catherine de Marneffe, Eric Fosler-Lussier, and Albert M. Lai. 2015. Corpus-based discovery of semantic intensity scales. In *Proceedings of NAACL-HLT*, pages 483–493.

Kristina Toutanova, Dan Klein, Christopher D Manning, and Yoram Singer. 2003. Feature-rich part-of-speech tagging with a cyclic dependency network. In *Proceedings of NAACL-HLT*, pages 173–180. Association for Computational Linguistics.

Robert van Rooij. 2011. Vagueness and Linguistics. In Giuseppina Ronzitti, editor, *Vagueness: A Guide*, chapter Vagueness, pages 123–170. Springer Netherlands.

Graph-based Semi-supervised Gene Mention Tagging

Golnar Sheikhshab[1,2], Elizabeth Starks[2], Aly Karsan[2], Anoop Sarkar[1], Inanc Birol[1,2]

`gsheikhs@sfu.ca,{lstarks, akarsan}@bcgsc.ca,anoop@sfu.ca,ibirol@bcgsc.ca`

[1] School of Computing Science, Simon Fraser University, Burnaby, BC, Canada
[2] Canada's Michael Smith Genome Sciences Centre, British Columbia Cancer Agency, Vancouver, BC, Canada

Abstract

The rapidly growing biomedical literature has been a challenging target for natural language processing algorithms. One of the tasks these algorithms focus on is called named entity recognition (NER), often employed to tag gene mentions. Here we describe a new approach for this task, an approach that uses graph-based semi-supervised learning to train a Conditional Random Field (CRF) model. Benchmarking it on the BioCreative II Gene Mention tagging task, we achieved statistically significant improvements in F-measure over BANNER, a widely used biomedical NER system. We note that our tool is transductive and modular in nature, and can be integrated with other CRF-based supervised NER tools.

1 Introduction

Detecting biomedical named entities such as genes and proteins is one of the first steps in many natural language processing systems that analyze biomedical text. Finding relations between entities, and expanding knowledge bases are examples of research that highly depend on the accuracy of gene and protein mention tagging.

Named entity recognition is typically modelled as a sequence tagging problem (Sha and Pereira, 2003). One of the most commonly used models for sequence tagging is a Conditional Random Field (CRF) (Lafferty et al., 2001; Sha and Pereira, 2003).

Many popular and best performing biomedical named entity recognition systems, such as BANNER (Leaman et al., 2008), Gimli (Campos et al., 2013) and BANNER-CHEMDNER (Munkhdalai et al., 2015) use CRF as their core machine learning model built on the MALLET toolkit (McCallum, 2002).

Inspired by the success of graph-based semi-supervised learning methods in other NLP tasks (Subramanya et al., 2010; Zhu et al., 2003; Subramanya and Bilmes, 2009; Alexandrescu and Kirchhoff, 2009; Liu et al., 2012; Saluja et al., 2014; Tamura et al., 2012; Talukdar et al., 2008; Das and Petrov, 2011), we integrated the graph based semi-supervised algorithm of Subramanya et al. (2010) and adapted their approach to improve on the results from BANNER. We show that our approach achieves a statistically significant improvement in terms of F-measure on the BioCreative II dataset for gene mention tagging.

Semi-supervised learning for gene mention tagging is not without precedent. There has been several semi-supervised approaches for the gene mention task and they have always been more successful than fully supervised approaches (Jiao et al., 2006; Ando, 2007; Campos et al., 2013; Munkhdalai et al., 2015).

Ando (2007) used a semi-supervised approach, Alternative Structure Optimization or ASO, in the BioCreative II gene mention shared task along with other extensions, such as using a lexicon or combining several classifiers. ASO ranked first among all competitors in the shared task competition 2007. Ando reported usage of unlabeled data as the most useful part of his system improving the F-measure of the baseline by 2.09 points where the complete (winning) system had a total improvement of 3.23 points over the baseline CRF (Ando, 2007). Jiao et al. (2006) used conditional entropy over the unlabeled data combined with the conditional likelihood over the labeled data in the objective function of CRF (Jiao et al., 2006). Munkhdalai et al. (2015) trained word representations using Brown clustering (Brown et al., 1992) and word2vec (Mikolov et al., 2013) on MEDLINE and PMC document collections and used them as features along with traditional features in a CRF. Like many of these approaches we

Proceedings of the 15th Workshop on Biomedical Natural Language Processing, pages 27–35,
Berlin, Germany, August 12, 2016. ©2016 Association for Computational Linguistics

also use unlabeled data to augment our baseline CRF model. In all these previous studies the unlabelled data was orders of magnitude more than labelled data and distinct from the test data.

In this paper we take a transductive approach and use the test set as our unlabelled data. Moreover, our approach is orthogonal to all these approaches and can be used to augment many of them. This approach can be easily implemented as a post-processing step in any system that uses a CRF model. Examples of such systems include Gimli (Campos et al., 2013) and BANNER-CHEMNDNER (Munkhdalai et al., 2015). These tools have achieved the highest F-scores in the literature after ASO (Ando, 2007). Our approach relies on the extraction of label distributions from the CRF and augments the decoding algorithm to incorporate the new information about gene mentions from the graph-based learning approach we describe in this paper.

2 Method

Like many previous studies (Leaman et al., 2008; Munkhdalai et al., 2015; Campos et al., 2013), we formulate the gene mention tagging problem as a word level sequence prediction problem, where labels for each word in the input are either Gene-Beginning, Gene-Inside, and Outside (not a gene). This representation is called IOB (for inside-outside-beginning). We applied a graph-based semi-supervised learning (SSL) approach, previously shown effective on a similar labelling task, part-of-speech tagging, for gene mention tagging. (Subramanya et al., 2010)

In graph-based SSL, a graph is constructed to represent partially labelled data. Each node in the graph represents a single word-level gene mention tagging decision and the edges between the nodes represent similarity between the nodes. The goal is to associate probability distributions over the IOB tags to all vertices. Label distributions for vertices that appear in labelled data are estimated based on the reference labels and propagate to vertices for unlabelled data in the graph. These label distributions are combined with the CRF decoding algorithm used for labelling the test data. Graph-based SSL is categorized into inductive and transductive approaches. In inductive settings (e.g. Subramanya et al. (2010)), a model is trained and can be used as-is for unseen data. In transductive settings however, the unlabelled data includes test data. We took a transductive approach in constructing our graph on the union of train set and test set as labelled and unlabelled data.

Since the graph is the cornerstone of the algorithm, let us describe its construction and usage before the overall algorithm.

2.1 Graph Construction

We use the following steps for constructing the graph for the gene mention tagging task adapted from the graph construction for part-of-speech tagging described in Subramanya et al. (2010):

1. Each vertex represents a 3-gram type and the middle word of this 3-gram is the word which is tagged as a gene mention using the IOB tags. The label distribution for this middle word is learned during graph propagation and subsequently combined with the CRF model at test time.

2. A vertex is represented by a vector of pointwise mutual information values between feature instances and its 3-gram type.

3. Edge weights represent the similarity between vertices and are obtained by computing the cosine similarity of feature vectors of their two end vertices.

4. For each vertex only the K nearest neighbours are kept (default = 10).

We considered several feature sets, namely contextual features (Table 1), simplified contextual features (Table 2), all features from the base CRF model, and the most informative features from the base CRF model. We picked the simplified contextual features based on preliminary results using cross-validation on our development set. To represent a vertex v with 3-gram $w_{-1}w_0w_1$, we look at all occurrences of its 3-gram in the text, consider the larger context $w_{-2}w_{-1}w_0w_1w_2$ and get the lemmas of these words. v is represented by a vector of point-wise mutual information values between all possible feature instances (e.g. all possible lemmas for w_{-2}) and $w_{-1}w_0w_1$.

We eliminated extremely frequent features (default $> 10,000$) to reduce the time complexity of graph construction. This should not affect the structure of the graph substantially because the point-wise mutual information between a feature and any given vertex decreases as the frequency

Description	Feature
3-gram + Context	$w_{-2}\ w_{-1}\ w_0\ w_1\ w_2$
3-gram	$w_{-1}\ w_0\ w_1$
Left Context	$w_{-1}\ w_{-2}$
Right Context	$w_1\ w_2$
Center Word	w_0
Trigram $-$ Center Word	$w_{-1}\ w_1$
Left Word + Right Context	$w_{-1}\ w_1\ w_2$
Left Context + Right Word	$w_{-2}\ w_{-1}\ w_1$

Table 1: Complete set of contextual features.

Description	Feature
Left Context Word	w_{-2}
Left Word	w_{-1}
Center Word	w_0
Right Word	w_1
Right Context Word	w_2

Table 2: Simplified set of contextual features.

of the feature increases leaving extremely frequent features with relatively small weights.

2.2 Graph Propagation

In graph propagation we associate any given vertex u with a label distribution X_u that represents how likely we think each label is for that vertex.

The goal of graph-based SSL is to propagate existing knowledge about the labels through the graph. The initial knowledge about graph nodes is provided by the labeled data and potentially some prior knowledge. Figure 1 shows how graph propagation can assign label distributions to unlabelled vertices and change the label distributions coming from labelled data.

Propagation is accomplished by optimizing an objective function over the label distributions at each node in the graph. The objective function consists of three types of constraints:

1. For any labeled vertex u, the associated label distribution X_u should be close to the reference distribution $\hat{X}_u$ (obtained from labeled data);

2. Adjacent vertices u and k should have similar label distributions X_u and X_k;

3. The label distributions of all vertices should comply with the prior knowledge, if such knowledge exists, or be uniformly distributed, otherwise.

The following objective function represents these three components:

$$
\begin{aligned}
C(X) \ = \ & \sum_{u \in L} ||X_u - \hat{X}_u||_2^2 \\
& + \mu \sum_{u \in V} \sum_{k \in N(u)} w_{u,k} ||X_u - X_k||_2^2 \\
& + \nu \sum_{u \in V} ||X_u - U||_2^2 \quad (1)
\end{aligned}
$$

where u and v are nodes in the graph, L is the set of labelled vertices, V is the set of all vertices, $N(u)$ is the set of neighbours of u, U is the uniform distribution over all labels, and μ and ν are weight constants for constraints 2 and 3, respectively. We used Euclidean distance as the distance metric.

While the first two terms in the objective function, and their corresponding constraints make intuitive sense, the uniformity constraint needs further explanation. The rationale behind using distance from uniform distribution is to avoid preferring a label over others in the absence of strong evidence.

The objective function is optimized using stochastic gradient descent. We implement the optimization algorithm for this as described in Subramanya et al. (2010):

$$
\begin{aligned}
X_i^{(m)}(y) =\ & \frac{\gamma_i(y)}{k_i} \\
\gamma_i(y) =\ & \hat{X}_i(y)\delta(i \in L) \\
& + \sum_{k \in N(i)} w_{i,k} X_k^{m-1}(y) + \nu\frac{1}{Y} \quad (2) \\
k_i =\ & \delta(i \in L) + \nu + \mu \sum_{k \in N(i)} w_{i,k}
\end{aligned}
$$

$X_i^{(m)}$ and $X_i^{(m-1)}$ denote the label distributions of vertex i in iterations m and $m - 1$, respectively, $\delta(i \in L)$ is 1 if and only if i is a labeled vertex, and Y is the number of labels.

2.3 Overall algorithm

Once propagated the label distributions through the graph, we would need to combine what we learned in the graph with the tagging results from the CRF model. For that we use a self-training algorithm, shown in Figure 2.

On an input of a partially-labeled corpus, we first train a CRF model in a supervised fashion on the labeled data (crf-train, line 1); we then use this trained CRF model to assign label probability distributions to each word in the entire (labeled + unlabeled) corpus (posterior decode, line 4). As a result, each n-gram token in the corpus has a label distribution (the posteriors). For each n-gram type u (a vertex in the graph), we find all instances (n-gram tokens) of u and average over the label distributions of these instances to get a label distribution for u (token to type, line 5). Next, we perform graph-propagation (i.e. we optimize the objective function in equation 1) to learn the label distributions for all vertices. Finally, we linearly interpolate the trained CRF model and the label distributions from the graph:

$$X_{int}(t) = \alpha X_{CRF}(t) + (1 - \alpha)X_{Graph}(t) \quad (3)$$

where t is a 3-gram token in a specific sentence, $X_{CRF}(t)$ denotes the posterior probability from the CRF model for the middle word in t, $X_{Graph}(t)$ denotes the label distribution of the 3-gram *type* t after graph propagationn, and $\alpha \in [0, 1]$ is the mixture parameter between the CRF and graph models. The best label for all words in the entire corpus is then found using Viterbi-decoding for the CRF using X_{int} instead of X_{CRF} (viterbi-decode, line 7). Viterbi decoding provides us with the best label for every n-gram token in the unlabeled corpus, which implies that our labeled set has grown to include the unlabeled corpus. We re-train the CRF on this expanded training set (crf-train, line 8); and iterate until convergence.

Note that the steps indicated by lines 1, 4, and

1. Λ^s =crf-train(D_l, Λ^0)
2. Set $\Lambda_0^{(t)} = \Lambda^{(s)}$
3. **while** not converged **do**
4. $\{p\}$ =posterior_decode(D_u, Λ_{old})
5. $\{q\}$ =token to type$(\{p\})$
6. $\{q'\}$ =graph propagate$(\{q\})$
7. $D_u^{(1)}$ =viterbi_decode$(\{q'\}, \Lambda_{old})$
8. $\Delta_{n+1}^{(t)}$ = crf-train$(D_l \cup D_u^{(1)}, \Delta_n^{(t)})$
9. return t

Figure 2: Iterative semi-supervised training of CRF with label distributions from the graph. (Subramanya et al., 2010).

8 work on the corpus whereas graph propagation in line 6 works on the graph. So, the step in line 5 takes us from corpus to the graph, and the step in line 7 takes us back from the graph to the corpus.

2.4 Integration with BANNER

BANNER (Leaman et al., 2008) is a well-known open-source biomedical named entity recognizer that is widely used. Many studies have used BANNER for gene mention tagging (Li et al., 2015; Hakala et al., 2015; Leaman et al., 2015; Pyysalo et al., 2015; Li et al., 2015; Lee et al., 2014; Leaman et al., 2013) and many have cited it as a biomedical NER system with good performance (Dai et al., 2015; Krallinger et al., 2015; Luo et al., 2016; Gonzalez et al., 2016; Hebbring et al., 2015).

BANNER uses CRF as its machine learning core, and we used it as our base CRF in lines 1 and 8 in Figure 2. We also modified BANNER's source code in order to extract the posterior proba-

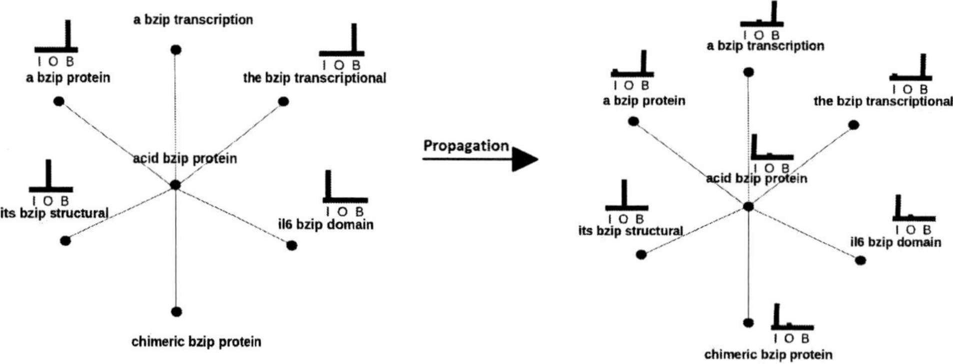

Figure 1: Neighbours of one vertex before and after Propagation. I,O,B stand for Inside-gene, Outside-gene, Beginning-gene.

Category	Method	Precision	Recall	F-Score
Baseline	BANNER	86.27	**85.57**	84.90
Our methods	Graph-based SSL	88.98	82.95	85.86
	Graph + postprocessing	**89.36**	82.95	**86.04**
More recent methods	BANNER-ChemDNER (2015)	88.02	86.08	87.04
	Gimli (2013)	90.22	84.32	87.17
Best performing methods in BioCreative II challenge	(Ando, 2007)	88.48	85.97	87.21
	(Kuo et al., 2007)	89.3	84.49	86.83
	(Huang et al., 2007)	84.93	88.28	86.57

Table 3: Graph-based SSL improves BANNER by increasing the precision.

bilities from the underlying MALLET CRF model (line 4). These probabilities were used in lines 5 through 7 in Figure 2.

Furthermore, the lemmas we used as features in our graph construction (see section 2.1) came from BANNER's lemmatizer.

BANNER also does some post-processing: it discards all the mentions that contain unmatched brackets. We ran our method with and without this post-processing step and verified its utility in our approach as well.

3 Experiments

We show improvements over BANNER on the dataset of BioCreative II Gene Mention Tagging Task. This data set contains 15,000 training sentences and 5,000 test sentences. Annotations are given by the starting character index and finishing character index of the gene in the sentence (space characters are ignored). Some sentences have alternative annotations presented in a separate file.

The upper part of Table 3 shows the results of BANNER; Graph-Based SSL without post-processing; and Graph-Based SSL with post-processing. The hyper-parameters of Graph-Based SSL were chosen by cross-validation over different train/test splits with different hyper-parameters tested for each split ($\alpha = 0.02$, $\mu = 10^{-6}$, $\nu = 10^{-4}$, and number of iterations = 2). Table 3 shows that the improvement we get in F-measure is due to better precision which is further boosted by dropping the candidates with unmatched parentheses (which is our only post-processing step).

The lower part of Table 3 puts our method in context. Although our method is competitive with these best performing methods in the literature, it has not outperformed any of them other than BANNER. Its precision however, is better than all other methods with the exception of Gimli. It would be interesting to integrate the graph-based approach to the ones with CRF as their machine

Type Of error	Number	Examples
FN in both BANNER and Graph	882	SST, R
FP in Graph	120	CD18, kinase, homeobox domain, transforming growth factor - beta, GRK6, POZ/Zn, HPR, E1B 19
FP in BANNER	337	oxidase, dose Ara C, mouse amino acid sequence, Ann Arbor, K1F, wild-type R. sphaeroides 2.4.1, SAS GLM, 1.6-kb cDNA, SH2, E3 ubiquitin, Xp22.3
FN in BANNER	158	LDL, bZIP protein, SL1, NF-kappaB, Ig-like domain, immunoglobulin genes, signal transducer and activator of transcription 1, bcr, ACTH, GFR, wnt
FN in Graph	197	SH3A, EGF, VA1, CBP, Decidual/trophoblast prolactin-related protein, CA 50

Table 4: Qualitative comparison by a human domain expert between BANNER and Graph Propagation. FN: false negatives. FP: false positives.

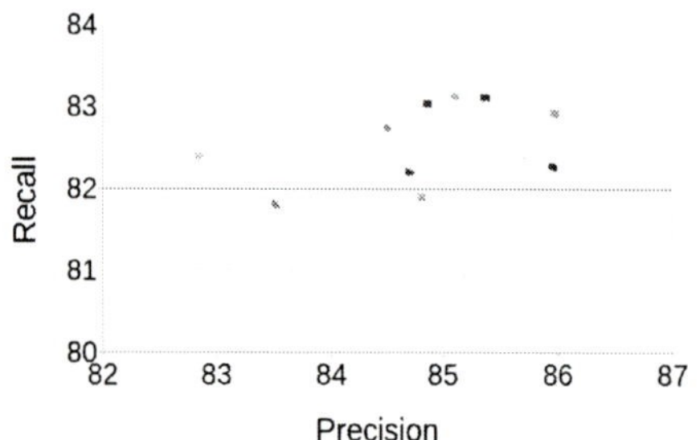

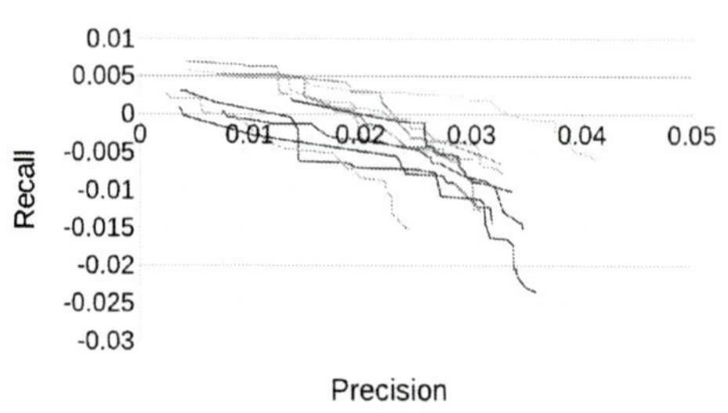

Figure 3: Precision and recall for different train/test splits and hyper-parameter choices. Each color represents a single train/test split. We include only the Pareto optimal points for each split.

Figure 4: The same points as in Figure 3 shown as the difference from the Banner scores for the same train/test split. The origin in this graph is the BANNER score. Each cluster of points in Figure 3 becomes a line in this graph.

learning core (BANNER-ChemdNER, Gimli, and the approach of (Kuo et al., 2007)) to further test the utility of the graph approach.

3.1 Qualitative analysis

To understand the differences between BANNER and the graph propagation results, a human domain expert compared the errors occurring in their respective outputs. Table 4 shows the number of these errors as well as some examples.

These examples illustrate two important observations. First, there are examples of categories more general than genes in both false positives and false negatives for both systems. For example *Kinase* is a functional group of proteins; *POZ/Zn, Ig-like domain*, and *SH2* are protein domains; and *E3 ubiquitin* and *NF-kappaB* are gene families. Anecdotal evidence suggests that this is due to presence of similar annotations in the training/test data set. For example the bZIP protein, a protein family, and Ig-like domain, a gene/protein functional domain were both annotated as genes. This calls for a better gene mention corpus annotated according to more recent gene annotation guidelines. Second, there are some hard to explain false positives in BANNER. Examples include *Ann Arbor*, a city in Michigan, *SAS GLM*, a type of statistical test, and *1.6-kb cDNA*, a molecular length. Our graph-based approach has eliminated these false positives.

3.2 Cross validation study

We conducted extensive cross-validation experiments using different train and test splits in order to explore the hyper-parameter values and to detect trends in the values that were optimal for this task. The results show that graph-propagation consistently improves results over BANNER.

Figures 3 and 4 were created by running graph-propagation over different train and test splits with different hyper-parameter values for each split. For each train/test split, we show only the Pareto optimal points (for each choice of hyper-parameters we include it in the graph only if the performance is not dominated by another choice in both recall and precision). Figure 3 illustrates two points: 1) the precision and recall for the different Pareto optimal points for each train/test split is very similar, and 2) overall the different train/test splits have similar precision and recall values. Figure 4 shows the performance for each train/test split shown as the difference from the BANNER scores for that split. It shows that the precision scores of graph-propagation is always better than the BANNER baseline, while recall is sometimes worse. The F-scores for all train/test splits and for all Pareto optimal points in each split is always better than the BANNER baseline.

We can collect useful statistics about which hyper-parameter values are the most useful in graph-propagation in this task from the extensive set of experiments described above: for different train/test splits and for each split with different hyper-parameter values. Figure 5 shows the number of times different hyper-parameter values have appeared in the set of Pareto optimal points over all the train/test splits.

The hyper-parameter α (see equation 3) controls the interpolation between the BANNER posterior probability over labels and the label distribution from the graph-propagation step. Higher α values would prefer BANNER over graph-propagation. Figure 5 shows that smaller α values are preferred, which implies that the label distribution produced through graph-propagation is found to be more useful than the label distribution produced by BANNER. We also investigated the two extreme cases of $\alpha = 0$ (only graph) and $\alpha = 1.0$ (only BANNER followed by an extra Viterbi decoding step), and observed that both of these options were worse than the BANNER baseline.

In equation (1) higher ν values keep the label distribution at each vertex of the graph closer to the uniform distribution. Higher μ values would allow adjacent vertices to have a greater influence on the label distribution at the vertex. Figure 5 shows that, in our experiments, graph-propagation is sensitive to the values of μ. Lower μ values appear in Pareto optimal points more often. On the other hand, Figure 5 shows that graph-propagation is not as sensitive to different values of ν as long as it is not too high (10^{-1}). This might be due to our setting, where about 73% of vertices are labelled.

We looked for strong correlations between ν values, μ values, and number of iterations in graph propagation and found none.

Finally, for different iteration numbers of graph-propagation, we collected the frequency with which each number appeared in the Pareto optimal results. One iteration of graph-propagation produced 68 Pareto optimal points, two iterations produced 198 points, and three iterations produced 120 points in our experiments. This shows that having more than one iteration of graph-propagation can improve the results.

Our algorithm (Figure 2) has two levels of iterations. One outer iteration (the while loop) and one inner iteration in graph propagation. The numbers mentioned above refer to this inner iteration. All our results reported are for one outer iteration only. Our experiments in this paper were in a transductive setting where the graph was constructed over the test and training data. For this reason we did not experiment extensively with more than one outer iteration. In future work, we plan to experiment with increasing the amount of unlabeled data, and in this setting explore increasing the number of outer iterations.

3.3 A note on scalability

The most time consuming step in our approach was graph construction, where the bootleneck is to compute the edge weights between any possible vertex pairs. We experimented with a naive algorithm, where for every vertex pair the values of feature vectors for shared features were considered, and the cosine similarity was computed. We also implemented a variation on it, where the similarities between all pairs sharing a specific feature instance were computed, and the contributions of individual feature instances were summed to give the final similarity between any given pair. The first algorithm was too slow as expected due to its $O(|V|^2)$ time complexity; the second one was too slow due to high frequency features. This is an important issue since the graph needs to be constructed for our approach to work on a new dataset.

Apart from the graph construction, the graph based approach is as scalable as CRF if a labeled train set is available for the new domain, as the CRF only needs to be trained on the new labelled set. If we wish to adapt the method in a domain where there is no labelled data in the target domain, there is no need for any training.

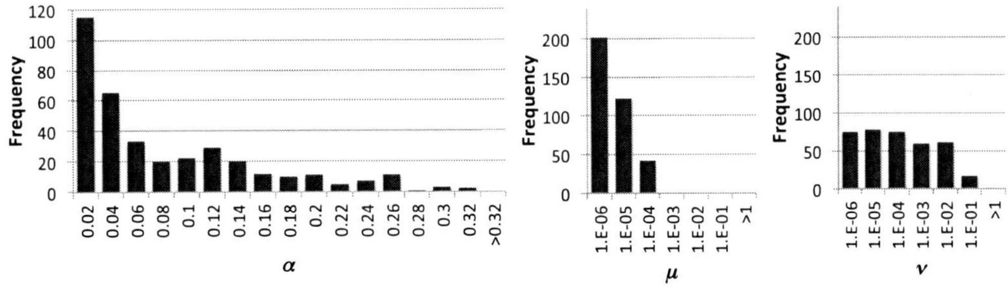

Figure 5: These graphs show the number of times specific hyper-parameter values α, μ and ν appeared in Pareto optimal points over all train/test splits.

4 Conclusion and future directions

Our results show that propagating labels from 3-grams present in training set to 3-grams only appearing in the test set can significantly improve BANNER, a well-known frequently used biomedical named entity recognition system for the gene mention tagging task. Our cross-validation study shows the robustness of this improvement. We also presented qualitative comparison by a human domain expert. Our ideas for future work are categorized into three groups:

1. **Adding more unlabelled data**: The only unlabelled data we included in the graph were the test data. Since the success of semi-supervised learning methods is usually due to huge amount of unlabelled data, we plan to use many more PubMed abstracts to construct the graph. This however will be challenging because the graph construction can be time consuming as it was in our case due to high frequency features.

2. **Constructing a better graph**: Contextual features we used to construct our graph are only one of the feature sets that have been shown useful in gene mention tagging task. Other feature sets include orthographic features, contextual features learnt from neural networks, features from parse trees. These features may also prove useful in constructing a graph that represents the similarity between gene mentions. Also, we can preprocess the raw sentences to collapse some collocations into one word so that the middle word in the 3-gram vertices are more meaningful.

3. **Improving the latest approach**: Although BANNER is one of the most frequently used biomedical named entity recognition system, it is not one with the best performance ever. Previous approaches have improved BANNER in a variety of ways, including semi-supervised learning. In particular, Munkhdalia et al. have achieved an F-measure of 87.04 by including word representations learnt from massive unlabelled data as features (Munkhdalai et al., 2015) . We plan to test our approach on their freely available system.

5 Acknowledgements

The authors thank the funding organizations, Genome Canada, British Columbia Cancer Foundation, and Genome British Columbia for their partial support. The research was also partially supported by the Natural Sciences and Engineering Research Council of Canada (NSERC RGPIN 262313 and RGPAS 446348) to the fourth author.

References

Andrei Alexandrescu and Katrin Kirchhoff. 2009. Graph-based learning for statistical machine translation. In *NAACL 2009*.

Rie Kubota Ando. 2007. BioCreative II gene mention tagging system at IBM Watson. In *Proceedings of the Second BioCreative Challenge Evaluation Workshop*, volume 23, pages 101–103. Centro Nacional de Investigaciones Oncologicas (CNIO) Madrid, Spain.

Peter F Brown, Peter V Desouza, Robert L Mercer, Vincent J Della Pietra, and Jenifer C Lai. 1992. Class-based n-gram models of natural language. *Computational linguistics*, 18(4):467–479.

David Campos, Sérgio Matos, and José Luís Oliveira. 2013. Gimli: open source and high-performance biomedical name recognition. *BMC bioinformatics*, 14(1):54.

Hong-Jie Dai, Po-Ting Lai, Yung-Chun Chang, and Richard Tzong-Han Tsai. 2015. Enhancing of chemical compound and drug name recognition using representative tag scheme and fine-grained tokenization. *Journal of cheminformatics*, 7(S1):1–10.

Dipanjan Das and Slav Petrov. 2011. Unsupervised part-of-speech tagging with bilingual graph-based projections. In *Proceedings of the 49th Annual Meeting of the Association for Computational Linguistics: Human Language Technologies-Volume 1*, pages 600–609. Association for Computational Linguistics.

Graciela H Gonzalez, Tasnia Tahsin, Britton C Goodale, Anna C Greene, and Casey S Greene. 2016. Recent advances and emerging applications in text and data mining for biomedical discovery. *Briefings in bioinformatics*, 17(1):33–42.

Kai Hakala, Sofie Van Landeghem, Tapio Salakoski, Yves Van de Peer, and Filip Ginter. 2015. Application of the evex resource to event extraction and network construction: Shared task entry and result analysis. *BMC bioinformatics*, 16(Suppl 16):S3.

Scott J Hebbring, Majid Rastegar-Mojarad, Zhan Ye, John Mayer, Crystal Jacobson, and Simon Lin. 2015. Application of clinical text data for phenome-wide association studies (PheWASs). *Bioinformatics*, 31(12):1981–1987.

Han-Shen Huang, Yu-Shi Lin, Kuan-Ting Lin, Cheng-Ju Kuo, Yu-Ming Chang, Bo-Hou Yang, I-Fang Chung, and Chun-Nan Hsu. 2007. High-recall gene mention recognition by unification of multiple backward parsing models. In *Proceedings of the* second BioCreative challenge evaluation workshop*, volume 23, pages 109–111. Centro Nacional de Investigaciones Oncologicas (CNIO) Madrid, Spain.

Feng Jiao, Shaojun Wang, Chi-Hoon Lee, Russell Greiner, and Dale Schuurmans. 2006. Semi-supervised conditional random fields for improved sequence segmentation and labeling. In *Proceedings of the 21st International Conference on Computational Linguistics and the 44th annual meeting of the Association for Computational Linguistics*, pages 209–216. Association for Computational Linguistics.

Martin Krallinger, Obdulia Rabal, Florian Leitner, Miguel Vazquez, David Salgado, Zhiyong Lu, Robert Leaman, Yanan Lu, Donghong Ji, Daniel M Lowe, et al. 2015. The chemdner corpus of chemicals and drugs and its annotation principles. *Journal of cheminformatics*, 7(S1):1–17.

Cheng-Ju Kuo, Yu-Ming Chang, Han-Shen Huang, Kuan-Ting Lin, Bo-Hou Yang, Yu-Shi Lin, Chun-Nan Hsu, and I-Fang Chung. 2007. Rich feature set, unification of bidirectional parsing and dictionary filtering for high f-score gene mention tagging. In *Proceedings of the second BioCreative challenge evaluation workshop*, volume 23, pages 105–107. Centro Nacional de Investigaciones Oncologicas (CNIO) Madrid, Spain.

John Lafferty, Andrew McCallum, and Fernando Pereira. 2001. Conditional random fields: Probabilistic models for segmenting and labeling sequence data. In *ICML*.

Robert Leaman, Graciela Gonzalez, et al. 2008. Banner: an executable survey of advances in biomedical named entity recognition. In *Pacific Symposium on Biocomputing*, volume 13, pages 652–663. Citeseer.

Robert Leaman, Rezarta Islamaj Doğan, and Zhiyong Lu. 2013. Dnorm: disease name normalization with pairwise learning to rank. *Bioinformatics*, 29(22):2909–2917.

Robert Leaman, Chih-Hsuan Wei, and Zhiyong Lu. 2015. tmchem: a high performance approach for chemical named entity recognition and normalization. *J. Cheminformatics*, 7(S-1):S3.

Hee-Jin Lee, Tien Cuong Dang, Hyunju Lee, and Jong C Park. 2014. Oncosearch: cancer gene search engine with literature evidence. *Nucleic acids research*, page gku368.

Gang Li, Karen E Ross, Cecilia N Arighi, Yifan Peng, Cathy H Wu, and K Vijay-Shanker. 2015. mirtex: A text mining system for mirna-gene relation extraction. *PLoS Comput Biol*, 11(9):e1004391.

Shujie Liu, Chi-Ho Li, Mu Li, and Ming Zhou. 2012. Learning translation consensus with structured label propagation. In *ACL 2012*.

Yuan Luo, Özlem Uzuner, and Peter Szolovits. 2016. Bridging semantics and syntax with graph algorithmsstate-of-the-art of extracting biomedical relations. *Briefings in bioinformatics*, page bbw001.

Andrew Kachites McCallum. 2002. Mallet: A machine learning for language toolkit. http://mallet.cs.umass.edu.

Tomas Mikolov, Kai Chen, Greg Corrado, and Jeffrey Dean. 2013. Efficient estimation of word representations in vector space. *arXiv preprint arXiv:1301.3781*.

Tsendsuren Munkhdalai, Meijing Li, Khuyagbaatar Batsuren, Hyeon Park, Nak Choi, and Keun Ho Ryu. 2015. Incorporating domain knowledge in chemical and biomedical named entity recognition with word representations. *J. Cheminformatics*, 7(S-1):S9.

Sampo Pyysalo, Tomoko Ohta, Rafal Rak, Andrew Rowley, Hong-Woo Chun, Sung-Jae Jung, Sung-Pil Choi, Jun'ichi Tsujii, and Sophia Ananiadou. 2015. Overview of the cancer genetics and pathway curation tasks of bionlp shared task 2013. *BMC bioinformatics*, 16(Suppl 10):S2.

Avneesh Saluja, Hany Hassan, Kristina Toutanova, and Chris Quirk. 2014. Graph-based semi-supervised learning of translation models from monolingual data. In *ACL 2014*.

Fei Sha and Fernando Pereira. 2003. Shallow parsing with conditional random fields. In *NAACL*.

Amarnag Subramanya and Jeff A Bilmes. 2009. Entropic graph regularization in non-parametric semi-supervised classification. In *Advances in Neural Information Processing Systems*, pages 1803–1811.

Amarnag Subramanya, Slav Petrov, and Fernando Pereira. 2010. Efficient graph-based semi-supervised learning of structured tagging models. In *Proceedings of the 2010 Conference on Empirical Methods in Natural Language Processing*, pages 167–176. Association for Computational Linguistics.

Partha Pratim Talukdar, Joseph Reisinger, Marius Paşca, Deepak Ravichandran, Rahul Bhagat, and Fernando Pereira. 2008. Weakly-supervised acquisition of labeled class instances using graph random walks. In *EMNLP 2008*.

Akihiro Tamura, Taro Watanabe, and Eiichiro Sumita. 2012. Bilingual lexicon extraction from comparable corpora using label propagation. In *EMNLP-CoNLL 2012*.

Xiaojin Zhu, Zoubin Ghahramani, John Lafferty, et al. 2003. Semi-supervised learning using gaussian fields and harmonic functions. In *ICML*, volume 3, pages 912–919.

Feature Derivation for Exploitation of Distant Annotation via Pattern Induction against Dependency Parses

Dayne Freitag and John Niekrasz
SRI International
9988 Hibert Street, Suite 203
San Diego, CA 92130
{freitag,niekrasz}@ai.sri.com

Abstract

We consider the use of distant supervision for biological information extraction, and introduce two understudied corpora of this form, the Biological Expression Language (BEL) Large Corpus and the Pathway Logic (PL) Datum Corpus. Each resource eschews annotation at the sentence constituent level, and the PL corpus requires synthesis of information across multiple sentences to construct composite knowledge frames. Decomposing this problem into feature induction for slot-level attributes, followed by event assembly over this space of features, we introduce a novel, general-purpose pattern induction procedure, evaluating it against these two corpora, demonstrating its ability to induce effective detection against dependency parses.

1 Introduction

Biological event and relation extraction have been the focus of considerable study in recent years, resulting in the availability of annotated corpora (Kim et al., 2003; Pyysalo et al., 2007; Kim et al., 2008; Thompson et al., 2009). In the interest of replicability and progress on critical challenges, such resources typically decompose the hard problem of factual understanding into several simpler problems, such as entity recognition, binary relation detection, and co-reference resolution.

This methodology is subject to several criticisms. The reliance on thorough annotation imposes overheads that prevent rapid progress. The targeting of a fixed set of simplified, typically binary relations does justice neither to the complexity of information expressed in a typical sentence, nor to the biological processes under discussion.

And the methodology places a emphasis on pieces of information amenable to expression in individual sentences, leaving untouched information that can be assembled only through traversal of paragraphs or complete documents.

Some of these limitations can be mitigated through *distant supervision*, a technique deriving noisy annotation through the heuristic alignment of structured knowledge resources to texts (Craven et al., 1999). The biological domain affords a number of high-quality knowledge resources with good coverage, making possible strongly competitive distantly supervised solutions (Poon et al., 2015). However, the distance between resource and text is often not great in such work, which focuses on relations for which entity co-occurrence in a sentence is strong evidence that the sentence expresses the target relation.

In this paper we attempt to exploit two knowledge resources, neither of which has received much attention from the BioNLP community, that increase this distance in interesting and distinct ways. The Biological Expression Language (BEL) is a knowledge interchange format intended to encode qualitative causal and correlative relations that supports nested knowledge frames. One product of the OpenBEL initiative[1] is the "Large BEL Corpus," which explicitly pairs a large number of literature excerpts with the BEL assertions that each supports. The relation between sentence and BEL statement is many-to-many, with no provisions for aligning specific statement components with specific sentence constituents.

The Pathway Logic (PL) project pursues high-fidelity signaling pathway models centering on Ras (Eker et al., 2004). Part of the effort involves a manual curation of experimental results, which has resulted in approximately 40K records, each

---

[1] http://www.openbel.org/

Proceedings of the 15th Workshop on Biomedical Natural Language Processing, pages 36–45,
Berlin, Germany, August 12, 2016. ©2016 Association for Computational Linguistics

containing a detailed formal representation of an experiment and its outcomes. Such records, called *datums*, retain pointers to the papers and figures from which they were derived. In general, assembling the rich information contained in a datum requires traversing multiple sentences, both in figure captions and paper bodies.

We view the problem of extracting composite knowledge frames based on these attenuated supervisory signals as having two parts. First, we seek to generate a set of features highly indicative of various aspects of the target frame (its type, various attributes, etc.). Second, we view the problem of assembling frames from the resulting enriched feature space as one of structured classification. Recent work on structured classification lends confidence that such empirical assembly models are possible in principle (Daum III et al., 2009) and applicable to discourse-level event extraction (Reschke et al., 2014).

In this paper we address the first problem, the derivation of features for downstream extraction. We treat this problem as one of sentence classification via pattern (or rule) set induction against dependency parses. Compared with related work involving rules in the BioNLP literature (Bunescu et al., 2005; Bui et al., 2013; Huang et al., 2004; Liu et al., 2011; Hunter et al., 2008; Valenzuela-Escarcega et al., 2015; Peng et al., 2014), our approach exhibits some interesting features, particularly the eschewal of domain heuristics and the nonreliance on constituent-level annotations. Our work can be viewed as complementary to manual rule writing, and we present evidence that our learned patterns outperform rules written by hand.

Our contributions in this paper are twofold:

- We present and evaluate a novel, general-purpose approach to the induction of classification and extraction patterns from dependency parses.

- We evaluate this approach against two BioNLP corpora that have received little attention in the literature. Each corpus presents an extraction problem of greater complexity than can be addressed by current methods, providing avenues toward models of greater scope and biological fidelity.

The remainder of the paper is organized as follows. In the next section, we describe these two data sources and the problems they pose. In Section 3 we present our approach to pattern induction. Then, we describe and discuss our experiments in Section 4. Finally, we compare our approach in Section 5 to related work.

2 Data

In this section we describe the BEL Large Corpus and the PL Datum Corpus, against which we evaluate our approach.

2.1 OpenBEL

The Biological Expression Language (BEL) is designed to capture rich qualitative biological relationships in context. For example, the BEL statement $[p(\text{HGNC:CCND1}) \implies kin(p(\text{HGNC:CDK4}))]$ expresses that "increased abundance of the protein HGNC:CCND1 directly increases the kinase activity of the abundance of the protein HGNC:CDK4." Here, the $\implies$ symbol expresses a causal *directly increases* relationship between two BEL functions. Relationship types include causal (e.g., increases) and correlative (e.g., association) relationships. BEL functions are defined for abundances (e.g., protein or rna abundances), modifications (e.g., phosphorylation), activities (e.g., kinase or catalytic activity), processes (e.g., angiogenesis), and transformations (e.g., translocation and cell secretion). Depending on their definition, functions can be nested, accepting entity or other functions as arguments.

The OpenBEL initiative distributes the 'large corpus."[2], a collection of ∼80k statements, ∼74k of which are associated with natural language evidence passages and a PubMed article ID. The relationship between BEL statements and supporting sentences is many-to-many — some sentences are used to support multiple BEL statements, and whole paragraphs can be mustered in support of a given statement. After sentence segmentation and minor cleanup (e.g., removing inline comments from curators), we obtained a total of ∼40k unique supporting evidence sentences. In terms of biological content, the corpus contains independent observations (in human, mouse, and rat) not selected to represent any specific biological process. Given its size, this implies a lack of comprehensive coverage for any specific biological domain.

Figure 1: An example of a Pathway Logic datum.

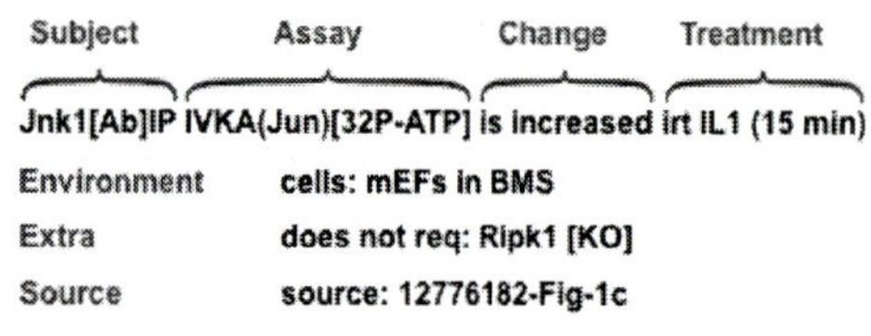

2.2 Pathway Logic

Pathway Logic (PL) is an approach to modeling biological entities and processes based on rewriting logic.[3] PL models can include specific facts and general principles relating entities and processes. PL is currently being used for the analysis of signal transduction and metabolic networks, including the STM model, a network of protein interactions and modifications used by the cell to transmit signals from its environment to the nucleus. Using STM, PL is able to predict and explain the effect of interventions (removal/inhibition/mutation of proteins) on downstream events.

In PL, reactions are curated by expert biologists from published experimental evidence (normally a Pubmed article). This experimental evidence is captured in formal expressions called "datums," encoded in a structured syntax over a controlled vocabulary, each representing one assay. An example datum is shown in Figure 1. Each datum captures, among other things, the protein(s) that are observed (Subject), the assay (Assay), the stimulus (Treatment), a result (Change), and the cells and culture conditions that were used in the assay (Environment).

Importantly, each datum also includes a reference to the figure in its source article containing the experimental result. These references allow us to link each datum to a small set of natural language sentences, namely those in the caption of the referenced figure or citing it in the paper body. This alignment, along with any mentions of entities listed in key datum roles, provides our supervisory signal. Note that the datum corpus consists primarily of PDF documents, necessitating a somewhat noisy conversion and alignment. We use a version of the PL knowledge base that contains ~39k unique datums sourced from figures in ~2,000 Pubmed articles.

In Section 4 we benchmark our pattern induc-

tion procedure against rule sets written by hand using the ODIN framework (Valenzuela-Escarcega et al., 2015). We created these rules over a period of several weeks while implementing a datum extraction system evaluated under DARPA Big Mechanism. This activity took place before the work described here had begun. Rule authors had full access to the datum corpus and possessed tools that exploited the same sentence-to-datum alignment heuristics used in this paper's experiments. Thus, although we cannot claim to have produced optimal manual rule sets, these sets can be viewed as characteristic of what can be achieved with reasonable effort. Of course, the two rule sources are not mutually exclusive. We are currently extending the datum extraction system to use *both* manual and automatically induced rules, and expect to see improvements in both precision and recall.

3 Approach

3.1 The Setting

We frame our approach as a problem of Boolean classification over dependency parses (more generally, graphs with multiply labeled nodes and edges), where the positive class typically reflects that a sentence communicates some information we seek to detect. For conciseness, in the remainder of the paper we will refer simply to "parses", leaving the "dependency" modifier implied.

Formally, we are given *data* as a set of examples from (X, Y), with $Y = \{0, 1\}$ reflecting class membership. Each member of X is a parse taking the form (V_d, E_d), with V_d a set of vertices and E_d a set of directed edges (v_i, v_j). In addition, we are given two feature spaces, $F_V : V_d \mapsto \{0, 1\}$ and $F_E : E_d \mapsto \{0, 1\}$, that range over vertices and edges, respectively, which represent things such as a vertex word or the dependency label of an edge.

We seek to classify such parses using *patterns*, which take the form $(r, V_p, E_p, T_V, T_E, D_E)$. As with parses, the components V_p and E_p define a tree, which in this case is rooted in the distinguished vertex r. D_E denotes the *direction* of an edge, taking the form $D_E : E_p \mapsto \{\uparrow, \downarrow\}$. As this implies, we allow pattern edges to traverse up or down a parse.

T_V and T_E represent the *types* of vertices and edges, respectively. The vertex type, $T_V : V_p \mapsto \{\lambda\} \cup F_V$, is intended to constrain compatibility

---

(call it C_V) with parse vertices:

$$C_V(v_d, v_p, T_V) = \begin{cases} 1 & : \quad T_V(v_p) = \lambda \; \vee \\ & \quad \;\; (T_V(v_p))(v_d) = 1 \\ 0 & : \quad \text{otherwise} \end{cases}$$

In other words, a *null* pattern vertex is compatible with any parse vertex, while a *feature* pattern vertex is compatible only with parse vertices for which the feature tests true, i.e., that *have* that feature. Thus, a feature pattern vertex with type f_{the} will match only parse vertices that correspond to the word "the", f_{NN} only to nouns, etc.

Edge types, $T_E : E_p \mapsto \{\lambda, *\} \cup F_E$, are analogous to vertex types. A pattern edge with type f_{amod} will match only parse edges having the "amod" dependency label. Edges with type "*", which we call *Kleene* edges, are explained in the next section.

3.2 Matching

We say that a pattern *matches* a parse if we can find a one-to-one alignment between pattern vertices and a subset of parse vertices proceeding recursively from the root node. This matching procedure is most easily described by means of a hypothetical Boolean function MATCH that returns true if a specified pattern vertex matches a specified parse vertex.

Algorithm 1 Procedure for matching patterns to parses.

```
 1: function MATCH(X, v_d, P, v_p)
 2:     (V_d, E_d) ← X
 3:     (r, V_p, E_p, T_V, T_E, D_E) ← P
 4:     if not C_V(v_d, v_p, T_V) then return false
 5:     for e_p in E_p s.t. e_p = (v_p, v'_p) do
 6:         Found = false
 7:         for e_d, v'_d in CandEdges(e_p, E_d, v_d) do
 8:             if T_E(e_p) = * then
 9:                 if KleeneMatch(X, e_d, v'_d, P, v'_p) then
10:                     Found = true
11:             else if T_E(e_p) = λ ∨ T_E(e_p)(e_d) = 1 then
12:                 if Match(X, v'_d, P, v'_p) then
13:                     Found = true
14:             if not Found then return false
15:     return true
```

MATCH, shown in Algorithm 1, can be broken into three parts: a check for vertex compatibility (Line 4); a check for edge (or edge-to-path) compatibility (Line 5–13); and a recursive call to align unmatched pattern vertices (Lines 9 and 12). For brevity, we assume the existence of two helper functions. CANDEDGES (Line 7) merely selects

and returns all edges (with destination vertex) attached to v_d that are compatible with the directional restriction of e_p. KLEENEMATCH (Line 9) enumerates all nodes on any path in the direction selected by e_p, internally calling MATCH on each until a match is found.

3.3 Induction

Linguistic variation usually ensures that no single pattern can adequately account for the ways in which target information is expressed. Therefore, our objective is to learn a set of patterns covering the forms observed in the training data. In pursuit of this objective we follow a top-down set covering procedure. At each step in this procedure, a single pattern is learned from the training data, all positive parses matching the new pattern are removed from the training set, and the process repeats. If no positive parses remain, or if the algorithm fails to induce a pattern, the process terminates.

Algorithm 2 Pattern induction procedure.

```
 1: function INDUCE(T, V, α)
 2:     P ← {}
 3:     p ← null vertex
 4:     while p' ← Specialize(T, p, α) do
 5:         p ← p'
 6:         s ← Score(p, V)
 7:         P ← P ∪ {(p, s)}
 8:         T ← T − {(x, y)|y = 1 ∧ Match(p, x)}
 9:     return (p, s) ∈ P with max s
10:
11: function SPECIALIZE(T, p, α)
12:     E ← {}
13:     for (x, y) ∈ T s.t. y = 1 do
14:         M ← Matches(p, x)
15:         for o_1 ⋯ o_k ∈ Extensions(M, x, α) do
16:             E ← E ∪ {o_1 ⋯ o_k ↦ (0, 0)}
17:     for (x, y) ∈ T do
18:         M ← Matches(p, x)
19:         for o_1 ⋯ o_k ∈ Extensions(M, x, α) do
20:             if o_1 ⋯ o_k ∈ E then
21:                 E[o_1 ⋯ o_k][y] += 1
22:     o_1 ⋯ o_k ← BestExtention(E)
23:     if no best extension found then
24:         return false
25:     return p extended with o_1
```

The procedure for inducing a single pattern is presented as Algorithm 2. The top-level function INDUCE (Line 1) subjects an initial pattern containing a single null vertex (i.e., a pattern matching any non-empty parse—Line 3) to a series of specializations selected by the function SPECIALIZE (Line 4), and scores them against hold-out training data V (Line 6). The score of a rule is its precision, or $(p + m)/(p + n + 2m)$, where p is the number of matching positive parses, n the number of negative, and $m > 0$ a smoothing parameter.

SPECIALIZE (Line 11) takes the training set T, the pattern p in its current form, and an integer "look-ahead" parameter α. The procedure involves two passes over the training data, one collecting a set of candidate extensions to p (Lines 13–16), the other accumulating statistics for those extensions (Lines 17–21), which, as implied by Lines 16 and 21, are simple counts of the number of positive and negative parses matching each extension. These two steps assume the existence of two procedures: MATCHES (a straightforward variant of MATCH, Lines 14 and 18) returns all alignments between p and x; and EXTENSIONS (Lines 15 and 19), the behavior of which will be described in the next section.

Extensions are sequences of specialization operations $o_1 \cdots o_k$, $1 \leq k \leq \alpha$. Once statistics have been collected, the extension that best favors positive examples at the expense of negative ones is selected (Line 22). We use the "FOIL gain" in making this determination (Quinlan, 1990):

$$y' \cdot (\log(\frac{y'}{y' + n'}) - \log(\frac{y}{y + n})) \qquad (1)$$

where y and n (respectively, y' and n') are the number of positive and negative parses matched by p (respectively, p' formed by extending p). [4]. Importantly, once the best extension is identified, only the first specialization operation in the sequence is applied (Line 25) In this way, some of the greediness of the extension search is mitigated.

3.4 Specialization Operations

When considering specializations of a pattern, we have as reference its alignment to some parse, each vertex to some parse vertex, and each non-Kleene edge to some parse edge. We generate extensions by iterating over this alignment and collecting all possible specialization operations supported by it. Let us use v_p (pattern) and v_d (data) to represent two vertices in an alignment (similarly e_p and e_d for edges). We consider the following specialization operations (the o_i in Algorithm 2):

- **Specialize a null vertex**. If v_p is null, change it to require a feature of v_d. Because parse

nodes may have multiple features, this operation in general generates multiple specializations.

- **Specialize a null edge**. Analogous to the previous item, but defined on edges. We currently only consider features based on dependency labels, but others are possible in principle.

- **Add a null edge**. If v_d has edges to unaligned parse vertices, add a null edge in the appropriate direction from v_p to a new null vertex.

- **Add a Kleene edge**. Except for the type, the conditions and effects of this operation are identical to the previous item.

Because these operations are typically considered in the context of a multi-step extension search, we can align any newly introduced vertices and edges to the parse, and consider further specializations either to the current element or to any newly introduced elements, up to the limit specified by α. Being a hyperparameter that inversely affects accuracy and running time, we set α manually to maximize accuracy given practical constraints on compute resources ($\alpha = 3$ for experiments in this paper).

Feature	Description
Word	Word associated with a vertex
POS	Part-of-speech tag
NER	The named entity type, if any
Cluster	Cluster of a word derived through distributional clustering

Table 1: Vertex feature types.

The specialization operations are defined in part by the feature spaces available, particularly by the vertex (word) feature space F_V. Conceptually, these features belong to an extensible set of types at the lexical level. The types used in our experiments are presented in Table 1.

Figure 2 shows an example pattern aligned to a matching sentence. In addition to the words in the sentence, the diagram also shows the associated POS and NER tags (clusters are not shown). An induced pattern matching the sentence is shown below the horizontal line. The pattern has three Kleene edges and one null edge. It has five nodes: two matching the NER=PRO (protein) feature, one matching the word "phosphorylation",

---

Figure 2: An example of an automatically induced pattern matching a sentence.

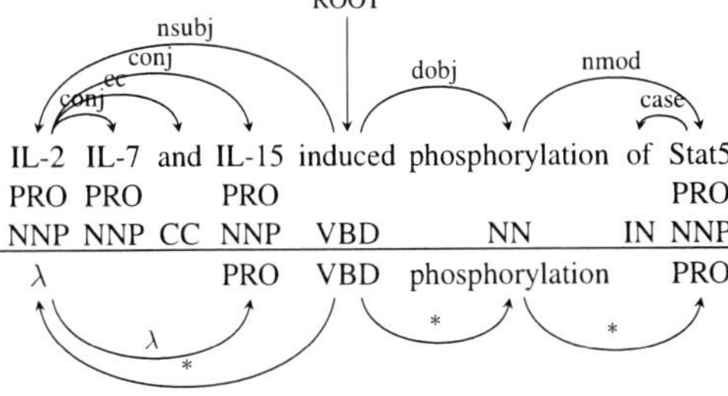

another matching POS=VBD, and another matching any word.

3.5 Pattern Application

The output of the induction process is a set of patterns scored for precision against hold-out data. Our primary interest in these patterns is as feature detectors used by some downstream process that assembles composite events, but the set of patterns can be used and evaluated as a stand-alone classification model. In this mode, the scores attached to the rules enable us to estimate the precision of matches against novel parses, and therefore to control precision and recall.

Multiple patterns may and often do match an individual parse. In such cases we estimate the precision of the ensemble match M as:

$$1 - \prod_{p \in M} (1 - s_p) \qquad (2)$$

where p is a pattern and s_p is its estimated precision. This estimate essentially treats the individual estimates as mutually independent.

4 Experiments

We evaluated our pattern induction procedure on the BEL and PL corpora for its effectiveness in detecting sentences expressing information needed for composite knowledge frames.

4.1 BEL evaluation

We converted BEL statements into a set of overlapping binary distinctions, called *fragments*, each a possible abstraction of the statement. Our objective is to convert each BEL extraction into a large set of redundant simpler problems, from which the original statement might be reconstituted. For example, the BEL statement

"$p(\mathrm{HGNC:CCND1}) \implies kin(p(\mathrm{HGNC:CDK4}))$" yields (among other fragments) "kin" (describes kinase activity), "$kin(p)$" (kinase activity of a protein), and "$\implies kin(p)$" (kinase activity of a protein resulting from unspecified cause). Generating BEL fragments from fully specified BEL statements proceeds by first abstracting away any entity or numeric function arguments. Then, fragments are generated for every subtree in the abstract syntax tree of the statement. (We distinguish between functions occurring in the statement's subject and object position, in other words treating "subject" and "object" as a named element of the syntax tree.) Additionally, a fragment is generated for the relationship type and all functions occurring anywhere in the statement. Table 2 lists examples of BEL fragments with the number of positive training set sentences associated with each of them in the corpus.

BEL fragment	no. pos. sentences
$p \implies ?$	8728
r	6355
$p \implies bp$	1737
$c(p, p, p)$	127
$trans(p) \implies r$	22

Table 2: Example BEL statement fragments ($p = proteinAbundance$; $r = rnaAbundance$; $bp = biologicalProcess$; $trans = translocation$; $c = complexAbundance$).

From the ~74,000 BEL statements associated with validated evidence, this process generates ~3k unique fragments, of which we retained those associated with at least 20 sentences (resulting in ~400 unique fragments). For each unique fragment, all sentences associated with the fragment were labeled as positive, and all other sentences were labeled as negative. Patterns were then induced and evaluated on a 60/20/20% train/validation/test split defined on each of the sets. Note that the corpus sometimes associates BEL statements with *multiple* sentences, e.g., and entire paragraph, which means our labeling procedure sometimes treats proximal sentences as positive, even though they may not directly instantiate a target statement. The inductive procedure, which assumes that the target class is a disjunction of mutually exclusive cases, handles such "noise" well, essentially failing to derive patterns from (i.e., ignoring) the superfluous "positive" examples.

Figure 3: Scatterplot showing results of the ~400 sentence classification experiments run on the BEL corpus.

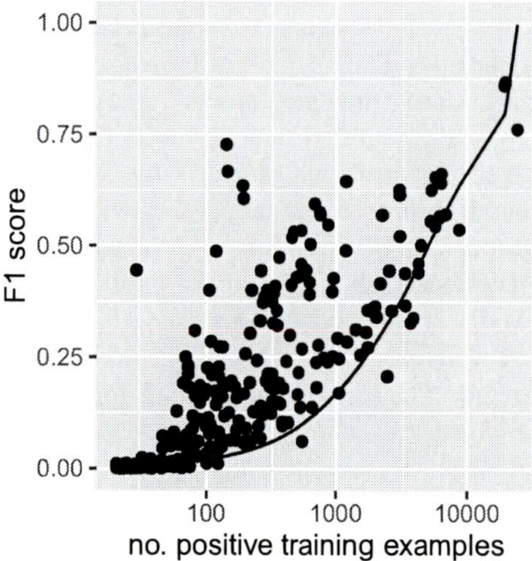

PL datum fragment	Learned	Written
phos & subject	0.54	0.37
ubiq & subject	0.53	0.41
GTP-assoc & subject	0.60	0.20
phos & treatment	0.48	0.35
ubiq & treatment	0.29	0.17
GTP-assoc & treatment	0.32	0.05

Table 3: F1 performance in the extraction of datum fields by learned and hand-written rules.

We evaluate the induced rule ensembles by calculating an idealized F1 score, identifying a threshold for classification based upon the validation estimated precision of the ensemble match (Eq. 2) that maximizes F1 on the test set. We acknowledge that finding this threshold using test data produces an overly-optimistic result, but doing so provides us with an informative upper bound. In practice, the threshld would be tuned using an alternative validation set to prevent overfitting. The mean cardinality of the induced rule ensembles was 7, with a total of 2991 rules (2402 *unique*) induced across the ~400 fragments.

Results are plotted in Figure 3, with one dot per classification experiment, each being an application of the rule induction approach against a single unique BEL statement fragment. For each experiment, the F1 classification result is plotted against the size of the positive training set. The plot also contains a line showing the F1 results of a random chance baseline. To calculate the baseline, we assume a classifier that randomly labels sentences as positive or negative with the same marginal probabilities as observed in the training set.

4.2 Pathway Logic evaluation

We next conducted a set of experiments targeting classification of sentences associated with various PL datum fragments. In this case, we employed named entity resolution for proteins, label-

ing parse nodes as to whether they refer to the value of associated Subject or Treatment fields, and restricting rules to necessarily include nodes that match the protein mention. Results are shown in Table 3. These experiments targeted the two datum fields (*subject* and *treatment*) that correspond to extractible entities, and focus on the three important assay types (*phos*, *ubiq*, and *GTP-assoc*) for which we had written ODIN rules while implementing our heuristic datum extractor.

4.3 Discussion

The results presented in Figure 3 clearly establish the effectiveness of the induced rule ensembles at detecting the information conveyed in the BEL statements. Not surprisingly, this effectiveness increases with increased training data, though there is considerable variation. We attribute this variation to the indiscriminate way in which fragments were generated. Presumably, some kinds of information correspond more strongly to detectible linguistic regularities than others. A brief investigation informally confirmed our intuitions. For example, a common but difficult fragment is "$complex(p, p) \implies ?$" (about 0.17 F1), or, roughly, "a complex between two proteins increases some effect." A key insight is that " $\implies$?" corresponds to a large range of effects, and these effects govern the form a sentence takes. In contrast, the fact that a complex is the agent is often expressed subtly, often requiring inference over multiple sentences. Note that the set of classes we target deliberately *overdetermine* the typical BEL statement, making it possible in principle to reassemble many statements.

The results presented in Table 3 promise immediate practical value. As described in Section 2.2, the hand-written rules were used in a system extracting simplified datums. Although this system produced very noisy outputs, we were able to show in an official evaluation that the extractions could be used to corroborate mechanistic assertions extracted by other systems, increasing baseline precisions of 50% to 80% at the strictest corroboration levels. Based on these results, we intend to supplement or replace the hand-written rules with learned ones.

In the slightly longer term, it remains to validate the second part of our hypothesis: that *assembly* of information captured by these induced patterns into composite frames (BEL statements or PL datums) similarly can be realized. To this end, we plan to explore the selection and use of learned patterns as features, as well as alternative approaches to induction, such as boosting. Experience has shown that contextualizing symbolic learning methods in this way yields performance as strong as other leading learning paradigm (Caruana and Niculescu-Mizil, 2006).

5   Related Work

Progress in biological information extraction (BioIE) is measured against shared annotated corpora that decompose the problem into entity extraction and sentence-level relation detection (Kim et al., 2003; Pyysalo et al., 2007; Kim et al., 2008; Thompson et al., 2009). The BEL corpus has recently joined the ranks of these shared corpora as part of BioCreative, where early F1 scores on the task of assembling complete BEL statements average about 0.2, reflecting the difficulty of the task (Fluck et al., 2015).

Corpora annotated for entities and pairwise relations enable the application of machine learning, which has been shown to be as effective for such problems (Bunescu et al., 2005). Perhaps because of a relative wealth of structured knowledge resources, there are several competitive rule-based approaches to BioIE, involving both hand-written rules (Hunter et al., 2008; Valenzuela-Escarcega et al., 2015; Peng et al., 2014) and rules induced or tuned from data. For the most part, these rule learning approaches introduce domain-specific heuristics that limit their generality. Bui et al (2013) begin with a pre-specified library of syntactic patterns, which are instantiated from training data through a domain-specific procedure. Huang et al (2004) heuristically simplify training sentences, align them in pairs using a specialized edit distance, derive a pattern from the alignment, applying a series of heuristic checks to discard problematic patterns. Liu et al (2011) presents what is in some ways a generalization of this procedure, deriving a graph structure that is the union of individual simplified parses. It is unclear how the resulting extractor controls for overgeneration.

In contrast to these rather specific solutions, Bunescu et al (2005) evaluate a number of machine learning approaches to BioIE problems, finding them generally viable and noting that the rule-based learners yield high precision. These rule learners, which were drawn from a tradition of general-purpose rule induction for IE (Ciravegna and others, 2001; Soderland, 1997; Califf and Mooney, 1999; Freitag, 2000; Freitag and Kushmerick, 2000), notably make no or relatively modest assumptions about syntax. We are aware of no prior work applying such techniques to full dependency or constituent parses.

The relaxations of the annotation requirement that we explore in this paper (absence of phrase-level annotations, distant supervision) have been thoroughly studied in other contexts. An early instance of the IE-as-classification idea was the AutoSlog system (Riloff, 1996), which gave birth to bootstrapping techniques commonly used for many NLP problems (Riloff et al., 1999). Similarly, distant supervision, pioneered in the biological domain (Craven et al., 1999), has matured toward yielding performance comparable to complete supervision on certain problems (Poon et al., 2015). In contrast with such work, which focuses on sentence-local targets, relatively little work has been done on *discourse-level* distant supervision. A counter-example is Reschke et al (2014), which addresses event extraction at the document level, showing promising results but leaving many unanswered questions.

6   Conclusion

We have presented two understudied corpora providing distant annotation for the extraction of composite frames constituted from multiple sentences, none of which are annotated at the constituent level. We have argued for a two-phase approach to the exploitation of these resources—feature derivation and frame assembly—and presented a novel pattern induction procedure applicable to the first phase. Experiments with the two corpora demonstrate the procedure is effective, yielding patterns superior to those authored by humans in a comparable pattern language.

References

Quoc-Chinh Bui, David Campos, Erik van Mulligen, and Jan Kors. 2013. A fast rule-based approach for biomedical event extraction. In *Proceedings of the BioNLP Shared Task 2013 Workshop*, pages 104–108. Association for Computational Linguistics.

Razvan Bunescu, Ruifang Ge, Rohit J. Kate, Edward M. Marcotte, Raymond J. Mooney, Arun K. Ramani, and Yuk Wah Wong. 2005. Comparative Experiments on Learning Information Extractors for Proteins and their Interactions. *Artificial Intelligence in Medicine: Special Issue on Summarization and Information Extraction from Medical Documents*.

M.E. Califf and R.J. Mooney. 1999. Relational learning of pattern-match rules for information extraction. In *Proceedings of the Sixteenth National Conference on Artificial Intelligence*, volume 334.

Rich Caruana and Alexandru Niculescu-Mizil. 2006. An empirical comparison of supervised learning algorithms. In *Proceedings of the 23rd international conference on Machine learning*, pages 161–168. ACM.

Fabio Ciravegna and others. 2001. Adaptive information extraction from text by rule induction and generalisation. In *International joint conference on artificial intelligence*, volume 17, pages 1251–1256. LAWRENCE ERLBAUM ASSOCIATES LTD.

Mark Craven, Johan Kumlien, and others. 1999. Constructing biological knowledge bases by extracting information from text sources. In *ISMB*, volume 1999, pages 77–86.

Hal Daum III, John Langford, and Daniel Marcu. 2009. Search-based structured prediction. *Machine learning*, 75(3):297–325.

Steven Eker, Merrill Knapp, Keith Laderoute, Patrick Lincoln, and Carolyn Talcott. 2004. Pathway logic: Executable models of biological networks. *Electronic Notes in Theoretical Computer Science*, 71:144–161.

Juliane Fluck, Sumit Madan, Tilia Renate Ellendorff, Theo Mevissen, Simon Clematide, Adrian van der Lek, and Fabio Rinaldi. 2015. Track 4 Overview: Extraction of causal network information in biological expression language (bel). In *Fifth BioCreative Challenge Evaluation Workshop*, pages 333–346.

Dayne Freitag and Nicholas Kushmerick. 2000. Boosted wrapper induction. In *AAAI/IAAI*, pages 577–583.

Dayne Freitag. 2000. Machine learning for information extraction in informal domains. *Machine learning*, 39(2-3):169–202.

Minlie Huang, Xiaoyan Zhu, Yu Hao, Donald G. Payan, Kunbin Qu, and Ming Li. 2004. Discovering patterns to extract proteinprotein interactions from full texts. *Bioinformatics*, 20(18):3604–3612.

Lawrence Hunter, Zhiyong Lu, James Firby, William A. Baumgartner, Helen L. Johnson, Philip V. Ogren, and K. Bretonnel Cohen. 2008. OpenDMAP: an open source, ontology-driven concept analysis engine, with applications to capturing knowledge regarding protein transport, protein interactions and cell-type-specific gene expression. *BMC bioinformatics*, 9(1):78.

J.-D. Kim, Tomoko Ohta, Yuka Tateisi, and Junichi Tsujii. 2003. GENIA corpusa semantically annotated corpus for bio-textmining. *Bioinformatics*, 19(suppl 1):i180–i182.

Jin-Dong Kim, Tomoko Ohta, and Jun'ichi Tsujii. 2008. Corpus annotation for mining biomedical events from literature. *BMC bioinformatics*, 9(1):1.

Haibin Liu, Ravikumar Komandur, and Karin Verspoor. 2011. From graphs to events: A subgraph matching approach for information extraction from biomedical text. In *Proceedings of the BioNLP Shared Task 2011 Workshop*, pages 164–172. Association for Computational Linguistics.

Yifan Peng, Manabu Torii, Cathy H. Wu, and K. Vijay-Shanker. 2014. A generalizable NLP framework for fast development of pattern-based biomedical relation extraction systems. *BMC bioinformatics*, 15(1):1.

Hoifung Poon, Kristina Toutanova, and Chris Quirk. 2015. Distant supervision for cancer pathway extraction from text. In *Pac. Symp. Biocomput.*, pages 120–131.

Sampo Pyysalo, Filip Ginter, Juho Heimonen, Jari Bjrne, Jorma Boberg, Jouni Jrvinen, and Tapio Salakoski. 2007. BioInfer: a corpus for information extraction in the biomedical domain. *BMC bioinformatics*, 8(1):50.

J. Ross Quinlan. 1990. Learning logical definitions from relations. *Machine learning*, 5(3):239–266.

Kevin Reschke, Martin Jankowiak, Mihai Surdeanu, Christopher D. Manning, and Daniel Jurafsky. 2014. Event Extraction Using Distant Supervision. In *LREC*, pages 4527–4531.

Ellen Riloff, Rosie Jones, and others. 1999. Learning dictionaries for information extraction by multi-level bootstrapping. In *AAAI/IAAI*, pages 474–479.

Ellen Riloff. 1996. Automatically generating extraction patterns from untagged text. In *Proceedings of the national conference on artificial intelligence*, pages 1044–1049.

Stephen G. Soderland. 1997. *Learning text analysis rules for domain-specific natural language processing*. Ph.D. thesis, Citeseer.

Paul Thompson, Syed A. Iqbal, John McNaught, and Sophia Ananiadou. 2009. Construction of an annotated corpus to support biomedical information extraction. *BMC bioinformatics*, 10(1):1.

Marco A. Valenzuela-Escarcega, Gus Hahn-Powell, Thomas Hicks, and Mihai Surdeanu. 2015. A Domain-independent Rule-based Framework for Event Extraction. In *ACL-IJCNLP 2015 System Demonstrations*.

Inferring Implicit Causal Relationships in Biomedical Literature

Halil Kilicoglu
Lister Hill National Center for Biomedical Communications
National Library of Medicine
Bethesda, MD, 20894, USA
kilicogluh@mail.nih.gov

Abstract

Biomedical relations are often expressed between entities occurring within the same sentence through syntactic means. However, a significant portion of such relations (in particular, causal relations) are expressed implicitly across sentence boundaries. Inferring these discourse-level relations can be challenging in the absence of syntactic clues. In this paper, we present a study of textual characteristics that contribute to expression of implicit causal relations across sentence boundaries. Focusing on a chemical-disease relationship corpus, we identify and investigate the contribution of various features that can assist in identifying such inter-sentential relations. Using these features for supervised learning, we were able to improve previously reported best results by more than 13%. Our results demonstrate the usefulness of the proposed features and the importance of using a balanced dataset for this task.

1 Introduction

Causal associations between entities, events, and processes are central to biomedical knowledge (Mihăilă et al., 2013). Such associations extend from physical causation, such as gene-disease relationships and adverse drug reactions, to rhetorical causation between claims and their justifications. Detecting causal associations in biomedical literature can assist in biocuration of pathways and databases, such as the Comparative Toxicogenomics Database (CTD)[1], and support tasks such as drug discovery and pharmacovigilance. Recognizing this need, the recent BioCreative V challenge included a task (CID) on extraction of

chemical-induced disease relationships from Medline abstracts (Wei et al., 2016).

Chemical-disease relationships that the CID task focuses on are causal relationships in which a chemical acts as the *cause* and a disease or an adverse effect acts as the *effect*. In the simplest case, these relationships can be expressed intra-sententially through syntactic means. For example, in the sentence below (taken from the CDR corpus used in the CID task), the causal relationship between the drug *tacrolimus* and the disease *myocardial hypertrophy* is expressed explicitly with the causal trigger *induce*, which has the drug mention as its subject and the disease mention as its direct object.

(1) *Thus, we conclude that <u>tacrolimus</u> **induces** reversible <u>myocardial hypertrophy</u>.*

Assuming that the named entities have been successfully recognized by a named entity recognition (NER) system, lexical clues (the causal trigger *induce*) and syntactic dependency path between the entities and the trigger can be used to establish a causal link. However, not all causal relationships are expressed intra-sententially, and crucial information may be missed if the implicit, discourse-level relationships are simply ignored. For illustration, consider the discourse fragment below.

(2) *We investigated the efficacy and toxicity of a 3-hour <u>paclitaxel</u> infusion in a phase II trial in patients with inoperable stage IIIB or IV NSCLC. ... Hematologic toxicities were mild: only one patient (2%) developed grade 3 or 4 <u>neutropenia</u>, while 29% had grade 1 or 2. Grade 1 or 2 <u>polyneuropathy</u> affected 56% of patients while only one (2%) experienced severe <u>polyneuropathy</u>. Similarly, grade 1 or 2 <u>myalgia/arthralgia</u> was observed in*

---

[1] http://ctdbase.org/

Proceedings of the 15th Workshop on Biomedical Natural Language Processing, pages 46–55,
Berlin, Germany, August 12, 2016. ©2016 Association for Computational Linguistics

Limiting relation extraction to sentence level, we would miss the causal relationships between the drug *paclitaxel* and the adverse effects (*neutropenia, polyneuropathy, myalgia, arthralgia, nausea,* and *vomiting*). The difficulty of extracting implicit, discourse-level relationships is due to several factors. First, the role of syntax in expressing relationships is limited; no syntactic dependency exists between the entities. Secondly, discourse-level phenomena, such as coreference, implicit argumentation, and rhetorical relations between sentences play a larger role. Resolving such phenomena could aid in identifying implicit relationships; however, these are all challenging NLP tasks in their own right. Thirdly, potential relationships between all entities occurring in the document may need to be considered, which can lead to a data sparsity/imbalance problem due to the smaller number of relations expressed across sentence boundaries.

In the biomedical domain, to our knowledge, there is little research specifically focusing on implicit, inter-sentential relations. In the GENIA event corpus (Kim et al., 2008), one of the major corpora for biomedical relation extraction, 7.8% of all events cross sentence boundaries and the majority of these events (4.8%) are causal. In contrast to the text-bound and linguistically-motivated annotation in the GENIA event corpus, the CDR corpus annotation is not concerned with explicit event triggers and implicit causal inferences are annotated much more frequently, as illustrated in the example above. 27.2% of all relations in the corpus are expressed only at the discourse level; that is, their arguments never co-occur within the same sentence. Therefore, the CDR corpus provides a good opportunity to study implicit causal relationships. While systems participating in the CID task have addressed discourse-level relations to some extent, only a few have explicitly reported results on discourse-level relations. Among these, the top-ranked system (CD-REST) (Xu et al., 2016) incorporated a document-level classifier, which uses entity and context-based features as well as knowledge-based features. Knowledge-based features, particularly those extracted from the CTD database, proved to be the difference, since this database provides manually curated relationships between chemical and diseases.

In this paper, we aim to elucidate the textual characteristics that play a role in implicit, discourse-level relations. While the CD-REST system (Xu et al., 2016) demonstrates that curated knowledge about chemical-disease relationships in structured resources can be used to great advantage, we approach the problem purely as a natural language processing task and specifically focus on characteristics that can be derived from the text, since presence of curated relationships cannot be assumed for all relation extraction tasks and therefore such an approach may not be generalizable. Based on the characteristics that are discussed, we propose specific features that can play a role in recognizing implicit relations, use these features for supervised learning and investigate their effect. To address the imbalance of the data, we also experiment with different training sizes. Our results show that the features we propose aided by a balanced training set can provide state-of-the-art performance in recovering implicit causal relationships and indicate that named entity recognition has a significant impact on the performance.

2 Related Work

In the general domain, Swampillai and Stevenson (2011) used an SVM-based approach to address inter-sentential relations in the MUC6 dataset. Adapting structural features used for intra-sentential relation extraction (e.g., parse trees) to the inter-sentential case and addressing the data sparsity problem by hyperplane adjustment, they were able to obtain comparable performance to intra-sentential relation extraction. A relevant research thread in semantic role labeling (SRL) is concerned with implicit arguments of predicates. Gerber and Chai (2010) studied implicit arguments of a small number of nominal predicates, such as *price* and *shipping*. Their model used a variety of features such as VerbNet classes and semantic roles for predicates and arguments, sentence distance, predicate frequency, and pointwise mutual information between arguments to identify implicit arguments. The SemEval-2010 Task 10: Linking Events and their Participants in Discourse (Ruppenhofer et al., 2010) addressed the same problem on a larger set of event predicates. The participating systems performed very poorly; however, more recent studies were able to improve results, by casting the problem as an anaphora resolution task (Silberer and Frank, 2012) and by using the previously identified explicit arguments of a given predicate in linking (Laparra and Rigau, 2013). Causal relations have also been studied in the general domain from a wide range of perspectives. For example, Girju (2003) learned patterns indicating causal relationships between noun phrases to improve question answering. Other research focused on causal relations between discourse segments (rather than individual entities) and generally reported poorer results on causal relations than other types of discourse relations (Subba and Di Eugenio, 2009). It should be noted that most research on implicit arguments and causal relations assume the presence of explicit triggers (e.g., *produce, as a result*).

In the biomedical domain, there is little work that specifically addresses implicit arguments. Focusing on consumer health questions, Kilicoglu et al. (2013) incorporated resolution of anaphora and ellipsis to their question frame extraction pipeline and reported an 18 point improvement in F_1 score due to implicit argument resolution. Coreference resolution has been studied as a strategy to recover implicit arguments and improve event extraction and varying degrees of improvement due to coreference resolution have been reported (Yoshikawa et al., 2011; Miwa et al., 2012; Kilicoglu and Bergler, 2012; Lavergne et al., 2015; Kilicoglu et al., 2016).

Regardless of whether they are expressed implicitly, a wide range of causal relations have also been addressed in biomedical text. GENIA event corpus (Kim et al., 2008) and BioInfer corpus (Pyysalo et al., 2007) contain causal relationships between genes/proteins (e.g., REGULATION, POSITIVE_REGULATION, and NEGATIVE_REGULATION), in addition to other relation types. Causal relations in these corpora were often found to be more challenging to identify than other relation types (Kim et al., 2012). In the BioCause corpus (Mihăilă et al., 2013), causality was addressed as a discourse coherence relation and 850 causal discourse relations from full-text journal articles on infectious diseases (94% of which have explicit causal triggers) were annotated. In the BioDRB corpus (Prasad et al., 2011), a larger number of discourse relation types were annotated, one of which is causality. Mihăilă and

Ananiadou (2014) focused on discourse causality in BioCause and used a semi-supervised method to recognize causal triggers and their arguments in biomedical discourse. They did not address implicit discourse causality.

BioCreative V CID task involved chemical-disease relationships at the discourse level, even though they were often not specifically addressed. The top-ranked system (CD-REST) (Xu et al., 2016) incorporated a discourse-level classifier, which interestingly performed better than the sentence-level classifier; however, most of the performance gain was due to features extracted from curated resources, particularly CTD. Similarly, the next best system (Pons et al., 2016) used domain knowledge from various databases, and one of better performing systems, UET-CAM (Le et al., 2015), incorporated features from coreference resolution into an intra-sentential relation classifier. The present study diverges from these studies by specifically addressing implicit, discourse-level causality and focusing on textual characteristics.

3   Methods

In this section, we first describe the corpus we used for analysis and experiments. Next, we discuss the linguistic characteristics of inter-sentential, implicit causal relationships. In the following subsection, we describe our supervised learning approach and features that we developed. Finally, we discuss our evaluation.

3.1   CDR Corpus

For our analysis and experiments, we used the CDR corpus that was used in the BioCreative V CID task (Wei et al., 2016). This corpus consists of 1,500 Medline abstracts, annotated with chemical and disease mentions, normalized to MeSH identifiers, and the abstract-level chemical-disease causal relationships between the normalized entities. The corpus is split into three, one-third is used for training, one-third for development, and the rest for testing. Causal triggers have not been annotated in the corpus. The distribution of chemical/disease entities as well as that of the relations are given in Table 1. For our experiments, we focused on relations that are solely expressed across sentences (i.e., entity pairs co-occurring in the same sentences are ignored). The statistics for these relations are also given in Table 1. We did not perform any named entity recognition or nor-

Dataset	# Diseases	# Chemicals	# Relations	# Discourse-level Relations
Training	4,182 (1,965)	5,203 (1,467)	1,038	283
Development	4,244 (1,865)	5,347 (1,507)	1,012	246
Testing	4,244 (1,988)	5,385 (1,435)	1,066	320
TOTAL	12,670 (5,818)	15,935 (4,409)	3,116	849

Table 1: CDR corpus characteristics

malization and conducted our analysis and experiments using the gold entities. For comparison, we also used DNorm (Leaman et al., 2013) for disease and tmChem (Leaman et al., 2015) for chemical name recognition/normalization. On the test portion of the corpus, DNorm achieves 81% F_1 score and tmChem achieves 91% F_1 score.

3.2 Characteristics of implicit causal relations

Focusing only on inter-sentential relations in the training set, we examined the linguistic characteristics that play a role in expressing them. We examine and exemplify some of the important characteristics below.

3.2.1 Causal ordering of events

A significant portion of the implicit chemical-disease relationships can be seen as inferences, rather than explicit assertions. One minimal condition for such causal inference is temporality: if a chemical causes a disease in a patient, then the chemical administration has to occur before the manifestation of the disease. In biomedical abstracts, language describing such event ordering is present, particularly in descriptions of experiments. An example, shortened from the original text, is shown below, with relevant chemical and disease mentions underlined.

(3) *We report on a combination of everolimus and tacrolimus in 24 patients ...with either myelodysplastic syndrome ...or acute myeloid leukemia .... All patients engrafted, and only 1 patient experienced grade IV mucositis. ...Transplantation-associated microangiopathy ...occurred in 7 patients ..., with 2 cases of acute renal failure.*

Similarly, case studies often involve language describing a sequence of events that lead to a medical problem. An example is given below.

(4) *We present a case of a 5-year-old child with cerebral palsy and seizure disorder, receiv-*

ing clonidine for restlessness, who presented for placement of a baclofen pump. Without the knowledge of the medical personnel, the patient's mother administered three doses of clonidine during the evening before and morning of surgery to reduce anxiety. During induction of anesthesia, the patient developed bradycardia and hypotension ...

3.2.2 Coreference

The role of coreference in expressing implicit arguments has been acknowledged (Silberer and Frank, 2012). Anaphora relations can create explicit links between sentences and assist in resolving implicit arguments. In the following example, the definite noun phrase *this regimen* and the personal pronoun *it* corefer with *combination therapy with pegylated interferon and ribavirin* in the previous sentence. If these anaphora relations are resolved, the anaphoric expressions can simply be substituted with the antecedent, simplifying the problem to sentence-bound relation extraction.

(5) *The current best treatment for HCV infection is combination therapy with pegylated interferon and ribavirin. Although **this regimen** produces sustained virologic responses (SVRs) in approximately 50% of patients, **it** can be associated with a potentially dose-limiting hemolytic anemia.*

Bridging (or associative) anaphora (Poesio et al., 1997), a type of indirect coreference that is distinguished by relations such as hypernymy (is-a) or meronymy (part-of), is also used considerably to indicate implicit causal relations. In the following example, the causal relation between *ventricular fibrillation* and the chemicals *sodium citrate* and *disodium edetate* can be identified, if we can recognize that there is a meronymic relationship between these chemicals and *Renografin*.

(6) *Renografin contains the chelating agents sodium citrate and disodium edetate,*

*while Hypaque contains calcium disodium edetate and no sodium citrate. Ventricular fibrillation occurred significantly more often with **Renografin**.*

3.2.3 Document Topic as Implicit Argument

Since abstracts are relatively short, it is common to have the main focus of the article mentioned only once and referred to implicitly throughout the abstract. For example, in an article investigating the side effects of a drug, the drug name is often mentioned early on (in some cases, only in the title), and the side effects of the drug are revealed later in the abstract. In the following example, the logical object argument of the predicate *treatment* is uninstantiated, and this implicit argument refers to the document topic, the drug *CCNU (lomustine)*.

(7) *<u>CCNU (lomustine)</u> toxicity in dogs: a retrospective study (2002-07) …CCNU was used most commonly in the treatment of lymphoma, mast cell tumour, …. Throughout **treatment**, 56.9% of dogs experienced <u>neutropenia</u>, 34.2% experienced <u>anaemia</u> and 14.2% experienced <u>thrombocytopenia</u>.*

3.2.4 Document Structure

The title and the abstract of an article need to convey the gist of the study in a small, often predetermined, number of words. To ensure that the content of the abstract is representative of the study, some journals require the abstracts to conform to a formal structure (structured abstracts), with sections such as Objective, Methods, and Results. Important findings are more likely to be reported in the Results section, and implicit causal relationships between entities in the Results section and the main topics of the articles are frequent. In the following example, *desipramine* is one of the main topics of the article and the only mention of *ventricular arrhythmias* is in the Results section.

(8) *Effect of calcium chloride and 4-aminopyridine therapy on <u>desipramine</u> toxicity in rats …The <u>incidence of ventricular arrhythmias</u> (p = 0.004) and seizures (p = 0.03) in the CaCl2 group was higher than the other groups.*

3.3 Supervised learning of implicit causal relationships

We formulate implicit causal relation extraction as a binary classification task, where examples consist of chemical-disease mention pairs whose corresponding normalized entities do not co-occur intra-sententially in the abstract. Positive examples are mention pairs that are causally related, and negative examples are those that are not. We used linear SVM (Fan et al., 2008) to train the binary classifier and empirically set the regularization parameter C to 0.1. To address the imbalance of the dataset (approximately 85% of all examples are negative), we trained the classifier with varying number of negative examples (undersampling). We selected negative examples from the documents in proportion to the number of all examples extracted from the document.

The classifier uses features developed based on the analysis presented in the previous section as well as standard n-gram (unigram and bigram) features. Features that proved predictive are provided in Table 2 and illustrated on the *desipramine:ventricular arrhythmias* pair from Example (8). In Table 2, we also indicate whether the feature or an approximation was used by the top-performing system (Xu et al., 2016) in the CID task. We distinguish between lexical, semantic, and discourse features.

Lexical features are simple n-gram features extracted from the sentences of the target mentions. We use unigrams and bigrams of the mentions as well as those of sentences that the mentions appear in.

Semantic features include conceptual knowledge about the entities (their MeSH identifiers and the MeSH identifiers of their ancestors in the MeSH hierarchy) as well as other semantic information that occur in the sentence context. For this purpose, we use an existing dictionary of causal predicates, previously compiled from several corpora. The list consists of 201 predicates and mainly includes triggers for regulatory events (e.g., *induce, effect, develop*) as well as discourse connectives that describe causal (e.g., *as a result*) or temporal relations (e.g., *before, after*). We also use a feature that indicates whether an experiencer (e.g., *patient, rats*) is mentioned in the sentence context. Finally, a binary feature indicates whether any mention belonging to the opposing semantic class occurs in the sentence (i.e., if the classified example includes a chemical mention in the current sentence, this feature is true if the current sentence contains a disease mention).

Discourse features are mainly features based on

Feature	Description	CD-REST
Lexical features		
F_1	Uncased unigrams of the mentions	✓
F_2	Uncased bigrams of the mentions	✓
F_3	Uncased unigrams of the mention sentence(s)	
F_4	Uncased bigrams of the mention sentence(s)	
Semantic features		
F_5	Uncased causal predicate lemmas preceding the chemical mention (\{*effect*\})	S
F_6	Uncased causal predicate lemmas following the chemical mention ($\emptyset$)	S
$F_7 - F_8$	Same as $F_5 - F_6$, for the disease mention (\{$\emptyset,\emptyset$\})	S
F_9	Whether the opposing semantic class in the mention pair exists in the sentence (*true*)	
F_{10}	Whether an experiencer trigger exists in either mention sentence (*true*)	✓
F_{11}	Disease MeSH identifier (*D001145*)	✓
F_{12}	chemical MeSH identifier (*D003891*)	✓
F_{13}	disease MeSH hypernyms (\{*D002318, D006331, D010335, D013568*\})	✓
F_{14}	chemical MeSH hypernyms (\{*D003984, D006571, D006575*\})	✓
Discourse features		
F_{15}	chemical in focus (*true*)	S
F_{16}	disease in focus (*false*)	
F_{17}	normalized section name of the chemical (*TITLE*)	
F_{18}	normalized section name of the disease (*RESULTS*)	
F_{19}	main verb POS sequence in target and intervening sentences (*NONE*)	
F_{20}	whether the sentences of the mentions are adjacent (*false*)	✓
F_{21}	the document contains sortal anaphors (*true*)	
F_{22}	MeSH descendant of the disease occurs in the document (*true*)	✓
F_{23}	MeSH ancestor of the disease occurs in the document (*true*)	✓

Table 2: The features used by the binary classifier (S: a similar feature is used)

our analysis. To address causal ordering of events by capturing tense information, we include a feature that concatenates the part-of-speech tags of the main verbs of the mention sentences and those of the sentences intervening between them (ignoring title sentences). Adjacent sentences are often implicitly related, and therefore, we include a binary feature that indicates whether the mention sentences are adjacent. To address anaphora, we include a binary feature that indicates whether the abstract contains any sortal anaphors that can refer to chemical or disease mentions (e.g., *this drug, the condition*). We extracted this information using the Bio-SCoRes tool (Kilicoglu and Demner-Fushman, 2016). With regards to bridging anaphora, we use binary features that indicate whether a MeSH ancestor or descendant of one of the entities in the pair appear in the abstract, addressing hypernymy. Whether the chemical en-

tity and the disease entity may be document topics are also included as features. We simply included all entities that appear in the title of the article as document topics. To capture document structure, we normalized the section names in structured abstracts using the mappings curated at the NLM[2]. If the abstracts are not structured, we simply used TITLE or ABSTRACT as the normalized section name.

Feature extraction presupposes a standard linguistic processing pipeline (i.e., tokenization, part-of-speech tagging, syntactic parsing). We performed this processing using the Stanford CoreNLP toolkit (Manning et al., 2014).

---

[2]http://structuredabstracts.nlm.nih.gov/Downloads/Structured-Abstracts-Labels-110613.txt.

3.4 Evaluation

In separate experiments, we used gold standard entities and those recognized by DNorm (Leaman et al., 2013) and tmChem (Leaman et al., 2015) as the basis for relation extraction. Following the CID baseline, we took simple abstract-level entity co-occurrence as the baseline method. We also compared our results to those reported with CD-REST (Xu et al., 2016). This comparison is made somewhat difficult by the fact that their discourse-level classifier considers entities co-occurring in the same sentences as candidates, as well. Another complicating factor in comparison is their classifier's use of curated knowledge-base features. In particular, two features from CTD provide more than 18% improvement in their overall F_1 score using gold standard entities (from 56.7% to 67.1%). For a fair comparison, we implemented these CTD features and incorporated them into our best model.

- CTD relation between the chemical and the disease: *null, inferred-association, therapeutic,* or *marker/mechanism*

- Whether the disease has a marker/mechanism association with any chemical in CTD

In addition, we performed an ablation study to better understand the contribution of various feature sets. We used the standard evaluation metrics, precision, recall, and F_1 score, to assess relation extraction performance.

4   Results and Discussion

The results of implicit causal relation extraction on the test set using the gold standard entities and DNorm/tmChem entities are provided in Table 3. The effect of CTD features on classification performance is also shown. We obtained the highest F_1 score and recall when we undersampled negative examples to yield a 1:1 positive/negative sample ratio (balanced training)[3]. The highest precision was obtained in both cases when all available data are used for training.

The improvement due to CTD features was less dramatic than that found by Xu et al. (2016) but still significant (more than 11% improvement with the gold entities, from 66.1% to 73.7%). However, we believe that the results obtained without CTD features are a better representation of the

state-of-the-art for implicit causal relation extraction from a purely NLP perspective. In this setting, we obtained 66.1% F_1 score with gold entities and 48.6% F_1 score with DNorm/tmChem entities (in italics).

In comparing our performance to that of CD-REST, we find that our approach overall outperforms CD-REST. Using gold entities, CTD features, and a balanced training set, we outperformed their system by more than 9% (67.3% vs. 73.7%). They have not used a balanced training set, so the difference with their reported system is even wider (56.7% to 73.7%). Without CTD features, we slightly outperformed their reported results (66.1% vs. 64.9%), indicating that our approach in some sense compensates for the CTD knowledge and suggesting that it could support biocuration of these relationships. The performance they reported with gold entities is somewhat higher than what we obtained with our implementation of their features (64.9% vs. 56.7%); however, it is worth pointing out that their classifier takes into account mention pairs that co-occur in the same sentences, as well, which can explain the difference to some extent. The small differences in our implementation of their features could also account for some of the difference. CD-REST uses its own named entity recognition tool, which outperforms the DNorm/tmChem combination, and this is partly reflected in the performance difference between using DNorm/tmChem entities with their features and their reported end-to-end performance (50.2% vs. 56.8%).

To better understand the contribution of features, we performed an ablation study in which we removed a set of features, retrained our classifier, and assessed the performance. The results of this evaluation are shown in Table 4. In these experiments, we used gold entities and a balanced training set and did not include CTD features. The results show that lexical and discourse features contribute similarly to implicit causal relation extraction, while the contribution of semantic features is much smaller. We observe that the effect of lexical features is to improve precision, whereas discourse features contribute significantly to recall, with a minor degradation in precision.

While the discourse features we used were overall successful, our attempts at using more sophisticated discourse features have often resulted in performance loss. For example, coref-

---

[3]Not all the ratios we experimented with are shown.

Experiment	Precision	Recall	F_1
Using DNorm/tmChem entities			
Baseline	20.3	67.3	31.2
Balanced training	*46.9*	*50.5*	*48.6*
Balanced + CTD features	56.4	54.9	55.6
Unbalanced	56.4	36.2	44.1
Using gold entities			
Balanced	*59.7*	*74.0*	*66.1*
Balanced + CTD	68.0	80.3	73.7
Unbalanced	67.6	52.4	59.0
Our CD-REST implementation			
Balanced + CTD w/ gold entities	70.1	64.8	67.3
Unbalanced + CTD w/ gold entities	79.4	44.1	56.7
Balanced + CTD w/ tmChem/DNorm entities	60.5	42.9	50.2
Reported CD-REST performance (Xu et al., 2016)			
Using gold entities	68.4	61.8	64.9
End-to-end results	64.1	50.5	56.8

Table 3: Evaluation results

Experiment	Precision	Recall	F_1
All	59.7	74.0	66.1
-Lexical features	42.8	91.4	58.3
-Semantic features	58.6	73.7	65.3
-Discourse features	60.7	55.9	58.2

Table 4: Feature ablation results

erence emerged as an important aspect of implicit causal relations, and it seemed that fully resolving disease/chemical coreference in the abstract could improve the performance. We adapted the Bio-SCoRes framework (Kilicoglu and Demner-Fushman, 2016) to extract anaphora relations and incorporated more sophisticated features based on these relations into our classifier, such as whether a mention corefers with an anaphor in the sentence of the other mention in the pair (Example 5). While this improved precision (59.7% to 66.2%), the recall loss was more significant (74.0% to 62.9%), leading to a lower F_1 score (64.5%). Similar, unsuccessful features include a binary feature indicating whether there is a potential bridging anaphora that involves the chemical or the disease mention itself. On the other hand, a simplistic discourse feature that indicates whether the document contains any sortal anaphor at all improved the F_1 score from 65.1% to 66.1%. Along the same lines, using normalized structured abstract section labels improved the classification performance. How-ever, most abstracts are not structured, and our attempts to automatically assign section labels using the sentence position in the abstract in such cases did not improve results.

The named entity recognition tools we used have reported relatively high performance on the test set (81% and 91% F_1 scores for DNorm and tmChem, respectively). However, the performance difference when using these tools in comparison to using gold entities is relatively large; gold entities yield more than 30% higher F_1 on average. This indicates that the relation extraction performance is highly sensitive to entity recognition and normalization, and that even small performance drop in this task can cause a major performance drop in relation extraction.

Data sparsity is a well-known problem for inter-sentential relation extraction (Swampillai and Stevenson, 2011). To deal with this problem, we experimented with training various positive/negative sample ratios, and found that a balanced training set led to superior overall performance, at the expense of loss of precision. This result is similar to that of Swampillai and Stevenson (2011), which they achieved with hyperplane adjustment.

There are several limitations to the study presented. First, we have not investigated the generalizability of the approach to other relation types expressed implicitly. The GENIA event corpus, with its text-bound event triggers, presents an opportunity to study implicit argumentation more widely from a semantic role labeling perspective, even though the number of relevant events in the corpus is relatively small. Secondly, whether the method can be extended to extracting relations from full-text articles remains to be seen. Thirdly, there are NLP methods that can provide more predictive features that we have not attempted to incorporate into our models. For example, temporal ordering of events have been the subject of much research recently, in both general (Chambers et al., 2014) and clinical (Bethard et al., 2015) domains, and tools based on these methods can provide useful features to detect causal ordering of events. Similarly, while our simple sentence position-based heuristics to assign sections to unstructured abstract sentences did not yield predictive features, more advanced methods to classify sentences into rhetorical categories (Agarwal and Yu, 2009) could be beneficial.

5   Conclusion

We presented a method to extract implicit, inter-sentential causal relationships from Medline abstracts. The method incorporates lexical, semantic, and discourse features and a simple undersampling approach for data sparsity to achieve state-of-the-art results. In this study, we specifically focused on implicit relationships across sentences, since they are more challenging from an NLP perspective, and future work involves combining the proposed method with methods that extracts sentence-bound, mostly explicit relationships. Improving feature extraction and named entity recognition/normalization are likely to be beneficial in further improving the state-of-the-art in causal relationship extraction. Joint learning of named entities and causal relationships could further improve performance by preventing, to some extent, the propagation of named entity recognition errors to relation extraction step.

Acknowledgments

This work was supported by the intramural research program at the U.S. National Library of Medicine, National Institutes of Health.

References

Shashank Agarwal and Hong Yu. 2009. Automatically classifying sentences in full-text biomedical articles into Introduction, Methods, Results and Discussion. *Bioinformatics*, 25(23):3174–3180.

Steven Bethard, Leon Derczynski, Guergana Savova, James Pustejovsky, and Marc Verhagen. 2015. SemEval-2015 Task 6: Clinical TempEval. In *Proceedings of the 9th International Workshop on Semantic Evaluation (SemEval 2015)*, pages 806–814.

Nathanael Chambers, Taylor Cassidy, Bill McDowell, and Steven Bethard. 2014. Dense event ordering with a multi-pass architecture. *Transactions of the Association for Computational Linguistics*, 2:273–284.

Rong-En Fan, Kai-Wei Chang, Cho-Jui Hsieh, Xiang-Rui Wang, and Chih-Jen Lin. 2008. LIBLINEAR: A library for large linear classification. *Journal of Machine Learning Research*, 9:1871–1874.

Matthew Gerber and Joyce Chai. 2010. Beyond Nom-Bank: A Study of Implicit Arguments for Nominal Predicates. In *Proceedings of the 48th Annual Meeting of the Association for Computational Linguistics*, pages 1583–1592, Uppsala, Sweden.

Roxana Girju. 2003. Automatic detection of causal relations for question answering. In *Proceedings of the ACL 2003 Workshop on Multilingual Summarization and Question Answering*, pages 76–83.

Halil Kilicoglu and Sabine Bergler. 2012. Biological Event Composition. *BMC Bioinformatics*, 13 (Suppl 11):S7.

Halil Kilicoglu and Dina Demner-Fushman. 2016. Bio-SCoRes: A Smorgasbord Architecture for Coreference Resolution in Biomedical Text. *PLoS ONE*, 11(3):e0118538.

Halil Kilicoglu, Marcelo Fiszman, and Dina Demner-Fushman. 2013. Interpreting consumer health questions: The role of anaphora and ellipsis. In *Proceedings of the 2013 Workshop on Biomedical Natural Language Processing*, pages 54–62.

Halil Kilicoglu, Graciela Rosemblat, Marcelo Fiszman, and Thomas C. Rindflesch. 2016. Sortal anaphora resolution to enhance relation extraction from biomedical literature. *BMC Bioinformatics*, 17:163.

Jin-Dong Kim, Tomoko Ohta, and Jun'ichi Tsujii. 2008. Corpus annotation for mining biomedical events from literature. *BMC Bioinformatics*, 9:10.

Jin-Dong Kim, Ngan Nguyen, Yue Wang, Jun'ichi Tsujii, Toshihisa Takagi, and Akinori Yonezawa. 2012. The Genia Event and Protein Coreference tasks of the BioNLP Shared Task 2011. *BMC Bioinformatics*, 13 Suppl 11:S1.

Egoitz Laparra and German Rigau. 2013. ImpAr: A Deterministic Algorithm for Implicit Semantic Role Labelling. In *Proceedings of the 51st Annual Meeting of the Association for Computational Linguistics*, pages 1180–1189.

Thomas Lavergne, Cyril Grouin, and Pierre Zweigenbaum. 2015. The contribution of co-reference resolution to supervised relation detection between bacteria and biotopes entities. *BMC Bioinformatics*, 16 (Suppl 10):S6.

Hoang-Quynh Le, Mai-Vu Tran, Thanh Hai Dang, and Nigel Collier. 2015. The UET-CAM System in the BioCreAtIvE V CDR Task. In *Fifth BioCreative challenge evaluation workshop*, pages 208–213.

Robert Leaman, Rezarta Islamaj Dogan, and Zhiyong Lu. 2013. DNorm: disease name normalization with pairwise learning to rank. *Bioinformatics*, 29(22):2909–2917.

Robert Leaman, Chih-Hsuan Wei, and Zhiyong Lu. 2015. tmChem: a high performance approach for chemical named entity recognition and normalization. *Journal of Cheminformatics*, 7(S-1):S3.

Christopher D. Manning, Mihai Surdeanu, John Bauer, Jenny Finkel, Steven J. Bethard, and David McClosky. 2014. The Stanford CoreNLP natural language processing toolkit. In *Proceedings of 52nd Annual Meeting of the Association for Computational Linguistics: System Demonstrations*, pages 55–60.

Claudiu Mihăilă and Sophia Ananiadou. 2014. Semi-supervised learning of causal relations in biomedical scientific discourse. *BioMedical Engineering Online*, 13(2):1–24.

Claudiu Mihăilă, Tomoko Ohta, Sampo Pyysalo, and Sophia Ananiadou. 2013. BioCause: Annotating and analysing causality in the biomedical domain. *BMC Bioinformatics*, 14(1).

Makoto Miwa, Paul Thompson, and Sophia Ananiadou. 2012. Boosting automatic event extraction from the literature using domain adaptation and coreference resolution. *Bioinformatics*, 28(13):1759–1765.

Massimo Poesio, Renata Vieira, and Simone Teufel. 1997. Resolving bridging references in unrestricted text. In *Proceedings of a Workshop on Operational Factors in Practical, Robust Anaphora Resolution for Unrestricted Texts*, pages 1–6.

Ewoud Pons, Benedikt F.H. Becker, Saber A. Akhondi, Zubair Afzal, Erik M. van Mulligen, and Jan A. Kors. 2016. Extraction of chemical-induced diseases using prior knowledge and textual information. *Database*, 2016.

Rashmi Prasad, Susan McRoy, Nadya Frid, Aravind Joshi, and Hong Yu. 2011. The biomedical discourse relation bank. *BMC Bioinformatics*, 12:188+.

Sampo Pyysalo, Filip Ginter, Juho Heimonen, Jari Björne, Jorma Boberg, Jouni Järvinen, and Tapio Salakoski. 2007. BioInfer: a corpus for information extraction in the biomedical domain. *BMC Bioinformatics*, 8:50.

Josef Ruppenhofer, Caroline Sporleder, Roser Morante, Collin Baker, and Martha Palmer. 2010. SemEval-2010 Task 10: Linking Events and Their Participants in Discourse. In *Proceedings of the 5th International Workshop on Semantic Evaluation*, pages 45–50.

Carina Silberer and Anette Frank. 2012. Casting implicit role linking as an anaphora resolution task. In **SEM 2012: The First Joint Conference on Lexical and Computational Semantics*, pages 1–10.

Rajen Subba and Barbara Di Eugenio. 2009. An effective discourse parser that uses rich linguistic information. In *Proceedings of Human Language Technologies: The 2009 Annual Conference of the North American Chapter of the Association for Computational Linguistics*, pages 566–574.

Kumutha Swampillai and Mark Stevenson. 2011. Extracting Relations Within and Across Sentences. In *Proceedings of RANLP*, pages 25–32.

Chih-Hsuan Wei, Yifan Peng, Robert Leaman, Allan Peter Davis, Carolyn J. Mattingly, Jiao Li, Thomas C. Wiegers, and Zhiyong Lu. 2016. Assessing the state of the art in biomedical relation extraction: overview of the BioCreative V chemical-disease relation (CDR) task. *Database*, 2016.

Jun Xu, Yonghui Wu, Yaoyun Zhang, Jingqi Wang, Hee-Jin Lee, and Hua Xu. 2016. CD-REST: a system for extracting chemical-induced disease relation in literature. *Database*, 2016

Katsumasa Yoshikawa, Sebastian Riedel, Tsutomu Hirao, Masayuki Asahara, and Yuji Matsumoto. 2011. Coreference Based Event-Argument Relation Extraction on Biomedical Text. *Journal of Biomedical Semantics*, 2 (Suppl 5):S6.

SnapToGrid: From Statistical to Interpretable Models for Biomedical Information Extraction

Marco A. Valenzuela-Escárcega, Gus Hahn-Powell, Dane Bell, Mihai Surdeanu
University of Arizona
Tucson, AZ 85721, USA
{marcov, hahnpowell, dane, msurdeanu}@email.arizona.edu

Abstract

We propose an approach for biomedical information extraction that marries the advantages of machine learning models, e.g., learning directly from data, with the benefits of rule-based approaches, e.g., interpretability. Our approach starts by training a feature-based statistical model, then converts this model to a rule-based variant by converting its features to rules, and "snapping to grid" the feature weights to discrete votes. In doing so, our proposal takes advantage of the large body of work in machine learning, but it produces an interpretable model, which can be directly edited by experts. We evaluate our approach on the BioNLP 2009 event extraction task. Our results show that there is a small performance penalty when converting the statistical model to rules, but the gain in interpretability compensates for that: with minimal effort, human experts improve this model to have similar performance to the statistical model that served as starting point.

1 Introduction

Due to the deluge of unstructured data, information extraction (IE) systems, which aim to translate this data to structured information, have become ubiquitous. For example, applications of IE range from parsing literature (Iyyer et al., 2016) to converting thousands of cancer research publications into complex proteins signaling pathways (Cohen, 2015).

By and large, in academia most of these approaches are implemented using machine learning (ML). This choice is warranted: generally, ML approaches, where the machine learns directly from the data, perform better than approaches where human domain experts encode the structure to be extracted manually. For example, the top systems in the BioNLP event extraction shared tasks have consistently been ML-based approaches (Kim et al., 2009; Kim et al., 2013). However, this is only part of the story: most of these models cannot be easily understood by their users, and, by and large, cannot be modified without retraining. This "technical debt" of ML (Sculley et al., 2014) is better understood in industry: Chiticariu et al. (2013) report that 67% of large commercial vendors of natural language processing (NLP) software focus on rule-based IE, and an additional 17% on hybrid systems that combine rule-based and ML approaches.

In this paper we focus on *interpretable models* for information extraction, i.e., models that: (a) can be understood by human users, and (b) can be directly edited and improved by these users. In particular, we focus on deterministic, rule-based models. Here, we introduce a novel approach to generate such models, which maintains both the advantages of ML such as learning from data, and the benefits of interpretability such as allowing human domain experts to directly edit and improve these models. Specifically, our contributions are:

(1) We introduce a simple strategy that converts statistical models for IE to rule-based models. We call the proposed algorithm SnapToGrid. Our approach works in three steps. First, we train a statistical model for the task at hand. Here we experiment with logistic regression, but the proposed method is, in principle, independent of the underlying statistical model. Further, our strategy can operate over multiple classifiers that are part of the same IE system (e.g., one classifier to identify event triggers, and another to identify event arguments). Second, we convert features

Proceedings of the 15th Workshop on Biomedical Natural Language Processing, pages 56–65,
Berlin, Germany, August 12, 2016. ©2016 Association for Computational Linguistics

to rules implemented in Odin, a modern declarative rule language (Valenzuela-Escarcega et al., 2016; Valenzuela-Escarcega et al., 2015). We also discard most of the statistical information acquired previously, by converting feature weights to discrete votes, which guarantees interpretability (hence the SnapToGrid name). Third, human domain experts inspect and manually improve the generated model, under certain time constraints.

(1) We evaluate our approach on the BioNLP 2009 core event extraction task, and demonstrate that the resulting interpretable model has similar performance to the statistical model that served as starting point.

2 Approach

Our motivation for this work is to keep the human domain expert in the loop when building IE systems. We show in Section 3 that this is beneficial, even when the domain experts have limited time to work on the task and no access to data other than the model itself. To achieve this "human in the loop" goal we propose the following three-step algorithm:

1. Train a statistical model for the IE task at hand (Section 2.1). The model may consist of several statistical classifiers. For example, for the BioNLP event extraction task, the most common approach involves two classifiers: one to identify event triggers, and a following classifier to identify event participants. One restriction is that these classifiers be feature-based classifiers, e.g., logistic regression, rather than the classifiers based on latent representations, e.g., neural networks.

2. Convert the statistical model into an interpretable, rule-based model (Section 2.2):

   (a) First, we convert the features to rules in the Odin language.
   (b) Then, we assign to each rules "votes" for a given class, by "snapping to grid", i.e., converting to discrete values, the weights computed by the above statistical model.

3. Domain experts edit the produced rule-based model directly, aiming to improve its quality with respect to both coverage and precision (Section 2.3).

We detail this process in the rest of this section, focusing on the BioNLP core event extraction task as the domain of interest.

2.1 Step 1: Build Statistical Model

Our statistical model is inspired by the top performing approach at the 2009 evaluation (Björne et al., 2009). The approach is summarized in Figure 1. Similar to (Björne et al., 2009), our approach consists of two classifiers: the first classifier detects and labels event trigger words in the input text; the second classifiers extracts and labels relations between event triggers and potential event participants, which can be either Protein entities or other event triggers. Both classifiers are implemented using multi-class logistic regression (LR), but our conversion process (Steps 2 and 3) is independent of the underlying statistical model, so, in principle, other feature-based classifiers that assign explicit weights to features could be used, e.g., perceptron, or linear support vector machines.

The Trigger Classifier

The first classifier sequentially labels each word in the input text as a trigger for a specific BioNLP event class, or as Nil otherwise. We implemented the following features:

Surface features: These features include the original and lemmatized words, and the presence of the word in a gazetteer of known event triggers (constructed automatically from the training data). These features are generated for the word being classified, as well as the words surrounding it inside a window of n tokens. We used two windows in our experiments, with $n = 1$ and $n = 4$. Further, bag-of-words features are generated for the windows and for the sentence as a whole.

Syntactic features: These features capture the syntactic dependencies (both incoming and outgoing) directly connected to the token. All syntactic information was represented using Stanford dependencies (De Marneffe and Manning, 2008), and was generated using the CoreNLP toolkit (Manning et al., 2014). For each of these paths, we generate two different versions: one containing just the label and direction of the syntactic dependencies, and another including also the destination words.

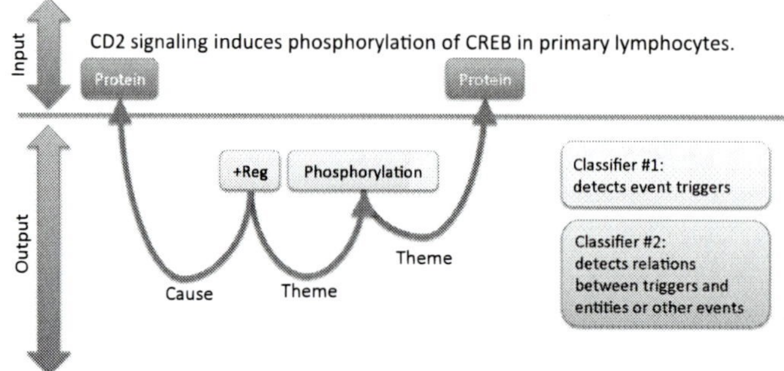

Figure 1: Architecture of the statistical model for the BioNLP core event extraction task.

Entity features: These features encode the number of other entities surrounding the token, both inside a window and in the sentence as a whole.

The Event Participant Classifier

This classifier pairs all the triggers detected by the previous classifier with other named entities (Proteins in this case) or event triggers that occur in the same sentence. These pairs are then classified into one of the possible participant relations, or `Nil` indicating that there is no relation between the pair. This classifier uses the following features:

Syntactic features: These features are based on the shortest path connecting the two mentions (trigger and candidate participant) in the Stanford syntactic dependency graph. Two versions of the shortest path are used: a lexicalized one (capturing the words along the path), and an unlexicalized one.

Surface features: These features include: the order of the two mentions in text, their distance in terms of tokens, the number of entities and triggers in the sentence, the parts of speech and words of the mentions, and the number of triggers and entities between the mentions.

Consistency features: These features encode the labels of the two mentions jointly, as well as the labels of their superclasses. For example, the features <Regulation, Phosphorylation> and <Regulation, Event> are generated for a relation between a Regulation event trigger and a Phosphorylation trigger as its theme. These feature capture selectional preferences for arguments, e.g., the Theme of a regulation event should be another event.

Graph features: The parent, children, and siblings of the mentions in the syntactic dependency graph.

Limitations

Not all of the above features can be represented as rules in the current implementation of the chosen rule language. Currently[1], Odin rules capture paths (over sequences or directed graphs) that are anchored at both ends (e.g., from an event trigger to an event argument) (Valenzuela-Escarcega et al., 2015; Valenzuela-Escarcega et al., 2016). Because of this, Odin cannot represent the following information: bag-of-word features, syntactic paths that are not anchored at both ends (such as dependencies connected only to event trigger candidates), and features that count occurrences of tokens or entities in text. In Section 3 we analyze the performance drop when such features are removed from the model.

2.2 Step 2: Convert the Statistical Model to a Rule-based Model

Once the statistical model is constructed, we employ the lossy process below to convert it to an interpretable one.

Converting Features to Rules

First, we convert the features encoded in the statistical model to rules in the Odin language (Valenzuela-Escarcega et al., 2015; Valenzuela-Escarcega et al., 2016). In general, the features previously introduced consist of conjunctions of information bits, each of which corresponds to a different rule fragment. For example, for the classification of event participants, one such conjunction captures the type of the expected trigger (e.g., Phosphorylation), combined

---

[1] As of June 2016

```
-   name: phospho_event
    label: Phosphorylation
    pattern: |
        trigger:Phosphorylation
        theme:Protein = >nsubjpass
```

Figure 2: Example of a rule for event participant classification that is built from a single feature. The feature captures the passive nominal subject (nsubjpass) outgoing (>) from a Phosphorylation trigger and landing on a Protein. The bold font indicates the rule output, i.e., the nominal subject is the **theme** of a **Phosphorylation** event.

with the syntactic path that connects the trigger with the participant candidate (e.g., an outgoing passive nominal subject – nsubjpass), and a semantic constraint for the type of named entity of the participant (e.g., Protein). These are immediately translatable to Odin rules, as illustrated in Figure 2.

Importantly, the rules encode output information as well, e.g., the recognized event participant serves as a theme for a Phosphorylation event in Figure 2. At this stage, this information is exhaustively generated from all possible classifier labels (e.g., for the classification of event participants these labels are the cartesian product of {theme, cause} and possible event labels {Phosphorylation, Binding, ...}). Of course, some of these outputs do not apply. For example, it is highly unlikely that the rule shown in Figure 2 produces the cause of a Regulation event. We quantify the confidence in these outputs in the next stage of the algorithm.

Converting Weights to Votes

Feature weights are unbounded continuous values that are difficult to interpret and manually modify. For this reason, we would ideally prefer to exclude them completely from the interpretable model. Conceptually, this is simple: we could use the weights to choose the most likely output label for a rule (from the options generated previously), and discard them afterwards. However, our early experiments demonstrated that this performs poorly, because it forces the algorithm to ignore the inherent ambiguity of language, which is captured by the statistical model through weights. For example, the trigger classifier learns that "recruits" serves as trigger for two different events, Binding and Localization, and, consequently, assigns different weights to the two labels based on

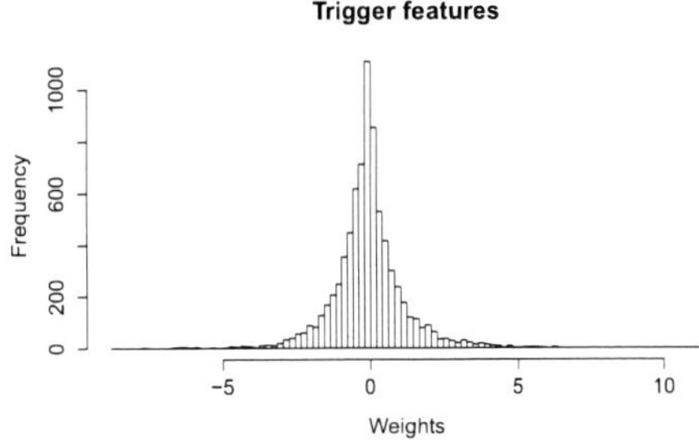

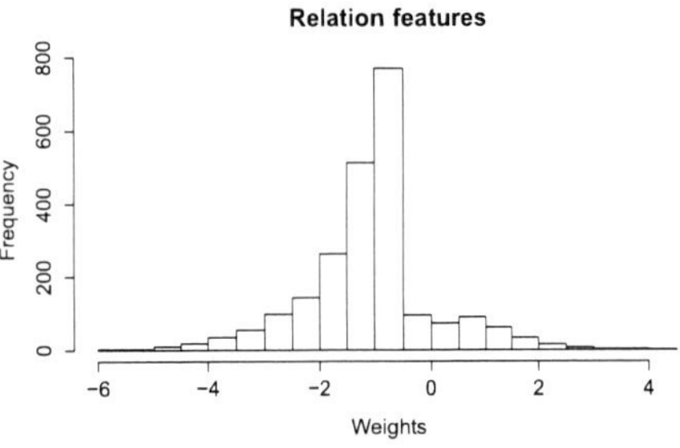

Figure 3: Weights of the two classifiers converted to votes (trigger classifier – top, participant classifier – bottom). Each histogram bin receives a number of votes (positive or negative) equal to its offset from 0.

the amount of evidence seen in training. During inference, the most likely class is chosen by aggregating the weights of all features that apply.

Given this observation, we chose to preserve the weights, but convert them from the original unbounded continuous values to discrete "votes" (positive or negative) that are then used during inference to resolve conflicts. This achieves two things. First, we increase the interpretability of the model: humans can now interpret these discrete votes, which mimic a Likert scale (Likert, 1932). Second, by keeping and using these discretized votes, we preserve some of the statistical power of the model. We show in Section 3 that some performance is indeed lost in this conversion, but the loss is small and the gain in interpretability compensates for that.

The conversion from continuous weights to discrete votes is a process similar to choosing the bins in a histogram. In our case, we first construct a histogram of all feature weights. Then, each histogram bin receives a number of votes equal to its offset (positive or negative) from 0. For example, all the weights in the second bin to the left of 0 receive two negative votes. Several methods have been proposed for selecting the number of bins in a histogram, for example (Sturges, 1926; Doane, 1976; Freedman and Diaconis, 1981). Here, we use the formula proposed by (Scott, 1979):

$$h = 3.5\hat{\sigma}n^{-1/3} \qquad (1)$$

where h is the estimated bin width, n is the sample size, and $\hat{\sigma}$ is the estimated standard deviation. We chose this formula because it gives a good compromise between retaining most of the information in the weights while minimizing the number of bins. The resulting binned weights for trigger and relation features (generated using the BioNLP 2009 training corpus) are shown in Figure 3.

2.3 Step 3: Edit the Rule-based Model

The output of the previous two steps is a model consisting of a set of rules. The association between rules and output classes is measured through votes that each matching rule gives to each output label. The last step in our proposed approach is to let human domain experts improve this model by directly editing it. The experts had complete freedom in the operations they were allowed to do. For example, they could improve the syntactic paths captured by the rules, or increase/decrease the number of votes assigned to a specific rule. The only constraints were: (a) they were not allowed to look only at the learned rules and not at the training data, and (b) they had to complete the process within one hour. This setting is of course artificial and unrealistic. We enforced it in this work to demonstrate the interpretability of the generated model.

3 Empirical Results

We analyze the performance of our approach on the core event extraction dataset from the BioNLP 2009 shared task (Kim et al., 2009). All the results reported in this section were measured on the *development* partition of the dataset, which was not used at all during training.[2] To minimize overfitting, we did not implement any feature selection or other hyper parameter tuning process.

Table 2 lists the results of the complete statistical model, i.e., using all features introduced in Section 2.1, trained using L_2-regularized LR. This configuration generated 1,190,029 features with non-zero weights. The table shows that this model

Event Class	Recall	Precision	F1
Gene_expression	67.70	68.08	67.89
Transcription	57.32	50.00	53.41
Protein_catabolism	71.43	68.18	69.77
Phosphorylation	68.09	68.09	68.09
Localization	69.81	74.00	71.84
Binding	31.85	25.57	28.37
Event Total	55.89	51.48	53.59
Regulation	17.16	33.33	22.66
Positive_regulation	19.45	41.67	26.52
Negative_regulation	14.29	36.36	20.51
Regulation Total	18.02	39.16	24.69
All Total	35.10	47.29	40.30

Table 1: Performance of the statistical model using L_2-regularized LR, and all available features.

Event Class	Recall	Precision	F1
Gene_expression	57.58	74.28	64.87
Transcription	40.24	57.89	47.48
Protein_catabolism	61.90	86.67	72.22
Phosphorylation	51.06	82.76	63.16
Localization	47.17	92.59	62.50
Binding	18.15	34.62	23.81
Event Total	42.75	64.61	51.45
Regulation	8.28	40.00	13.73
Positive_regulation	17.18	42.74	24.51
Negative_regulation	7.14	40.00	12.12
Regulation Total	13.65	42.14	20.62
All Total	26.77	56.22	36.27

Table 2: Performance of the statistical model with L_2-regularized LR, using only features that can be converted to rules.

achieved an overall F1 score of over 40 points, which likely puts it in the top 5 or 6 (out of 24) systems that participated in the actual challenge.[3] The performance of this system could be further improved by adding more features proposed in other event extraction approaches (Miwa et al., 2010), feature selection, hyper parameter tuning, etc.

For a fair comparison, we next trained the same model but using only features that can be converted to rules. As discussed, the features that were removed include bag-of-word features and features that count occurrences of tokens or entities in text. These results, summarized in Table 2, show that the overall F1 score drops 4 points. This suggests that rule languages need to be extended if they are to have the same representational power as feature-based models. Given that the focus of

---

[2]The online scoring website, which would have allowed us to also obtain scores on the official test partition, was down due to updates during the development of this work.

[3](Kim et al., 2009) report results on the official test partition, which are not directly comparable with our results. However, in the authors' experience, the difference in scores between the development and test partitions in this dataset tend to be small. Since the 2009 evaluation, several works have improved upon these results, with performance reaching 58 F1 points, but using more complex methods, including joint inference, coreference resolution, and domain adaptation (Miwa et al., 2012; Bui and Sloot, 2012; Venugopal et al., 2014).

Event Class	Recall	Precision	F1
Gene_expression	58.71	78.28	67.09
Transcription	37.80	55.36	44.93
Protein_catabolism	61.90	86.67	72.22
Phosphorylation	46.81	84.62	60.27
Localization	56.60	88.24	68.97
Binding	16.13	33.33	21.74
Event Total	42.75	66.60	52.08
Regulation	8.88	65.22	15.62
Positive_regulation	13.13	40.50	19.83
Negative_regulation	8.16	55.17	14.22
Regulation Total	11.41	44.44	18.15
All Total	25.54	59.35	35.72

Table 3: Performance of the statistical model with L_1-regularized LR, using only features that can be converted to rules.

Event Class	Recall	Precision	F1
Gene_expression	55.34	76.95	64.38
Transcription	28.05	53.49	36.80
Protein_catabolism	57.14	85.71	68.57
Phosphorylation	40.43	90.48	55.88
Localization	45.28	88.89	60.00
Binding	12.90	33.33	18.60
Event Total	38.04	67.18	48.58
Regulation	5.33	75.00	9.94
Positive_regulation	10.70	48.89	17.55
Negative_regulation	5.61	55.00	10.19
Regulation Total	8.76	51.50	14.97
All Total	21.97	62.98	32.57

Table 4: Performance of the rule-based model before expert intervention.

this work is not on the design of rule-based languages for IE, we will use this latter model as the starting point of our approach, ignoring (for now) the performance penalty observed above.

Importantly, a system with more than 1 million features is not interpretable. To address this, we trained the same system using L_1 regularization as a form of feature selection. This reduced the number of features with non-zero weights by two orders of magnitude: from over 1 million to 10,926. The performance of this model is shown in Table 3. The results demonstrate that this drastic reduction in the number of useful features came with a small performance cost, of less than 1 F1 point.

Given this successful compression of the feature space, we next convert this L_1-regularized model to rules, using the approach discussed in Section 2.2. The performance of the rule-based model (before expert intervention!) is summarized in Table 4. The table shows that the overall cost of "snapping to grid" the statistical model is approximately 3 F1 points, which come from a drop in recall. This happens because many feature weights associated with specific labels (such as specific event triggers) have low values (due to sparsity), and, after the discretization process, the model can no longer prioritize these labels over the `Nil` class. Interestingly, the same process yielded a small increase in precision from 59% to 62%.

All in all, we consider a drop of 3 F1 points for the gain of interpretability an acceptable trade-off. To empirically demonstrate the value of interpretability, we let two Linguistics PhD students edit the generated rule-based model for one hour, aiming to improve its generalization, robustness to syntactic errors, and readability. The students were familiar with the Odin language (Valenzuela-Escarcega et al., 2015) so they could "read" the

model, and had a high-level understanding of the BioNLP shared task (although they did not participate in it). To guarantee that their recommendations came from understanding the model rather than other external factors, they were not given access to the BioNLP dataset. Given the large number of rules at this point, the students tended to randomly sample the rules in the model attempting to find repeated mistakes, rather than linearly inspect the list of rules. Table 5 summarizes the experts' recommendations. As shown, several of the experts' suggestions involved removing or collapsing rules, which reduced the number of rules from 10,926 to 8,868.

Table 6 lists the performance of the resulting model, after implementing the experts' recommendations. The table shows that most of the F1 loss has been recovered: the overall F1 score for this system approaches 35 F1 points, and is less than 1 F1 point behind the $L1$-regularized LR statistical model. In addition of reducing the number of rules in the model, the experts' recommendations increased recall by over 4%, which is more than what was lost during the conversion to rules. However, the precision of this configuration decreased by 11%, which we blame on the experts' limited familiarity with the BioNLP task, and the strict settings of the experiment (no access to data, limited time). However, all in all, this experiment demonstrates that the rule-based model produced by the proposed approach is interpretable: the experts understood the model, and were able to improve it, both with respect to its generalization power and its readability.

Lastly, Figure 4 shows a learning curve for the statistical model and the corresponding rule-based model (before expert intervention). The curve shows that the rule-based model follows closely

Suggested Change	Description
Generalization	
Add `/conj_(and|or|nor)|dep|cc|nn|prep_of/{,2}` to the end of Theme paths.	This transformation adds an optional modifier dependency to capture event participants when they appear either as nominal heads or modifiers. For example, because of this transformation, the model handles both these phrases similarly: "phosphorylation of MEK" and "phosphorylation of the MEK protein"
Ensure that all syntactic paths end in `appos?`.	This change handles optional apposition to increase rule coverage. For example, in the sentence "we found that A20 binds to a novel protein, ABIN", the word ABIN is an appositive for the word protein, so ABIN can serve as an argument in the binding event.
Replace all specific named entities with their label.	For example, in rules such as `[word=phosphorylates] (?=MEK)` that reference a specific protein, this replaces the specific protein (MEK) with the label `Protein`. This improves rule generalization and, at the same time, reduces the total number of rules.
Make the `>nn` dependency optional in `Theme:Protein = >nsubjpass >nn.`	The output of this transformation is similar to the first suggested change, i.e., the same rule captures event participants when they appear either as nominal heads or modifiers.
Robustness	
Replace `agent` with `/^(agent|prep_by)$/`.	This modification is designed to account for a common parsing error of passive sentences, where `agent` dependencies are incorrectly parsed as `prep_by`.
Change `ccomp` to `/(c|x)comp/` and `acomp` to `/(a|x)comp/`.	Parsers often confuse clausal and adjectival complements with open clausal complements. This transformation allows the rules to be robust to these errors.
Readability	
Merge rules when possible, e.g. `prep_of, prep_of nn, prep_of appos` become `prep_of (nn? appos | nn appos? nn)?`.	This transformation collapses rules to improve readability.
Eliminate trigger rules that are not sufficiently discriminative (e.g., `(?<=[lemma="be"]) [tag=/^(V|N|J)/]`).	Some uninformative rules survived feature regularization but should be removed, as with the example rule which looks for any verb, noun, or adjective preceded by any conjugation of the verb "be". These rules inflate the grammar without adding discriminative power.
Do not use `word` constraints. Only use `lemma` and `tag` features in trigger rules for simple events (other than transcription and binding).	This modification prefers lexical constraints on lemmas, because they generalize better than constraints on actual words.
Remove redundant constraints.	For example, in patterns like `[incoming=nsubj & tag=/^N/]` the POS tag is redundant because it is implicitly defined through the incoming dependency (nominal subject).

Table 5: Representative examples of the rule changes suggested by linguistic experts.

the behavior of its statistical counterpart, with a small penalty of 1-2 F1 points throughout. As discussed before, this performance loss can be mitigated through interventions by domain experts.

4   Related Work

Most of the biomedical IE systems in academia rely on supervised machine learning. This includes the top performing system at the BioNLP 2009 shared task (Björne et al., 2009), as well as several following approaches that improve upon its performance (Miwa et al., 2010; McClosky et al., 2012; Miwa et al., 2012; Bui and Sloot, 2012; Venugopal et al., 2014).

However, rule-based approaches (Appelt et al., 1993; Cunningham et al., 2002; Piskorski et al., 2004; Li et al., 2011; Chang and Manning, 2014) are preferable when the corresponding systems have to be deployed for long periods of time, during which they have to be maintained and improved. This has been recognized in industry (Chiticariu et al., 2013).

We bring together these two diverging directions by combining the advantages of ML with the interpretability of rule-based approaches. By representing the model as a collection of declarative rules, experts can directly edit the model, thus guaranteeing that the desired changes are actually applied. This is in contrast with methods such as active learning, in which the learning algorithm

Event Class	Recall	Precision	F1
Gene_expression	60.39	70.49	65.05
Transcription	31.71	57.78	40.94
Protein_catabolism	61.90	81.25	70.27
Phosphorylation	42.55	86.96	57.14
Localization	45.28	88.89	60.00
Binding	22.18	23.50	22.82
Event Total	43.74	54.31	48.46
Regulation	10.06	40.48	16.11
Positive_regulation	12.80	44.89	19.92
Negative_regulation	10.71	51.22	17.72
Regulation Total	11.91	45.17	18.86
All Total	26.27	51.71	34.84

Table 6: Performance of the rule-based model, after expert intervention.

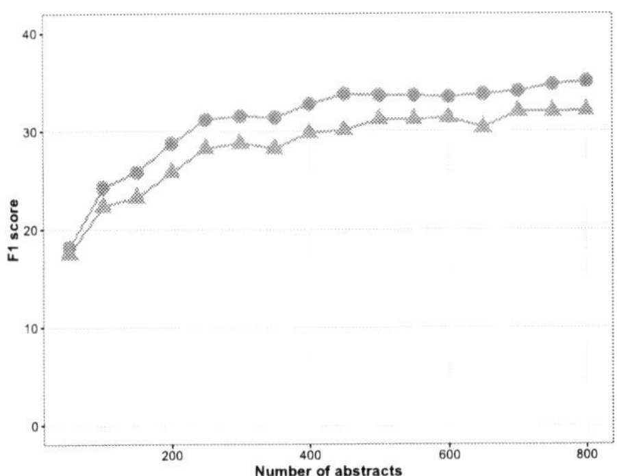

Figure 4: Learning curve showing the change in F1 performance as a function of the amount of training data. We compare the performance of the L_1-regularized logistic regression (shown using circles) with the rule-based model prior to the expert intervention (shown using triangles).

presents the "human in the loop" with new examples to annotate (Thompson et al., 1999). Although active learning may require less domain expertise than our proposal, it generally does not guarantee that the examples provided are actually propagated in the model (the learning algorithm may choose to override them with other data).

5 Conclusion and Future Work

We have proposed a simple approach that marries the advantages of machine learning models for information extraction (such as learning directly from data) with the benefits of rule-based approaches (interpretability, easier maintainability). Our approach starts by training a feature-based statistical model, then converts this model to a rule-based variant by converting its features to rules and its feature weights to discrete votes. In doing so, our proposal learns from data similar to other machine learning approaches, but produces an interpretable rule-based model that can be directly edited by experts. Using the BioNLP 2009 event extraction task as a test bed, we show that while there is a small performance penalty when converting the statistical model to rules, the gain in interpretability compensates for that.

In this work, we focused on building upon feature-based classifiers, in particular logistic regression, due to their potential extensions to distant supervision (DS), where training data is generated automatically by aligning a knowledge base (KB) of known examples (e.g., known drug-gene interactions) with text (e.g., scientific publications). Distant supervision has obvious applications to bioinformatics (Craven et al., 1999), but it generally suffers from noise in the automatically-generated annotations (Riedel et al., 2010). In future work, we plan to combine our work with distant supervision by adapting our proposal to logistic regression variants that are robust to the noise introduced in DS (Surdeanu et al., 2012). This extension would make it possible to generate rules even when no annotated examples are available, as long as a suitable KB of known examples exists.

Another planned extension of this work focuses on reducing the number of generated rules by merging/collapsing similar paths into a single pattern. This can be achieved by constructing a minimal deterministic acyclic finite-state automaton (DAFSA) (Daciuk et al., 2000) with the paths that are similar, and then converting the DAFSA into a single pattern (Neumann, 2005). For example, such approaches would collapse the two patterns: `dobj` and `dobj nn`, into a single one: `dobj nn?`. This is fundamental for the long-term maintainability of the rule-based model, because the human experts would have to maintain considerably fewer rules.

Lastly, we plan to improve the "snap to grid" algorithm. Currently, the conversion of weights to votes is implemented using Scott's rule (Scott, 1979), which is one method among several available to choose a histogram's bin size. Scott's method assumes that all bins have the same size, which may not be the best solution if interpretability is the goal. A potentially better approach is to select the bin divisions in a way that retains as much of the information contained in the weights as possible, while minimizing the number of bins.

Acknowledgments

This work was funded by the Defense Advanced Research Projects Agency (DARPA) Big Mechanism program under ARO contract W911NF-14-1-0395.

References

Douglas E. Appelt, Jerry R. Hobbs, John Bear, David Israel, and Mabry Tyson. 1993. Fastus: A finite-state processor for information extraction from real-world text. In *Proceedings of the International Conferences on Artificial Intelligence (IJCAI)*.

Jari Björne, Juho Heimonen, Filip Ginter, Antti Airola, Tapio Pahikkala, and Tapio Salakoski. 2009. Extracting complex biological events with rich graph-based feature sets. In *Proceedings of the Workshop on Current Trends in Biomedical Natural Language Processing: Shared Task*, pages 10–18. Association for Computational Linguistics.

Quoc-Chinh Bui and Peter MA Sloot. 2012. A robust approach to extract biomedical events from literature. *Bioinformatics*, 28(20):2654–2661.

Angel X. Chang and Christopher D. Manning. 2014. TokensRegex: Defining cascaded regular expressions over tokens. Technical Report CSTR 2014-02, Computer Science, Stanford.

Laura Chiticariu, Yunyao Li, and R. Reiss, Frederick. 2013. Rule-based Information Extraction is Dead! Long Live Rule-based Information Extraction Systems! In *Proceedings of 2013 Conference on Empirical Methods in Natural Language Processing (EMNLP)*.

Paul R. Cohen. 2015. DARPA's Big Mechanism program. *Physical Biology*, 12(4):045008.

Mark Craven, Johan Kumlien, et al. 1999. Constructing biological knowledge bases by extracting information from text sources. In *ISMB*, volume 1999, pages 77–86.

Hamish Cunningham, Diana Maynard, Kalina Bontcheva, and Valentin Tablan. 2002. A framework and graphical development environment for robust nlp tools and applications. In *ACL*, pages 168–175.

Jan Daciuk, Stoyan Mihov, Bruce W Watson, and Richard E Watson. 2000. Incremental construction of minimal acyclic finite-state automata. *Computational linguistics*, 26(1):3–16.

Marie-Catherine De Marneffe and Christopher D Manning. 2008. Stanford typed dependencies manual. Technical report, Stanford University.

David P Doane. 1976. Aesthetic frequency classifications. *The American Statistician*, 30(4):181–183.

David Freedman and Persi Diaconis. 1981. On the histogram as a density estimator: l_2 theory. *Probability theory and related fields*, 57(4):453–476.

Mohit Iyyer, Anupam Guha, and Jordan Boyd-Graber. 2016. Feuding Families and Former Friends: Unsupervised Learning for Dynamic Fictional Relationships. In *Proceedings of the 15th Annual Conference of the North American Chapter of the Association for Computational Linguistics: Human Language Technologies (NAACL-HLT)*.

Jin-Dong Kim, Tomoko Ohta, Sampo Pyysalo, Yoshinobu Kano, and Junichi Tsujii. 2009. Overview of BioNLP09 Shared Task on Event Extraction. In *Proceedings of the Workshop on BioNLP: Shared Task*.

Jin-Dong Kim, Yue Wang, and Yamamoto Yasunori. 2013. The Genia Event Extraction Shared Task, 2013 Edition - Overview. In *Proceedings of the BioNLP Shared Task 2013 Workshop*.

Yunyao Li, Frederick R Reiss, and Laura Chiticariu. 2011. Systemt: A declarative information extraction system. In *Proceedings of the 49th Annual Meeting of the Association for Computational Linguistics: Human Language Technologies: Systems Demonstrations*, pages 109–114. Association for Computational Linguistics.

Rensis Likert. 1932. A technique for the measurement of attitudes. *Archives of psychology*.

Christopher D Manning, Mihai Surdeanu, John Bauer, Jenny Rose Finkel, Steven Bethard, and David McClosky. 2014. The stanford corenlp natural language processing toolkit. In *ACL (System Demonstrations)*, pages 55–60.

David McClosky, Sebastian Riedel, Mihai Surdeanu, Andrew McCallum, and Christopher D. Manning. 2012. Combining joint models for biomedical event extraction. *BMC Bioinformatics*, 13(Suppl 11)(S9).

Makoto Miwa, Rune Sætre, Jin-Dong Kim, and Jun'ichi Tsujii. 2010. Event extraction with complex event classification using rich features. *Journal of bioinformatics and computational biology*, 8(01):131–146.

Makoto Miwa, Paul Thompson, and Sophia Ananiadou. 2012. Boosting automatic event extraction from the literature using domain adaptation and coreference resolution. *Bioinformatics*, 28(13):1759–1765.

Christoph Neumann. 2005. Converting deterministic finite automata to regular expressions. http://neumannhaus.com/christoph/papers/2005-03-16.DFA_to_RegEx.pdf.

Jakub Piskorski, Ulrich Schäfer, and Feiyu Xu. 2004. Shallow processing with unification and typed feature structures–foundations and applications. *Knstliche Intelligenz*, 1(1).

Sebastian Riedel, Limin Yao, and Andrew McCallum. 2010. Modeling relations and their mentions without labeled text. In *Machine Learning and Knowledge Discovery in Databases*, pages 148–163. Springer.

David W Scott. 1979. On optimal and data-based histograms. *Biometrika*, 66(3):605–610.

D. Sculley, Gary Holt, Daniel Golovin, Eugene Davydov, Todd Phillips, Dietmar Ebner, Vinay Chaudhary, Michael Young, Jean-Francois Crespo, and Dan Dennison. 2014. Hidden Technical Debt in Machine Learning Systems. In *Proceedings of SE4ML: Software Engineering for Machine Learning (NIPS 2014 Workshop)*.

Herbert A Sturges. 1926. The choice of a class interval. *Journal of the american statistical association*, 21(153):65–66.

Mihai Surdeanu, Julie Tibshirani, Ramesh Nallapati, and Christopher D Manning. 2012. Multi-instance multi-label learning for relation extraction. In *Proceedings of the 2012 Joint Conference on Empirical Methods in Natural Language Processing and Computational Natural Language Learning*, pages 455–465. Association for Computational Linguistics.

Cynthia A Thompson, Mary Elaine Califf, and Raymond J Mooney. 1999. Active learning for natural language parsing and information extraction. In *ICML*, pages 406–414. Citeseer.

Marco A. Valenzuela-Escarcega, Gustave Hahn-Powell, and Mihai Surdeanu. 2015. Description of the Odin Event Extraction Framework and Rule Language. In *arXiv:1509.07513*.

Marco A. Valenzuela-Escarcega, Gustave Hahn-Powell, and Mihai Surdeanu. 2016. Odin's Runes: A Rule Language for Information Extraction. In *Proceedings of the 10th edition of the Language Resources and Evaluation Conference (LREC)*.

Deepak Venugopal, Chen Chen, Vibhav Gogate, and Vincent Ng. 2014. Relieving the computational bottleneck: Joint inference for event extraction with high-dimensional features. In *EMNLP*, pages 831–843.

Character based String Kernels for Bio-Entity Relation Detection

Ritambhara Singh
Department of Computer Science
University of Virginia
Charlottesville
rs3zz@virginia.edu

Yanjun Qi
Department of Computer Science
University of Virginia
Charlottesville
yanjun@virginia.edu

Abstract

Extracting bio-entity relations has emerged as an important task due to the ever-growing number of bio-medical documents. In this paper, we present a simple and novel representation for extracting bio-entity relationships. The state-of-the-art systems for such tasks rely on word based representations and variations of linguistic driven features. In contrast, we model bio-text by the most basic character based string representation with a family of string kernels. This eliminates time consuming parsing, issue of rare words and domain specific pre-processing. This simple representation makes our approach fast and flexible for any bio-NLP dataset. We demonstrate comparable performance and faster computation time of our approach versus previous state-of-the-art kernel methods.

1 Introduction

Relation extraction from biomedical documents is an important task in knowledge representation and inference. It helps to construct and enhance structured knowledge-bases and in turn support automatic question answering and decision making. In today's era of vast amount of information collection and retrieval, the task of naming and identifying the relations between annotated bio-entities can become complex and time consuming. This can be deducted from the fact that the MED-LINE database has more than 22 million journal articles related to biomedicine. Many state-of-the art methods have been applied for the popular tasks of extracting protein-protein interaction (PPI) and drug-drug interaction (DDI) as a part of BioCreative shared task challenges (Segura Bedmar et al., 2011; Segura Bedmar et al., 2013;

Krallinger et al., 2008). While these methods have achieved good performance, they mostly rely on word-level features, are dependent on time-consuming parsers or require domain knowledge for pre-processing.

This paper uses characters instead of words for bio-entity relation extraction. Characters are the most fundamental building blocks in any language. We propose to model bio-text using its most basic character-based string representation. Through a string kernel implementation, in the framework of support vector machine (SVM), we separate positive and negative interaction instances to detect bio-entity relationships. This basic representation is independent of parsers, does not require domain-related pre-processing and eliminates the rare words problem. It not only performs comparable to other state-of-the-art methods but also provides an exploration of new and simple feature sets (complementary to existing features) that have not been previously studied for bio-NLP shared tasks.

2 Related Work

Convolution-based kernel methods have been used extensively in the tasks of PPI and DDI extraction, and differ in the feature sets they explore. While the shallow linguistic (SL) kernel (Giuliano et al., 2006) uses simple linguistic features, others utilize more complex features. Constituent parse tree-based kernels, like subtree (ST) (Vishwanathan et al., 2004), subset tree (SST) (Collins and Duffy, 2001), partial tree (PT) (Moschitti, 2006) kernels, and spectrum tree (SpT) (Kuboyama et al., 2007) kernel, use subtree forms or path structures from constituent parse trees. Another category of methods use dependency parse tree-based features. This includes edit distance and cosine similarity kernels (using shortest paths) (Erkan et al., 2007), k-band shortest path spectrum (kBSPS) (Tikk et al., 2010) (a

Proceedings of the 15th Workshop on Biomedical Natural Language Processing, pages 66–71,
Berlin, Germany, August 12, 2016. ©2016 Association for Computational Linguistics

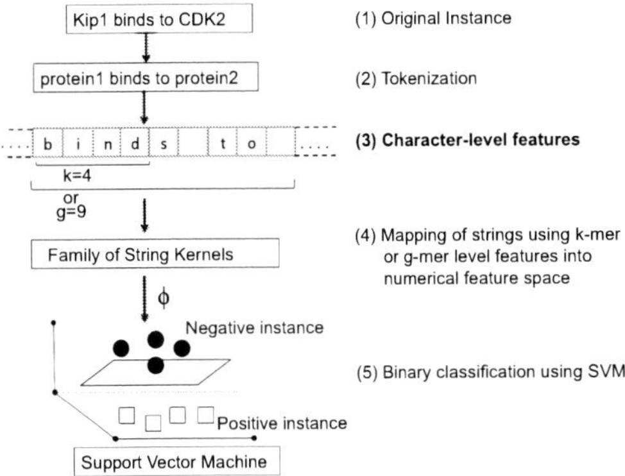

Figure 1: **End-to-end implementation of character-based string kernels for bio-entity relation detection.**

Corpus	Task	Sent.	Pos	Neg	Total
MEDLINE	DDI	1301	232	1555	1787
AIMed	PPI	1955	1000	4834	5834
LLL	PPI	77	164	166	330

Table 1: Statistics (number of sentences, positive, negative and total instances) of the MEDLINE corpus about DDI and, AIMed and LLL corpus about PPI extraction respectively.

k-band extension of shortest paths), all-path graph (APG) kernel (Airola et al., 2008) (weighing differently shortest paths), and Kim's kernels (Kim et al., 2008) (combines shortest path with different lexical, part-of-speech and syntactic features). Benchmark papers, such as (Tikk et al., 2010) and (Tikk et al., 2013), have performed thorough comparative and error analyses of all these different kernels. They concluded that APG, kBSPS and SL kernels give the best performance. Therefore, we use these three kernels as baselines in our experimental comparisons. Some studies include a combination of kernels and parsers for PPI extraction task, e.g. (Miwa et al., 2009). Similarly, (Thomas et al., 2013) implemented a two-step approach to first detect general DDIs and then classify detected DDIs into subtypes. For the general DDI task, they used voting to combine kernels including APG, subtree (ST), SST, SpT, and SL kernels.

All the above discussed methods suffer from the rare words problem, and require time consuming and domain specific pre-processing steps like parsing to obtain lexical features, constituent and dependency trees. Several recent studies have discovered that character-based representation provides simple and powerful models for sentiment classification (Zheng et al., 2015) and transition-based parsing (Ballesteros et al., 2015). (Lodhi et al., 2002) first used string kernels with character level features for text categorization. However, their kernel computation used dynamic programming which is computationally intensive. Over recent years, more efficient string kernel methods have been devised (Leslie and Kuang, 2004;

Kuksa et al., 2009). Therefore, we apply a family of state-of-the-art string kernels using simple character-based string representation for bio-entity relation detection in this work.

3 Approach

Figure 1 shows our end-to-end implementation of character-based string kernel approach for bio-entity relation detection.

3.1 Character-level features

Without relying on any pre-processing, we directly use the instance sentences of bio-NLP datasets as input to the string kernels. Here, each instance (whole sentence) is viewed as one long contiguous string comprised of characters. A string kernel is then used to convert these strings into a feature space (implicitly through kernel calculation) that can be used as input for support vector machine (SVM) classification algorithm.

3.2 Family of string kernels

The key idea of string kernels is to apply a function $\phi(\cdot)$, which maps strings of arbitrary length into a vectorial feature space of fixed length. In this space, a standard classifier such as SVM (Vapnik, 1998) can then be applied. Kernel-version of SVMs calculate the decision function for an input sample x :

$$f(x) = \sum_{i=1}^{N} \alpha_i K(x_i, x) + b \qquad (1)$$

where N is the total number of training samples. String kernels (Leslie and Kuang, 2004; Kuksa et al., 2009; Ghandi et al., 2014a), implicitly compute an inner product in the mapped feature space $\phi(x)$ as:

$$K(x, x') = \langle \phi(x), \phi(x') \rangle, \qquad (2)$$

where $x = (s_1, \ldots, s_{|x|})$. $x, x' \in \mathcal{S}$. $|x|$ denotes the length of the string x. $\mathcal{S}$ represents the set of all strings composed of dictionary Σ. $\phi : \mathcal{S} \to R^m$ defines the mapping from a sequence $x \in \mathcal{S}$ to a m-dimensional feature vector.

The feature representation $\phi(\cdot)$ plays a key role in the effectiveness of string analysis since strings cannot be readily described as feature vectors. We have implemented the following string kernels on the character representation.

Spectrum Kernel (SK): One classic representation is to represent a string as unordered set of k-mers, that is, combinations of k adjacent characters. A feature vector indexed by all k-mers records the number of occurrences of each k-mer in the current string. The string kernel using this representation is called spectrum kernel (Leslie et al., 2002), where the spectrum representation counts the occurrences of each k-mer in a string. Kernel scores between strings are then computed by taking an inner product between corresponding "k-mer - indexed" feature vectors:

$$K(x, x') = \sum_{\gamma \in \Gamma_k} c_x(\gamma) \cdot c_{x'}(\gamma) \qquad (3)$$

where γ represents a k-mer, Γ_k is the set of all possible k-mers, and $c_x(\gamma)$ is the number of occurrences (with normalization) of k-mer γ in string x. (Kuboyama et al., 2007) applied spectrum kernel on the constituent parse tree features.

Mismatch Kernel (MK): The spectrum kernel implementation is modified to include m number of mismatches in the k-mers (Leslie and Kuang, 2004; Kuksa et al., 2009). Thus, for a given k-mer γ in a string x, the (k, m)-neighborhood is generated. This consists of all k-length strings α from dictionary Σ^k such that they differ from original k-mer by at most m mismatches. The feature map of mismatch kernel can be defined as:

$$\phi_{(k,m)}(\gamma) = (\phi_\alpha(\gamma))_{\alpha \in \Sigma^k} \qquad (4)$$

where $\phi_\alpha(\gamma) = 1$ if $\alpha \in (k, m)$-neighborhood of γ, otherwise $\phi_\alpha(\gamma) = 0$. A mismatch kernel with $m = 0$ is essentially a spectrum kernel.

Wildcard Kernel (WK): For implementation of wildcard kernel, the dictionary Σ is augmented with a wildcard character $\star$ (Leslie and Kuang, 2004). Thus, the feature space consists of set of k-mers Γ_k obtained from $\Sigma \cup \{\star\}$ that consists of m occurrences of wildcard character ($\star$). The feature map of wildcard kernel can be defined as:

$$\phi_{(k,m,\lambda)}(\gamma) = \sum_{\Gamma_k \, in \, x} (\phi_\alpha(\gamma))_{\alpha \in (\Sigma \cup \{\star\})} \qquad (5)$$

where $\phi_\alpha(\gamma) = \lambda^m$ if γ matches α with m occurrences of character $\star$, otherwise $\phi_\alpha(\gamma) = 0$. Here $0 < \lambda \leq 1$.

Gapped k-mer based Kernel (GK): The previously described k-mer based string kernels generate extremely sparse feature vectors for even moderately sized values of k, resulting in overfitting. (Ghandi et al., 2014b) introduced a new feature set, called *gapped k-mers*, resolving the sparsity limitation with k-mer features. It is characterized by two parameters; (1) g, size of a gapped instance which is a segment of string including gaps and (2) k, the number of non-gapped k-mers or positions in each segment of size g. Thus, the number of gaps $d = g - k$. The inner product in equation 3 includes sum over all gapped k-mers features:

$$K(x, x') = \sum_{\gamma \in \Theta_g} c_x(\gamma) \cdot c_{x'}(\gamma) \qquad (6)$$

where γ represents a k-mer, Θ_g is the set of all possible g-mers in the given data.

3.3 Classification

Once the kernel matrix K is calculated, we input it into an SVM classifier as an empirical feature map using SVM^{light} (Joachims, 1999; Schölkopf and Burges, 1999). SVM maximizes the margin between the positive and negative instances of bio-entity interactions in the kernel defined feature space.

4 Experiment Setup

Datasets We demonstrate the benchmark implementations of our approach on three datasets with different sample sizes. They include MEDLINE corpus from the DDI extraction task (Segura Bedmar et al., 2011; Segura Bedmar et al., 2013), and the AIMed and LLL corpus from the PPI extraction task (Krallinger et al., 2008). [1]. The details of the datasets have been presented in Table 1.

Baselines We selected SL (Giuliano et al., 2006), APG (Airola et al., 2008), and kBSPS

---

[1] We use the same format as used in previous studies, that is, each interaction is represented as a separate input instance. Thus, a sentence about multiple interactions is represented as multiple instances. The protein name entities are replaced by special tokens. See details in (Tikk et al., 2010)

Corpus	Task	kBSPS	APG	SL	SK	(k)	MK	(k,m)	WK	(k,m)	GK	(g,k)
MEDLINE	DDI	-	82.3	78.9	82.1	(7)	82.7	(7,3)	**83**	(7,3)	82.4	(7,4)
AIMed	PPI	75.1	84.6	83.5	**75.6**	(8)	74.9	(10,5)	75.2	(10,5)	75.4	(8,6)
LLL	PPI	84.3	83.5	81.2	67.9	(7)	77.9	(7,3)	**78.4**	(8,5)	78.1	(7,5)

Table 2: Using AUC score to compare four character-based string kernels with APG, kBSPS and SL baselines. The best performing kernel parameters are also presented. AUC scores for APG and SL kernels for MEDLINE corpus have been reported in (Thomas et al., 2013), while scores of all three baseline kernels for AIMed and LLL corpus are reported in (Tikk et al., 2010).

Corpus	Task	kBSPS	APG	SL	SK	MK	WK	GK
MEDLINE	DDI	169.13	169.13	5.2	0.4	2.6	3.1	2.6
AIMed	PPI	254.15	254.14	7.82	76.8	79.5	78	41.3
LLL	PPI	10	10	0.3	0.2	1.3	1	0.2

Table 3: Comparing the kernel computation time (in seconds) for all four character-based string kernels versus the estimated parsing times of state-of-the-art baselines reported from (Tikk et al., 2010).

(Tikk et al., 2010) kernels as baselines for comparing with character-based string kernels. These kernels are the top-performing approaches, as reported by(Tikk et al., 2010; Tikk et al., 2013) and (Luo et al., 2016).

Parameters We ran all our string kernels across multiple kernel parameter settings as follows: (1) SK : $k = \{6,7,8,9,10\}$, (2) MK,WK : $k = \{6,7,8,9,10\}$ and $m = \{1,..,k-1\}$, and (3) GK : $g = \{6,7,8,9,10\}$ and $k = \{1,...,g-1\}$. These string kernels were implemented using *gkmsvm* (Ghandi et al., 2014a) tool. The character level dictionary, $\Sigma = \{a,...,z,0,1,...9\}$ (size=36), is consistent for all the datasets and kernels.

Evaluation Metrics We performed 10-fold document level cross-validation on each selected corpus and calculated the AUC score (area under the receiver operating characteristic curve) for performance evaluation. (Tikk et al., 2010) confirmed that AUC score is more stable to parameter modifications and less sensitive to the ratio of positive/negative pairs in the corpus than F-score. Hence, AUC score is our choice for performance metric. We also recorded the kernel calculation times (in seconds) for all four string kernels.

5 Results

Table 2 summarizes the performance evaluation. The AUC scores for APG and SL kernels for MedLine corpus have been reported in (Thomas et al., 2013), while scores of all baseline kernels for AIMed and LLL corpus are reported in (Tikk et al., 2010). Our string kernel approaches, with simple character features, (WK,MK, and GK), outperform the baseline kernels (Table 2) on the MedLine corpus. For the AIMed corpus, SK, GK, and WK give higher AUC score than the baseline kBSPS kernel. Our methods give reasonable performance for LLL corpus as well, however not as good as the three baseline kernels. The parameters giving the best AUC performance are also reported (Table 2). Our representation is complementary and can be plugged into state-of-the-art baselines to further improve their systems.

Table 3 presents the kernel computation time comparison of all four character-based string kernels versus the baselines. We compare this with the estimated parsing times. For the baseline kernels, these have been reported and calculated in (Tikk et al., 2010). Unlike baseline kernels, we use character level features directly and thus do not need the parsing step.

6 Discussion

We have proposed a simple and novel character representation for bio-entity relation detection task. We implement a family of string kernels on such simple features extracted directly from instances of PPI and DDI extraction task datasets. This eliminates time-consuming and domain-specific pre-processing steps, making our approach fast and flexible for any bio-NLP dataset. Hence, our work opens new avenues to explore different and simpler feature sets at the character level.

References

Antti Airola, Sampo Pyysalo, Jari Björne, Tapio Pahikkala, Filip Ginter, and Tapio Salakoski. 2008. All-paths graph kernel for protein-protein interaction extraction with evaluation of cross-corpus learning. *BMC bioinformatics*, 9(11):1.

Miguel Ballesteros, Chris Dyer, and Noah A Smith. 2015. Improved transition-based parsing by modeling characters instead of words with LSTMs. *arXiv preprint arXiv:1508.00657*.

Michael Collins and Nigel Duffy. 2001. Convolution kernels for natural language. In *Advances in neural information processing systems*, pages 625–632.

Günes Erkan, Arzucan Özgür, and Dragomir R Radev. 2007. Semi-supervised classification for extracting protein interaction sentences using dependency parsing. In *EMNLP-CoNLL*, volume 7, pages 228–237.

Mahmoud Ghandi, Dongwon Lee, Morteza Mohammad-Noori, and Michael A Beer. 2014a. Enhanced regulatory sequence prediction using gapped k-mer features. *PLoS Comput Biol*, 10(7):e1003711.

Mahmoud Ghandi, Morteza Mohammad-Noori, and Michael A Beer. 2014b. Robust k-mer frequency estimation using gapped k-mers. *Journal of mathematical biology*, 69(2):469–500.

Claudio Giuliano, Alberto Lavelli, and Lorenza Romano. 2006. Exploiting shallow linguistic information for relation extraction from biomedical literature. In *EACL*, volume 18, pages 401–408. Citeseer.

Thorsten Joachims. 1999. Making large scale SVM learning practical. Technical report, Universität Dortmund.

Seonho Kim, Juntae Yoon, and Jihoon Yang. 2008. Kernel approaches for genic interaction extraction. *Bioinformatics*, 24(1):118–126.

Martin Krallinger, Florian Leitner, Carlos Rodriguez-Penagos, Alfonso Valencia, et al. 2008. Overview of the protein-protein interaction annotation extraction task of BioCreative II. *Genome biology*, 9(Suppl 2):S4.

Tetsuji Kuboyama, Kouichi Hirata, Hisashi Kashima, Kiyoko F Aoki-Kinoshita, and Hiroshi Yasuda. 2007. A spectrum tree kernel. *Information and Media Technologies*, 2(1):292–299.

Pavel P Kuksa, Pai-Hsi Huang, and Vladimir Pavlovic. 2009. Scalable algorithms for string kernels with inexact matching. In *Advances in Neural Information Processing Systems*, pages 881–888.

Christina Leslie and Rui Kuang. 2004. Fast string kernels using inexact matching for protein sequences. *The Journal of Machine Learning Research*, 5:1435–1455.

Christina S Leslie, Eleazar Eskin, and William Stafford Noble. 2002. The spectrum kernel: A string kernel for SVM protein classification. In *Pacific symposium on biocomputing*, volume 7, pages 566–575.

Huma Lodhi, Craig Saunders, John Shawe-Taylor, Nello Cristianini, and Chris Watkins. 2002. Text classification using string kernels. *The Journal of Machine Learning Research*, 2:419–444.

Yuan Luo, Özlem Uzuner, and Peter Szolovits. 2016. Bridging semantics and syntax with graph algorithms–state-of-the-art of extracting biomedical relations. *Briefings in bioinformatics*, page bbw001.

Makoto Miwa, Rune Sætre, Yusuke Miyao, and Jun'ichi Tsujii. 2009. Protein–protein interaction extraction by leveraging multiple kernels and parsers. *International journal of medical informatics*, 78(12):e39–e46.

Alessandro Moschitti. 2006. Efficient convolution kernels for dependency and constituent syntactic trees. In *Machine Learning: ECML 2006*, pages 318–329. Springer.

Bernhard Schölkopf and Christopher JC Burges. 1999. *Advances in kernel methods: support vector learning*. MIT press.

Isabel Segura Bedmar, Paloma Martinez, and Daniel Sánchez Cisneros. 2011. The 1st DDIextraction-2011 challenge task: Extraction of drug-drug interactions from biomedical texts.

Isabel Segura Bedmar, Paloma Martínez, and María Herrero Zazo. 2013. Semeval-2013 task 9: Extraction of drug-drug interactions from biomedical texts (ddiextraction 2013). Association for Computational Linguistics.

Philippe Thomas, Mariana Neves, Tim Rocktäschel, and Ulf Leser. 2013. WDI-DDI. drug-drug interaction extraction using majority voting. In *Second Joint Conference on Lexical and Computational Semantics (* SEM)*, volume 2, pages 628–635.

Domonkos Tikk, Philippe Thomas, Peter Palaga, Jörg Hakenberg, and Ulf Leser. 2010. A comprehensive benchmark of kernel methods to extract protein–protein interactions from literature. *PLoS Comput Biol*, 6(7):e1000837.

Domonkos Tikk, Illés Solt, Philippe Thomas, and Ulf Leser. 2013. A detailed error analysis of 13 kernel methods for protein–protein interaction extraction. *BMC bioinformatics*, 14(1):1.

Vladimir N. Vapnik. 1998. *Statistical Learning Theory*. Wiley-Interscience, September.

SVN Vishwanathan, Alexander Johannes Smola, et al. 2004. Fast kernels for string and tree matching. *Kernel methods in computational biology*, pages 113–130.

Xiaoqing Zheng, Haoyuan Peng, Yi Chen, Pengjing Zhang, and Wenqiang Zhang. 2015. Character-based parsing with convolutional neural network. In *Proceedings of the 24th International Conference on Artificial Intelligence*, pages 1054–1060. AAAI Press.

Disambiguation of entities in MEDLINE abstracts
by combining MeSH terms with knowledge

Amy Siu, Patrick Ernst, Gerhard Weikum
Max Planck Institute for Informatics
66123 Saarbrücken, Germany
{siu, pernst, weikum}@mpi-inf.mpg.de

Abstract

Entity disambiguation in the biomedical domain is an essential task in any text mining pipeline. Much existing work shares one limitation, in that their model training prerequisite and/or runtime computation are too expensive to be applied to all ambiguous entities in real-time. We propose an automatic, light-weight method that processes MEDLINE abstracts at large-scale and with high-quality output. Our method exploits MeSH terms and knowledge in UMLS to first identify unambiguous anchor entities, and then disambiguate remaining entities via heuristics. Experiments showed that our method is 79.6% and 87.7% accurate under strict and relaxed rating schemes, respectively. When compared to MetaMap's disambiguation, our method is one order of magnitude faster with a slight advantage in accuracy.

1 Introduction

Motivation

The ever-growing volume of biomedical literature is published at a phenomenal pace. While the rich information buried in this literature can be extracted via text mining, entity recognition and entity disambiguation – both early tasks in a text mining pipeline – remain challenging. The ideal solution must not only address the quality of the results, but also cope with the sheer volume of textual input. Moreover, the solution should be able to tackle the full spectrum of entities without limiting its scope to narrow specializations such as genes, chemicals, and diseases. For information extraction tasks such as relation mining and knowledge base construction, it is crucial to go beyond merely recognizing entities and strive for precise entities via disambiguation. In this work, we focus on the entity disambiguation task, and propose a solution that attempts to balance quality with high throughput while addressing all entities.

Most existing biomedical entity disambiguation methods that do address all entities cannot be applied in practice to a large corpus for several reasons. The methods based on machine learning (such as Jimeno-Yepes (2016), Chen et al. (2013), Savova et al. (2008), Stevenson et al. (2008)) must identify in advance the exhaustive list of all ambiguous entity names. Where the training is supervised, labeled examples must be obtained, either by expensive manual annotation or by automatic curation (Jimeno-Yepes and Aronson, 2010). Finally, models – in general, one model per ambiguous entity name – must be trained prior to the disambiguation at runtime. All these setup costs render the methods impractical when all ambiguous entity names must be addressed. Alternative methods by Zheng et al. (2015) and Agirre et al. (2010) generate, at runtime, an entire instance of the problem customized per input text. MetaMap (Aronson and Lang, 2010), the de facto standard software tool, disambiguates amongst all entity types but the software is too slow for large-scale usage.

Approach and contributions

We present an automatic and light-weight method that disambiguates all entities in an indexed document by exploiting the indexing as well as domain knowledge. Specifically, the indexed documents are MEDLINE abstracts, which are the bulk of scientific literature in the biomedical domain. As for domain knowledge, the method draws upon UMLS (Unified Medical Language System). We choose MEDLINE abstracts and UMLS as our corpus and knowledge base for this work, respectively, because our method can then leverage the following unique characteristics of these biomedi-

Proceedings of the 15th Workshop on Biomedical Natural Language Processing, pages 72–76,
Berlin, Germany, August 12, 2016. ©2016 Association for Computational Linguistics

cal resources:

- MEDLINE abstracts are a large corpus indexed with rich, manually assigned MeSH (Medical Subject Heading) terms; we safely consider all MeSH terms to be accurate. In addition, since abstracts are very compactly written, their content rarely strays away from the biomedical domain. In other words, non-biomedical entities occur only rarely.

- UMLS is the authoritative and comprehensive knowledge base of the biomedical domain covering all aspects of the domain, with a vast collection of entities plus their lexical variations, semantic types, and inter-relationships.

- MeSH terms are themselves a crisp ontology that is already part of UMLS.

Putting these together: All the entities found in a MEDLINE abstract are of a biomedical nature, and all of them can be disambiguated to some canonical entity in UMLS. Therefore, given an abstract, its MeSH terms as ground truth, and all the text mentions in the abstract, the method first identifies unambiguous entities that we shall call *anchors*. The remaining text mentions are then disambiguated using heuristics based on linguistic-semantic patterns and knowledge base assets.

Under the best setting, our method achieves an average of 79.6% and 87.7% accuracy using the strict and relaxed rating schemes, respectively. To the best of our knowledge, this is the first work in the biomedical domain that evaluates all text mentions found in an abstract. In terms of throughput, our method processes 240k abstracts containing 24.5m text mentions in 400 minutes. We also present evaluations against established gold standards via a comparison to MetaMap.

The code is available as an open source project at `http://resources.mpi-inf.mpg.de/d5/bebe/`.

2 Related Work

In the biomedical domain, the terms entity disambiguation and word sense disambiguation are often used interchangeably, since the distinction between entity and sense is not always clear-cut. As mentioned in the Introduction, machine learning-based methods, both supervised and unsupervised, dominate existing works that address all entity types. Domain knowledge is a popular ingredient as well. The most recent work by Jimeno-Yepes (2016) combines word embeddings with long short term memory in a recurrent neural network model. The construction of a custom knowledge graph is the backbone of a collective inference approach by Zheng et al. (2015), where the approach disambiguates multiple entities simultaneously. Chen et al. (2013) applies active learning to support vector machine (SVM). Personalized PageRank is studied by Agirre et al. (2010), relying on and comparing different subsets of UMLS. Four further methods are compared by Jimeno-Yepes and Aronson (2010). In terms of evaluation, two gold standards, NLM WSD (Weeber et al., 2001) and MSH WSD (Jimeno-Yepes et al., 2011), are available.

When it comes to disambiguating only specific or highly specialized entities, a large body of work exists. To name a few representative specializations, there are works that disambiguate between species of genes (Harmston et al., 2012; Wang et al., 2010); chemicals (Batista-Navarro et al., 2015; Leaman et al., 2015); diseases (D'Souza and Ng, 2015); entities in clinical notes (Kang et al., 2012); and coarse entity types (Siu and Weikum, 2015; Jindal and Roth, 2013; Cohen et al., 2011).

3 Methodology

The input to the proposed method is a MEDLINE abstract and its MeSH terms. We use a fast dictionary-based entity recognition tool (Siu et al., 2013) to identify all longest text mentions that match UMLS entity names. (In this work, we use only the license level 0 subset of UMLS, but the proposed method works the same way for larger subsets.) Then the method proceeds in two phases: Phase 1 identifies unambiguous anchors amongst the text mentions. Phase 2 applies heuristics to disambiguate the remaining text mentions.

Unambiguous anchors

In phase 1, the method identifies *anchors* – given a text mention, the method determines if there is one UMLS entity that underlies this text mention unambiguously. A text mention may become an anchor in two ways:

- MeSH term (MESH): Recall that we assume MeSH terms are accurate ground truth. Following a strategy similar to Jimeno-Yepes et al. (2011), this heuristic identifies text mentions that are also MeSH terms for the abstract.

- Only one UMLS match (ONE): Recall that we assume UMLS has complete coverage of

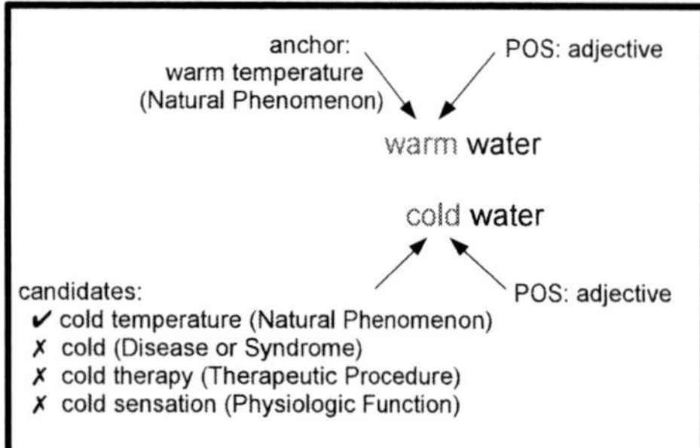

Figure 1: The linguistic-semantic pattern heuristic

all biomedical entities. A text mention that matches only a single UMLS entity is therefore considered unambiguous.

We pin down the anchors so that their underlying entities are considered correctly disambiguated.

Heuristics

In phase 2, the method disambiguates any remaining, non-anchor text mentions. Recall that the entity recognition tool already provides multiple matching UMLS entities to such a text mention. Taking these UMLS entities as candidates, select one candidate using one or more heuristics:

- Singular/plural (SP): Since abstracts are very short documents, we assume that, within one abstract, text mentions sharing the same surface string also share the same entity. Therefore, singular (e.g. *diet*) and plural (*diets*) forms of the same word should refer to the same entity. In UMLS, when the plural form is a unique entity (C0012155), that same entity is extended to the singular form, and vice versa.

- Linguistic-semantic pattern (PAT): Figure 1 depicts this heuristic via the example of two bigrams, *warm water* and *cold water*. When one word (*water*) appears in both bigrams in the same position, and when the other words (*warm* and *cold*) have the same part-of-speech, *warm* and *cold* ought to share the same linguistic function and some analogous meaning. Since *warm* is an anchor, take its UMLS semantic type (Natural Phenomenon), and pick for *cold* a candidate with the same type (cold temperature the Natural Phenomenon).

- Co-occurring semantic types (CO): The intuition behind this heuristic is that objects of the same semantic type often co-occur in the same

abstract. For instance, an abstract mentioning different fish species naturally also mentions the word *fish*. However, the candidates for *fish* belong to different UMLS semantic types (Fish, Gene or Genome, Organic Chemical (for fish extract), and Molecular Biology Research Technique (for Fluorescence in situ Hybridization)). When the entities in the abstract exhibit a predominant semantic type, pick the candidate with the same type.

- Ranked preferences of dictionary sources (RANK): UMLS comes with a pre-defined preference list of dictionary sources; more specifically, the source attributes have numerical ranks in the MRRANK table. When a text mention matches multiple candidates, each candidate's dictionary source leads to a corresponding rank number. This heuristic picks the candidate with the best rank. Under this heuristic, for instance, *HIV* the virus is preferred over *HIV* the vaccine.

- Prior probability (PRIOR): Thanks to the heterogeneous nature of UMLS, the listing of entity names contains much redundancy. Specifically, a single entity name is listed separately for each dictionary's contribution. A more popular meaning of the word (e.g. *cat* the animal) appears in more rows of the MRCONSO table than a less popular meaning (*CAT* the scan procedure). The prior probability distribution of candidates is thus estimated based on counts of entity name occurrences. Our prior work (Siu and Weikum, 2015) shows that estimated prior probabilities contribute to enriching disambiguation contexts 72% of the time. Here, the heuristic picks the candidate with the highest prior probability.

4   Results and Discussion

Ablation study of heuristics

We used disjoint sets of MEDLINE abstracts published in 2014 as the development and test datasets. The test dataset, in particular, consists of 20 randomly selected abstracts; in total, 2,549 text mentions were recognized. Two annotators evaluated all the recognized text mentions, including the anchors, rating the candidates as "completely correct", "partially correct", or "completely wrong". The inner-annotator agreement, calculated as Cohen's kappa, was 0.64, which in-

Heuristic(s)	Anchors		Non-anchors		All text mentions	
	Strict	Relaxed	Strict	Relaxed	Strict	Relaxed
MESH	16.0%	16.9%			8.9%	9.4%
ONE	83.3%	85.0%	not applicable		46.5%	47.5%
MESH + ONE	90.3%	93.0%			50.4%	51.9%
MESH + ONE + CO			47.2%	72.3%	71.3%	83.9%
MESH + ONE + PAT			7.8%	9.9%	53.9%	56.3%
MESH + ONE + PRIOR	remains at		53.9%	68.9%	74.2%	82.4%
MESH + ONE + RANK	90.3%	93.0%	63.2%	77.9%	78.5%	86.3%
MESH + ONE + SP			18.6%	22.1%	58.6%	61.7%
Successive filtering	remains at		66.2%	79.0%	**79.6%**	86.8%
Majority voting	90.3%	93.0%	64.3%	81.1%	78.8%	**87.7%**

Table 1: Contribution of different heuristics to accuracy

dicates mostly substantial agreement. The presence of fine shades of the same underlying entity in UMLS prompted the "partially correct" annotation choice. For instance, *children* exists as two separate entities with the semantic types Age Group and Family Group, and the exact distinction is difficult even for human judges. We therefore present results in two rating schemes: Under the strict rating scheme, only "completely correct" annotations count as correct; under the relaxed scheme, both "completely correct" and "partially correct" annotations count as correct.

Table 1 shows the accuracy and the contribution of each heuristic. We experimented with two types of ensembles, namely majority voting and applying heuristics as successive filters similar to D'Souza and Ng (2015). Under the relaxed rating scheme, majority voting consistently performed better. The best ensemble used, as expected, all heuristics to reach 87.7% accuracy. Under the strict rating scheme, on the other hand, successive heuristic filters consistently performed better. The best ensemble scored 79.6% accuracy using the following order of heuristics: MESH, ONE, SP, RANK, PRIOR, PAT, CO. On average, 56% of all text mentions in an abstract were anchors.

Comparison with MetaMap and other datasets

We compared the best setting of our method with MetaMap (version 2016 with disambiguation) using the aforementioned custom test dataset as well as 3 other datasets: NLM WSD (Weeber et al., 2001), EBI disease corpus (Jimeno et al., 2008), and a subset of the CRAFT corpus (Bada et al., 2012) that provides UMLS entity IDs in abstracts. Table 2 shows the accuracy for both systems. (The MSH WSD dataset (Jimeno-Yepes et al., 2011) was not used here because it was essentially constructed with the MESH heuristic; using the

	Custom strict	Custom relaxed	NLM WSD	EBI disease	CRAFT subset
Our method	79.6%	87.7%	39.9%	87.3%	38.8%
MetaMap	68.1%	76.1%	33.7%	78.4%	33.0%

Table 2: Comparison of accuracy between our method and MetaMap

dataset would not offer further insight.)

In terms of accuracy, both systems showed analogous trends for each dataset, though our proposed method outperformed MetaMap by 5% to 11%. Both systems performed poorly over the NLM WSD and CRAFT datasets due to their wide variety of highly ambiguous entity names. The disambiguation module in MetaMap is known to be a weaker module in the system (Aronson and Lang, 2010), while our method's heuristics are too simplistic for sophisticated cases. The same rationale explains why accuracy in EBI disease corpus was high, because disease names are much less ambiguous in general. In terms of speed, our system and MetaMap processed 600 and 11 abstracts per minute, respectively, on the same linux machine with 8 Intel Xeon 2.4GHz CPUs and 48GB RAM.

5 Conclusions and Future Work

We present a large-scale, high-quality, and automatic method that disambiguates entities in MEDLINE abstracts by exploiting MeSH terms as well as applying heuristics based on linguistic cues and knowledge assets in UMLS. Not only is the proposed method one order of magnitude faster than MetaMap, the overall accuracy is also slightly superior to that of MetaMap. Therefore we further propose our method as a viable alternative for real-time processing. We plan to harness the outputs of this work for future investigation on biomedical entity disambiguation.

References

Eneko Agirre, Aitor Soroa, and Mark Stevenson. 2010. Graph-based word sense disambiguation of biomedical documents. *Bioinformatics*, 26(22):2889–2896.

Alan R. Aronson and François-Michel Lang. 2010. An overview of MetaMap: historical perspective and recent advances. *Journal of the American Medical Informatics Association*, 17(3):229–236.

Michael Bada, Miriam Eckert, Donald Evans, Kristin Garcia, Krista Shipley, Dmitry Sitnikov, William A. Baumgartner, K. Bretonnel Cohen, Karin Verspoor, Judith A. Blake, and Lawrence E. Hunter. 2012. Concept annotation in the CRAFT corpus. *BMC Bioinformatics*, 13(1):1–20.

Riza Theresa Batista-Navarro, Rafal Rak, and Sophia Ananiadou. 2015. Optimising chemical named entity recognition with pre-processing analytics, knowledge-rich features and heuristics. *Journal of Cheminformatics*, 7(S-1):S6.

Yukun Chen, Hongxin Cao, Qiaozhu Mei, Kai Zheng, and Hua Xu. 2013. Applying active learning to supervised word sense disambiguation in MEDLINE. *Journal of the American Medical Informatics Association*, 20(5):1001–1006.

Raphael Cohen, Avitan Gefen, Michael Elhadad, and Ohad S. Birk. 2011. CSI-OMIM – Clinical synopsis search in OMIM. *BMC Bioinformatics*, 12:65.

Jennifer D'Souza and Vincent Ng. 2015. Sieve-based entity linking for the biomedical domain. In *Proceedings of the Annual Meeting of the Association for Computational Linguistics (ACL) (Volume 2: Short Papers)*, pages 297–302.

Nathan Harmston, Wendy Filsell, and Michael Stumpf. 2012. Which species is it? Species-driven gene name disambiguation using random walks over a mixture of adjacency matrices. *Bioinformatics*, 28(2):254–260.

Antonio Jimeno-Yepes, Ernesto Jimenez-Ruiz, Vivian Lee, Sylvain Gaudan, Rafael Berlanga, and Dietrich Rebholz-Schuhmann. 2008. Assessment of disease named entity recognition on a corpus of annotated sentences. *BMC Bioinformatics*, 9(3):1–10.

Antonio Jimeno-Yepes and Alan R. Aronson. 2010. Knowledge-based biomedical word sense disambiguation: comparison of approaches. *BMC Bioinformatics*, 11(1):1–12.

Antonio Jimeno-Yepes, Bridget McInnes, and Alan R. Aronson. 2011. Exploiting MeSH indexing in MEDLINE to generate a data set for word sense disambiguation. *BMC Bioinformatics*, 12:223.

Antonio Jimeno-Yepes. 2016. Higher order features and recurrent neural networks based on long-short term memory nodes in supervised biomedical word sense disambiguation. *arXiv preprint arXiv:1604.02506.*

Prateek Jindal and Dan Roth. 2013. Using soft constraints in joint inference for clinical concept recognition. In *Proceedings of the Conference on Empirical Methods in Natural Language Processing (EMNLP)*, pages 1808–1814.

Ning Kang, Zubair Afzal, Bharat Singh, Erik M. van Mulligen, and Jan A. Kors. 2012. Using an ensemble system to improve concept extraction from clinical records. *Journal of Biomedical Informatics*, 45(3):423 – 428.

Robert Leaman, Chih-Hsuan Wei, and Zhiyong Lu. 2015. tmChem: a high performance approach for chemical named entity recognition and normalization. *Journal of cheminformatics*, 7(supplement 1).

Guergana K. Savova, Anni R. Coden, Igor L. Sominsky, Rie Johnson, Philip V. Ogren, Piet C. de Groen, and Christopher G. Chute. 2008. Word sense disambiguation across two domains: biomedical literature and clinical notes. *Journal of Biomedical Informatics*, 41(6):1088–1100.

Amy Siu and Gerhard Weikum. 2015. Semantic type classification of common words in biomedical noun phrases. In *Proceedings of BioNLP 2015*, pages 98–103.

Amy Siu, Dat Ba Nguyen, and Gerhard Weikum. 2013. Fast entity recognition in biomedical text. In *Workshop on Data Mining for Healthcare (DMH) at Knowledge Discovery and Data Mining (KDD) 2013.*

Mark Stevenson, Yikun Guo, Robert Gaizauskas, and David Martinez. 2008. Disambiguation of biomedical text using diverse sources of information. *BMC Bioinformatics*, 9(Suppl 11):S7.

Xinglong Wang, Jun'ichi Tsujii, and Sophia Ananiadou. 2010. Disambiguating the species of biomedical named entities using natural language parsers. *Bioinformatics*, 26(5):661–667.

Marc Weeber, James G. Mork, and Alan R. Aronson. 2001. Developing a test collection for biomedical word sense disambiguation. In *Proceedings of the AMIA Symposium*, pages 746–750.

Jin G. Zheng, Daniel Howsmon, Boliang Zhang, Juergen Hahn, Deborah McGuinness, James Hendler, and Heng Ji. 2015. Entity linking for biomedical literature. *BMC Medical Informatics and Decision Making*, 15(1):1–9.

Using Distributed Representations to Disambiguate Biomedical and Clinical Concepts

Stéphan Tulkens[★] and **Simon Šuster**[★♠] and **Walter Daelemans**[★]

[★]CLiPS	[♠]CLCG
University of Antwerp	University of Groningen
Prinsstraat 13	Oude Kijk in 't Jatstr. 26
2000 Antwerpen	9700AS Groningen
Belgium	The Netherlands

{firstname.lastname}@uantwerpen.be

Abstract

In this paper, we report a knowledge-based method for Word Sense Disambiguation in the domains of biomedical and clinical text. We combine word representations created on large corpora with a small number of definitions from the UMLS to create concept representations, which we then compare to representations of the context of ambiguous terms. Using no relational information, we obtain comparable performance to previous approaches on the MSH-WSD dataset, which is a well-known dataset in the biomedical domain. Additionally, our method is fast and easy to set up and extend to other domains. Supplementary materials, including source code, can be found at `https://github.com/clips/yarn`

1 Introduction

Word Sense Disambiguation (WSD) is a procedure in which an ambiguous term or concept is assigned a single sense appropriate for that context, and is an important step in the creation of a semantic representation of a document (Ide and Véronis, 1998). While performing WSD will benefit most natural language processing applications, disambiguation of concepts is a critical component of applications operating on clinical and biomedical text, in which the same word can denote differing concepts, and may thus elicit radically different responses.

Compounding this problem of ambiguity is the fact that clinical text, in general, is noisier than other domains, and contains a large variety of abbreviations, some of which may be specific to a single hospital or physician. Additionally, there is a marked absence of large volumes of annotated clinical text, even for English, which presents a problem

for supervised approaches to Word Sense Disambiguation. For other languages, such as Dutch, there exist no freely available annotated corpora of clinical text.

A first step towards solving this problem could be the use of distributed representations. Where a more traditional word representation, such as a TF-IDF bag-of-words (BoW) representation, carries frequency information, distributed representations encode semantic information. A big advantage to using these representations is that they can be generated from large corpora of unlabeled text, and can be trained on very large corpora in a reasonable amount of time. These representations, especially when trained using neural architectures such as `word2vec` (Mikolov et al., 2013), have been shown to improve performance on a variety of tasks when compared to more traditional BoW representations.

We hypothesize that these kinds of distributional representations are well-suited for WSD in the clinical and biomedical domain because of the lack of training data, and the large terminological variety. We present a knowledge-based approach to Word Sense Disambiguation which creates concept representations by combining definitions from the Unified Medical Language System (UMLS) with distributed representations. We test our hypothesis on the MSH-WSD, which is a well-known dataset for WSD in the biomedical domain.

2 Related Research

All knowledge-based methods we review use the Unified Medical Language System® (UMLS) Metathesaurus® (Bodenreider, 2004) as a knowledge base, possibly augmented with external sources, such as MeSH®-indexed abstracts. Generally speaking, the UMLS contains two separate information sources that are suitable for use in dis-

77

Proceedings of the 15th Workshop on Biomedical Natural Language Processing, pages 77–82,
Berlin, Germany, August 12, 2016. ©2016 Association for Computational Linguistics

ambiguation: the concept unique identifier (CUI), which is a unique label for each concept, and the semantic type (ST), which is a set of 135 broad labels such as "Animal" or "Chemical". In general, a word is only considered disambiguated if the correct CUI can be selected; hence, as McInnes and Pedersen (2013) note, approaches based on semantic types are not able to disambiguate between approximately 12% of concepts, as some concepts with the same surface form have an identical ST, but a different CUI.

In terms of approaches using ST, Humphrey et al. (2006) create one vector for each semantic type by creating a BoW representation of all words that denote that semantic type. For each ambiguous term, a target word vector is created by taking a window of words from the right and left of the term. The concept which is associated with the ST with the lowest cosine distance is then taken to be the correct sense of the term. Similarly, Alexopoulou et al. (2009) create a method which finds the closest concept based on a combination of co-occurrence with other semantic types and ontological similarity through *is-a* relationships.

Closest to our approach is the machine readable dictionary (MRD) approach (McInnes, 2008; Jimeno-Yepes et al., 2011), which uses definitions from the UMLS to create concept vectors by creating BoW representations of concepts using all definitions of the concept and those of related concepts. This BoW representation contains TF-IDF values where D is the number of concepts in which a word appears, thereby reducing the influence of general words which occur in many concepts. These representations are then compared to the vectorized contexts of the ambiguous terms using cosine distance. A refinement of MRD, called second-order co-occurrence MRD (2-MRD) (McInnes, 2008), replaces each word in a definition by a vector which contains TF-IDF values of co-occurrence counts, thereby associating each word with a context.

McInnes and Pedersen (2013) introduce UMLS::SenseRelate, an approach which is based on Pedersen et al. (2004)'s WordNet::SenseRelate. In this system, each possible sense for an ambiguous term is assigned a distance-weighted score based on the *concepts* of the terms surrounding it, where the concepts of the surrounding terms are determined using UMLS::Similarity (McInnes et al., 2009).

Jimeno-Yepes and Berlanga (2015) present so-

	Medline	**Mimic-III**	**Bioasq**
Corpus size	920,081	13,097,844	-
Vocabulary	196,960	71,663	1,701,632
Dimension	320	320	200

Table 1: The number of words in the corpus, the resulting vocabulary size, and the dimension of the resulting vectors.

called step models, which calculate the probability of a word occurring with a certain concept by considering the number of times a word occurs in the definitions of that concept and its related concepts. It then steps through the UMLS-defined ontology of concepts, and refines the probabilities for each word and each concept based on the relations within the ontology.

Finally, Chen et al. (2014) present an approach for general WSD which uses word embeddings coupled with WordNet (Fellbaum, 1998) as a resource to perform sense disambiguation, and which creates sense-specific word embeddings from these sense-disambiguated word representations.

3   Materials

3.1   Test Corpus

We use the MSH-WSD corpus (Jimeno-Yepes et al., 2011), which consists of a set of 203 ambiguous terms, each associated with multiple concepts, to evaluate our approach. Of the 203 terms in the corpus, 106 are regular terms, 88 are acronyms, and 9 can be acronyms and regular terms. For each of these concepts, up to 100 MeSH abstracts were retrieved, resulting in a set of 37,888 abstracts. In our approach, all abstracts were pre-processed using the tokenizer from the Pattern package (De Smedt and Daelemans, 2012), and all stop words were removed using the English stop word list from `scikit-learn` (Pedregosa et al., 2011).

3.2   Word vectors

We evaluate our approach using three sets of vectors: The first set was trained on a small set of Medline abstracts[1], and a second set of vectors created on the entirety of the MIMIC-III corpus of clinical notes (Johnson et al., 2016). For both sets, we used the `word2vec` implementation from `gensim` (Řehůřek and Sojka, 2010), using skip-gram with negative sampling, a frequency cutoff

---

[1]The specific IDs of these abstracts are available in the online appendix.

of 5 and a negative sampling of 15. Additionally, we used a third set of vectors, available from the BioASQ organisers[2], which was trained on a much larger set of Medline abstracts.[3] The model statistics are visualized in Table 1.

4 Approach

Similar to the 2-MRD approach detailed above, our approach creates *concept vectors* by replacing each word in every definition by the vector representation of that word. This creates an $M \times n$ matrix for each definition, where M is the dimensionality of the word vectors, and n the number of words contained in that definition. Following this, for each definition, we then obtain a single vector of dimensionality M by applying a compositional function to the matrix, thereby obtaining so-called *definition vectors*, which represent the entire meaning of the definition in one vector. Each concept can then be represented by a $M \times d$ matrix, where d is the number of definitions that a concept has in the UMLS. Finally, we apply a second composition function to this matrix, thereby obtaining a single vector of dimensionality M which represents the combined meaning of all definitions for that concept, i.e. a *concept vector*.

For each abstract in the test corpus, we first locate each ambiguous term through a simple lookup. For each located term in the abstract we create a vector representation by retrieving all words in a window of size w surrounding the ambiguous term, and replacing the words by their vectors. Note that this window does not include the ambiguous term itself. These collections of vectors are then combined into M-dimensional vectors using the same composition function as above. This is done separately for each term occurrence within a single document, creating a $M \times x$ matrix, where x is the number of times the ambiguous term occurs in a single document. These are then combined in an M-dimensional *term vector* using the same composition we used for the concepts, above. A schematic representation of our model is given in Figure 1.

Because all concept and term vectors are created using the same distributed vectors and compositional functions, the vector space in which they are

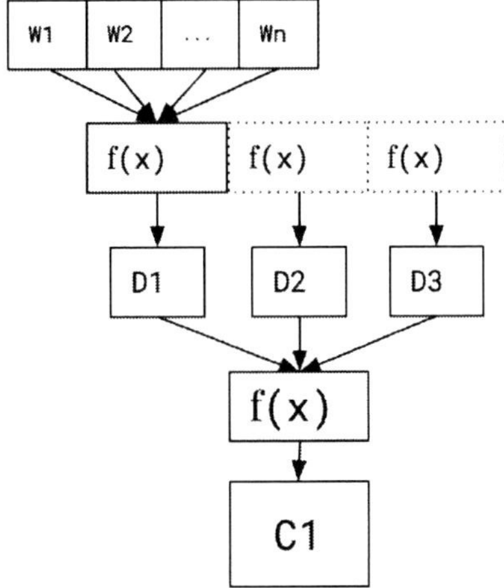

Figure 1: Our model represents a concept by replacing all words W in a definition D by their vectors, and then composing these into a definition vector with a function $f(x)$. For each concept, all definition vectors D are then composed into a concept vector C using a second composition function $f(x)$.

placed is also comparable. Hence, for each ambiguous word we encounter, we can use the cosine distance between the abstract vector of the ambiguous utterance and each possible sense of that word to determine the correct sense. This makes our approach very similar to the *Lesk* family of approaches (Lesk, 1986).

In terms of composition function we experimented with elementwise multiplication, averaging and summation, all of which are unordered compositional functions (Mitchell and Lapata, 2008). In addition, it is worth noting that there's still a lively debate whether ordered composition actually leads to better results for estimating document-, or sentence-level meaning, when compared to unordered composition (Iyyer et al., 2015; Socher et al., 2013).

5 Results

The accuracy scores obtained by our models using the different word vectors are displayed in Table 2. *med*, *mim* and *bio* denote the vectors created on the small Medline corpus, the Mimic-III corpus and the BioASQ vectors, respectively. We consider both a constrained and an unconstrained version of the task. For each word, the constrained version of the task only considers the senses present

---

[2] Available on the BioASQ website.

[3] While we concede that the BioASQ corpora might contain abstracts from the MSH dataset, it does not contain any explicit labeled information that might be used in disambiguation.

	med	mim	bio	MRD	2-MRD	0-step	2-step	r-step	UMLS::SenseRelate
Accuracy C	0.80	0.69	0.84	0.81	0.78	0.82	0.86	0.89	0.75
Accuracy U	0.72	0.63	0.75	-	-	-	-	-	-

Table 2: Results using constrained (C) and unconstrained (U) terms.

Term	Accuracy
DE	0.31
Hemlock	0.4
Brucella Abortus	0.46
WT1	0.46
Murine Sarcoma Virus	0.47

Table 3: The 5 lowest-performing terms.

in the MSH-WSD dataset as possible targets. The unconstrained version considers all concepts which are denoted by the ambiguous term in the 2015AB version of the UMLS as possible targets. The term cortex, for example, only has 2 concepts associated with it in the MSH-WSD dataset, while in the 2015AB UMLS release it can denote 5 separate concepts. Because the unconstrained version of the task considers all words, it therefore gives a better indication of real-life performance.

Accuracy C and U denote that the scores were obtained in the constrained settings and unconstrained setting, respectively. All reported scores use a window size of 6, which was optimized on a randomly selected set of 20 terms from the MSH-WSD set. Varying the window size had negligible results: all window sizes over 6 had comparable results, and increasing the window size over 30 causes a (small) decline in results. This is in line with McInnes and Pedersen (2013), who report a positive effect of window size that quickly tapers off for window sizes > 10. Concerning the composition functions, summation and averaging as first and second order composition function worked best, while using element-wise multiplication did not work well in any case. Where possible, we display the self-reported scores from the relevant papers on the same dataset.

A first thing to note is the large difference in accuracy when changing the set of word representations, especially the difference between the Medline vectors and the vectors derived from the Mimic-III corpus. It is currently unclear what causes these performance differences, although it is likely that the small vocabulary, caused by the noisiness of the clinical data in the MIMIC-III cor-

pus, reduces performance. Compared to previous approaches, our approach outperforms the MRD, 2-MRD, and UMLS::SenseRelate approaches, but does not manage to improve on the scores of the step models. Recall, however, that the step models largely rely on relationships in the UMLS ontology to estimate concept relatedness.

To compare how our models improved when including relation information, we also experimented with adding definitions of related concepts, i.e. concepts which had a sibling, parent or child relationship to each concept. In contrast to patterns observed in earlier work, this did not have a significant, and often a detrimental, effect on performance. Note that this makes our model entirely independent of the actual UMLS hierarchy, and more flexible as a result, as we only use the mappings from definition to CUI for disambiguation, and no other information, such as relations or semantic type. In addition, our system is also fast: on a consumer-grade laptop, our approach takes 10 seconds to vectorize and disambiguate all abstracts in the MSH dataset, not taking into account the time it takes to load the embeddings into memory.

Our approach obtains an accuracy of $> 90\%$ on 103 terms, showing that it is able to disambiguate a large variety of terms. For some terms, however, the performance was below random guessing. These are shown in Table 3. The pattern of errors is quite clear: Our approach has trouble with disambiguation if the definitions of the concepts themselves are lexically very similar. As an example, on the term Hemlock our approach performs below chance level because one of the concepts denotes a family of poisonous plants, while the other reports a tree, also called hemlock, the description of which mentions that it is explicitly *not* poisonous. We expect these kinds of problems to be alleviated with the addition of more data.

6   Conclusion and future work

In this paper we presented a novel approach to WSD in the biomedical domain which achieves comparable performance to existing methods without incorporating relational information from an ontology. This makes the approach easily transferable to other languages, for which such ontologies might not exist, and to other domains. The large variation in accuracy when changing sets of word embeddings also raises interesting prospects for improvement; better word representations will lead to an improvement in our approach without modifying the approach itself. Additionally, we would like to experiment with different composition functions for composing the definition and concept vectors.

Acknowledgments

Part of this research was carried out in the framework of the Accumulate IWT SBO project, funded by the government agency for Innovation by Science and Technology (IWT). We would also like to thank Elyne Scheurwegs for making the small set of Medline abstract available to us.

References

Dimitra Alexopoulou, Bill Andreopoulos, Heiko Dietze, Andreas Doms, Fabien Gandon, Jörg Hakenberg, Khaled Khelif, Michael Schroeder, and Thomas Wächter. 2009. Biomedical word sense disambiguation with ontologies and metadata: automation meets accuracy. *BMC bioinformatics*, 10(1):1.

Olivier Bodenreider. 2004. The unified medical language system (UMLS): integrating biomedical terminology. *Nucleic acids research*, 32(suppl 1):D267–D270.

Xinxiong Chen, Zhiyuan Liu, and Maosong Sun. 2014. A unified model for word sense representation and disambiguation. In *EMNLP*, pages 1025–1035. Citeseer.

Tom De Smedt and Walter Daelemans. 2012. Pattern for Python. *The Journal of Machine Learning Research*, 13(1):2063–2067.

Christiane Fellbaum. 1998. *WordNet*. Wiley Online Library.

Susanne M Humphrey, Willie J Rogers, Halil Kilicoglu, Dina Demner-Fushman, and Thomas C Rindflesch. 2006. Word sense disambiguation by selecting the best semantic type based on Journal Descriptor Indexing: Preliminary experiment. *Journal of the American Society for Information Science and Technology*, 57(1):96–113.

Nancy Ide and Jean Véronis. 1998. Introduction to the special issue on word sense disambiguation: the state of the art. *Computational linguistics*, 24(1):2–40.

Mohit Iyyer, Varun Manjunatha, Jordan Boyd-Graber, and Hal Daumé III. 2015. Deep unordered composition rivals syntactic methods for text classification. In *Association for Computational Linguistics*.

Antonio Jimeno-Yepes and Rafael Berlanga. 2015. Knowledge based word-concept model estimation and refinement for biomedical text mining. *Journal of biomedical informatics*, 53:300–307.

Antonio J Jimeno-Yepes, Bridget T McInnes, and Alan R Aronson. 2011. Exploiting MeSH indexing in MEDLINE to generate a data set for word sense disambiguation. *BMC bioinformatics*, 12(1):1.

AEW Johnson, TJ Pollard, L Shen, L Lehman, M Feng, M Ghassemi, B Moody, P Szolovits, LA Celi, and RG Mark. 2016. MIMIC-III, a freely accessible critical care database. *Scientific Data*.

Michael Lesk. 1986. Automatic sense disambiguation using machine readable dictionaries: how to tell a pine cone from an ice cream cone. In *Proceedings of the 5th annual international conference on Systems documentation*, pages 24–26. Association for Computing Machinery.

Bridget T McInnes and Ted Pedersen. 2013. Evaluating measures of semantic similarity and relatedness to disambiguate terms in biomedical text. *Journal of biomedical informatics*, 46(6):1116–1124.

Bridget T McInnes, Ted Pedersen, and Serguei VS Pakhomov. 2009. UMLS-Interface and UMLS-Similarity: open source software for measuring paths and semantic similarity. In *AMIA Annual Symposium Proceedings*, volume 2009, page 431. American Medical Informatics Association.

Bridget T McInnes. 2008. An unsupervised vector approach to biomedical term disambiguation: integrating UMLS and Medline. In *Proceedings of the 46th Annual Meeting of the Association for Computational Linguistics on Human Language Technologies: Student Research Workshop*, pages 49–54. Association for Computational Linguistics.

Tomas Mikolov, Ilya Sutskever, Kai Chen, Greg S Corrado, and Jeff Dean. 2013. Distributed representations of words and phrases and their compositionality. In *Advances in neural information processing systems*, pages 3111–3119.

Jeff Mitchell and Mirella Lapata. 2008. Vector-based models of semantic composition. In *Proceedings of the Association for Computational Linguistics*, pages 236–244.

Ted Pedersen, Siddharth Patwardhan, and Jason Michelizzi. 2004. WordNet:: Similarity: measuring the relatedness of concepts. In *Demonstration papers at HLT-NAACL 2004*, pages 38–41. Association for Computational Linguistics.

Fabian Pedregosa, Gaël Varoquaux, Alexandre Gramfort, Vincent Michel, Bertrand Thirion, Olivier Grisel, Mathieu Blondel, Peter Prettenhofer, Ron Weiss, Vincent Dubourg, et al. 2011. Scikit-learn: Machine learning in Python. *The Journal of Machine Learning Research*, 12:2825–2830.

Radim Řehůřek and Petr Sojka. 2010. Software Framework for Topic Modelling with Large Corpora. In *Proceedings of the LREC 2010 Workshop on New Challenges for NLP Frameworks*, pages 45–50, Valletta, Malta, May. ELRA. `http://is.muni.cz/publication/884893/en`.

Richard Socher, Alex Perelygin, Jean Y Wu, Jason Chuang, Christopher D Manning, Andrew Y Ng, and Christopher Potts. 2013. Recursive deep models for semantic compositionality over a sentiment treebank. In *Proceedings of the conference on empirical methods in natural language processing (EMNLP)*, volume 1631, page 1642.

Unsupervised Document Classification with Informed Topic Models

Timothy A. Miller[1] and Dmitriy Dligach[2] and Guergana K. Savova[1]

[1] Boston Children's Hospital Informatics Program, Harvard Medical School, Boston, MA 02115
{firstname.lastname}@childrens.harvard.edu

[2]Department of Computer Science, Loyola University Chicago, Chicago, IL 60611
ddligach@luc.edu

Abstract

Document classification is an important and common application in natural language processing. Scaling classification approaches to many targets faces a bottleneck in acquiring gold standard labels. In this work, we develop and evaluate a method for using informed topic models to noisily label documents, creating a noisy but usable set of labels for training discriminative classifiers. We investigate multiple ways to train this noisy classifier, and the best performing method uses Wikipedia-seeded topic models to approximately label training instances without any supervision. We evaluate these methods on the classification task as well as in an active learning setting, in which they are shown to improve learning rates over traditional active learning.

1 Introduction

Document classification is a standard task in machine learning and natural language processing which has been studied extensively (Joachims, 1999; Sebastiani, 2002). For many instances of this problem, standard supervised machine learning methods are now sufficient, so that any given document classification problem may be considered an application or engineering task rather than an interesting research problem. Recent work related to this problem has come mainly from the machine learning community and has focused on a generalization of the task called multi-label clas-

sification, in which each instance has multiple categories that must be predicted (Tsoumakas and Katakis, 2007; Read et al., 2011). That work has been concerned with the problem of how to best make use of correlations between the different labels, and using that information to perform the classifications non-independently.

In contrast, the work here is concerned with the more practical problem of obtaining these labels, and particularly the issue that ad hoc classification targets require obtaining supervised training data from scratch. This problem may arise in any application area of natural language processing, but in the clinical domain this problem is potentially more pressing because expert annotators (physicians) are expensive and traditional cost-saving approaches such as crowdsourcing are not always viable due to privacy concerns.

A common use case for clinical document classification is physicians mining patient notes for diseases, then using genetic samples of that "virtual cohort" to do phenotype-genotype correlation studies. Billing codes have high recall but varying precision depending on the disease. Thus, machine learning and NLP applied to the narrative text in the clinical record are now often used as a solution to this problem.

Our approach to this task is to use the unsupervised method of topic modeling, specifically Latent Dirichlet Allocation (LDA), which can learn word probabilities for semantically coherent topics, and by providing informed priors, we can steer topics to categories of interest and use these word lists like features in a classifier. As a first step, we take advantage of the crowd-sourced knowledge

Proceedings of the 15th Workshop on Biomedical Natural Language Processing, pages 83–91,
Berlin, Germany, August 12, 2016. ©2016 Association for Computational Linguistics

contained in Wikipedia to build a representation of the category of interest. We then use this category representation as an informed prior to LDA. This informed LDA algorithm then finds the topics that best satisfy the data given the priors, including both informed topics and traditional uninformed topics. In particular, we are able to guide the topic model to learn separate topics for similar categories if that is required by the categories we are interested in for classification.

The ability to extract pre-specified topics of varying granularity is interesting on its own, as it could be used for more guided data explorations of the kind that LDA is already in use for. But we can also use the output of this process to generate classifiers, by treating the occurrence of these topics in a document as a noisy label for that document. Given these noisy labels, we can immediately train a classifier, which performs much better than chance, without seeing a single gold standard training example.

Finally, we show that this has potential applications to active learning by using our noisy classifier's certainty estimates to select training examples, rather than first annotating a random seed set. This method results in faster learning rates than passive learning, standard active learning, and a baseline method that uses the Wikipedia-trained priors directly.

2 Background

2.1 Topic Modeling with Latent Dirichlet Allocation

Latent Dirichlet Allocation (LDA) (Blei et al., 2003) is a probabilistic unsupervised method for grouping tokens into a set of corpus-wide clusters. By setting parameters that constrain each document to use a subset of the clusters, frequently co-occurring words tend to get placed into the same clusters, and since distributionally similar words are often semantically similar, the result is that the clusters are often semantically coherent *topics*.

A document in LDA is represented as a bag of words. Each document has a probability distribution across K topic indices, and each topic is a global probability distribution across V words in the vocabulary. This leads to a generative story where the topic distribution for a document is drawn from a Dirichlet distribution, and each word is generated by first drawing a topic from the topic distribution, then drawing a word from the word

distribution indexed by that topic. One common inference method for LDA is to use Markov Chain Monte Carlo sampling, which is an iterative algorithm where each variable of interest is sampled probabilistically. In LDA, the standard sampling algorithm is derived by integrating out the topic and word distributions from the joint probability, so that the only random variable left to sample is the topic assignment for each word. Each topic assignment is typically randomly initialized, then at each iteration a topic is sampled from the sampling equation (from Griffiths and Steyvers (2004)):

$$p(z_i = j | \mathbf{z_{-i}}, \mathbf{w}) \propto \frac{n_{-i,j}^{w_i} + \beta}{n_{-i,j}^{(\cdot)} + |V|\beta} \cdot \frac{n_{-i,j}^{d_i} + \alpha}{n_{-i,\cdot}^{d_i} + T\alpha}$$

(1)

where i indexes words in the corpus and j is an index into K topics. The first factor represents the probability of the given word being selected for this topic ($n_{-i,j}^{w_i}$ is the count of the word at position i in topic j). The second factor represents the probability of topic j being selected for a word in this document ($n_{-i,j}^{d_i}$ is the count of words in the same document as w_i with topic j). α and β are the hyper-parameters from the Dirichlet priors used to draw the probability distributions. While the Dirichlet distribution accepts a vector of hyper-parameters the size of the output distribution, in most work these hyper-parameters are symmetrical, and are set using intuition or experimentation. Low values of these parameters (≤ 1) encourage sparse distributions, and sparsity constraints give rise to the clustering behavior typical of LDA.

One limitation of standard LDA in practice is that it will not always make fine-grained distinctions, even if they are known to exist in the data. For example, in the 20 Newsgroups data set (described in Section 4.2), there are different topics for baseball and hockey, which share quite a bit of terminology (teams, games, scores, etc.) but users may wish or expect them to be separate. Running standard topic modeling on this corpus with number of topics $K = 25$ using the Mallet topic modeling framework (McCallum, 2002), we observe that one topic seems to have merged baseball and hockey terminology (top words in that topic: *game, team, year, play, games, hockey, season, players, ca, win, league, baseball, nhl*). Simply increasing the number of topics may solve the problem but will also have the general effect of making categories more specific, which may adversely affect other topics. This problem can also be addressed by hierarchical models, in which topics that are higher in some hierarchy tend to model more general terms and lower topics are more specific. Hierarchical topic models (Blei et al., 2003) make use of a nested Chinese Restaurant Process where a word is a sample from a mixture between all the topics in a path from the root to a leaf node in a topic tree. Higher-level nodes will tend to be on more paths, and will thus be sampled more often and contain higher probability words. This method can then, for example, run on text without stop words removed and recover them as the top level of the hierarchy. One might imagine that for the baseball and hockey example, a hierarchical model would recover a higher-level sports topic with lower level topics specific to baseball and hockey.

Another method, Pachinko Allocation Model (Li and McCallum, 2006), generalized hierarchical topic models so that the topic hierarchy did not need to be a tree. This retains a hierarchy with higher and lower nodes corresponding to more or less general topics, but also allows for different words to be generated by different topic paths. While these hierarchical models have attractive properties, they are significantly more complex than standard LDA, which means they have more parameters, may take longer to train, and still may not recover topics of interest to a user.

Some relevant recent work in topic modeling has explored the importance of the prior values α and β. Wallach et al. (2009) developed optimization procedures for α and β and found that optimization of the document-topic prior α led to improved results, as measured by perplexity on held-out data. Jagarlamudi et al. (2012) found that priors on both α and β allowed them to incorporate information into the LDA inference, though they found that a more complex model structure was necessary to properly incorporate the information, which requires a more complex inference procedure.

Other relevant topic modeling work involves the augmentation of LDA-style models for labeling documents with multiple topics. Labeled LDA (Ramage et al., 2009) creates a topic for each label in a multi-label setting, and takes advantage of gold standard labels to learn topic distributions for each label. In the author-topic model (Rosen-Zvi et al., 2004), a document is generated by a set of authors, and an author is a distribution over topics. While both models are relevant to the multi-label classification problem, they both require gold standard labels, and we suspect that given gold standard labels discriminative classifiers will be superior.

3 Methods

Building on this existing work in topic modeling, we propose an extension to the LDA model that is able to find specific topics of interest, with minimal human effort. We call this method informed LDA, and the following sections will describe the method and how it can be used to train classifiers.

3.1 Building Informed Priors

We first build models for each of the target labels we are interested in. For this work, we use topics from two corpora, the 20 Newsgroups dataset mentioned above, as well as the 2008 i2b2 Challenge dataset[1], a set of 730 clinical discharge summaries labeled for multiple obesity-related diseases. Table 2 shows the 14 i2b2 labels we used for this work.

To build these models, we retrieved the Wikipedia article closest in meaning to each label. For most labels, there was an article with the exact title or a very similar title. We tokenized the articles and then TF-IDF (term-frequency/inverse document frequency) weighting was applied to these tokens (for the clinical articles we used an IDF derived from a sub-index of Wikipedia articles containing clinically relevant articles). The purpose of the TF-IDF reweighting is to downweight commonly occurring words like those representing broad terms (especially in the clinical data, terms like "disease" "surgery" are not as informative as they are generally).

While in the present work the step of identifying the relevant Wikipedia article required a small amount of manual effort, there are many ways that it could be automated – for example, by querying Wikipedia or the Web with the category name and performing token counts over multiple retrieved articles. Performing this step manually and obtaining high quality models of each category allows for a purer evaluation of the more technically challenging downstream steps.

---

[1] The i2b2 Challenge datasets are publicly available with a Data Use Agreement at https://i2b2.org/NLP/DataSets/.

3.2 Informed LDA

The standard LDA sampling equation, Equation 1, has a single value of α and β, assuming symmetric Dirichlet priors. A simple extension of the sampling equation for arbitrary priors can be obtained by vectorizing $\vec{\alpha}$ and $\vec{\beta}$:

$$p(z_i = j | \mathbf{z_{-i}}, \mathbf{w}) \propto \frac{n_{-i,j}^{w_i} + \beta^{w_i}}{n_{-i,j}^{(\cdot)} + \sum_{i'=1}^{|V|} \beta^{i'}} \cdot$$
$$\frac{n_{-i,j}^{d_i} + \alpha^j}{n_{-i,\cdot}^{d_i} + \sum_{j'=1}^{K} \alpha^{j'}} \quad (2)$$

Since each article has a different length, the prior vectors are first normalized so that all informed priors have equal strength. The $\vec{\beta}$ parameters are then filled in by these normalized weights. For token values that are not in the article, we use a default value of 0.01 in the prior.

To do inference using this model, we modified the source code of Mallet (McCallum, 2002), allowing for arrays of priors and modifying the sampling equation as described above. The number of topics K was set to 30, as this value gave reasonable results during preliminary experiments with standard LDA. This means that, in contrast to methods like Labeled LDA, not all topics are associated with a label – 16 of our topics were informed and the remaining 14 are uninformed, allowing the model to fit other topics in the data that we may not be currently interested in.

3.3 Creating a Bronze Standard

After running inference on the informed LDA model, the output of interest is the empirical estimates of document-topic probability – frequencies of each topic in each document. For example, the output may say that in document 0, the topic for *asthma* accounted for 3% of the tokens. Our goal is to use these values to assign noisy labels to each document for the value of that cluster category. We call this set of labels a *bronze standard*, in contrast to the *gold standard* of expert-generated labels.[2]

There are many ways one might go about converting topic frequencies to labels. For binary classification, as in the i2b2 data, one could set a threshold value and give all documents with topic frequencies above that threshold a positive label.

Possible thresholds include 0 and $1/K$. We found that thresholds allowed for too much variation, and led to some severely skewed label distributions, so that the next stage classifier may have only a few positive examples to work with. Even if this approximates a true distribution, it is probably not enough data points to find a signal in the features, and so the resulting classifier probably will not be useful.

Another option is, for each informed topic, sort all documents by that topic's frequency, and then split the data at the median frequency value into the *true* and *false* classes, so that the classifier gets training data with no skew in its distribution. We found that this method was the most reliable across labels and does not require fitting any parameter.

For multi-way classification, as in the 20 Newsgroups data, we use as the bronze label the topic whose document-topic probability was the maximum of all the informed topics. To simplify further, this is just the topic that accounts for the greatest number of words in the document.

3.4 Building a Classifier

However bronze labels are obtained, they can now be used in the typical way to train a classifier. The feature representation may also be varied. We will describe classifier settings in detail in the Evaluation, but we experimented with a variety of classifiers. The representation used here is bag of words for a document.

3.5 Active Learning with Bronze and Gold Labels

While this technique may have value as a low-cost low-accuracy classifier, we suspect that it might have additional value as an input to other systems. One such potential application is as an input to an active learning-based annotation system, as a way of obtaining gold standard labels to achieve optimal classification performance. Active learning is an annotation technique that has a classifier in the loop – instead of labeling examples randomly, examples are selected for labeling based on some notion of usefulness, such as classifier uncertainty (see Settles (2010) for an excellent overview of active learning). To initialize the active learning classifier, however, a small set of "seed" examples are randomly selected to be labeled.

Here, instead of using a seed set, we use our unsupervised classifiers from the start of the annotation process. Using this bronze-trained classifier,

---

[2] *Silver standard* is already used to describe huge automatically labeled datasets (Rebholz-Schuhmann et al., 2010).

we get a probability distribution across categories for every instance in the training data. We use uncertainty sampling to select the next instance, which in the two-class case means selecting the instance whose classification probability is closest to 0.5. The bronze standard label is then replaced with the gold standard label (simulating annotation of the instance) and the classifier is re-trained. This process repeats until every instance in the training data has its gold label uncovered. To test this method experimentally (as in Section 4), we would also have a set of held out data, and every time we train a classifier we would evaluate it on this held out data.

The active learning method just described differs from standard active learning in that there is no longer a breakdown into initial seed set and a pool set from which examples are drawn, but rather we have a mixed gold/bronze training set. Since the gold labels are more reliable, we give them a higher weight relative to the bronze labels, so the classifier can treat them differently.

4   Evaluation

To evaluate the effectiveness of this method, we will start with one brief qualitative evaluation to inspect the topics found, and then proceed to two quantitative evaluations. The first evaluation attempts to get a preliminary look at how the informed LDA method works on topics that are superficially similar, to gauge how adding information can guide the model to make difficult distinctions. The first quantitative evaluation is a simple set of unsupervised classification experiments. We build a bronze standard for the 20 Newsgroups and i2b2 data sets, then train classifiers for each category and evaluate the classifier. The second quantitative evaluation examines the use case of active learning. Our experiment uses the unsupervised classifiers from the previous experiment to evaluate whether active learning can be made even faster by using those classifiers to select examples at the start of the active learning procedure, when the gold standard training data is still quite small.

4.1   Qualitative Inspection of Similar Topics

Table 1 shows the results of inspecting a few sports-related topics from the 20 Newsgroups corpus. This is to simply see if this method can address the issue discussed in Section 2, the conflation of similar topics. The first column shows

LDA	Baseball	Hockey
game	year	team
team	baseball	game
year	hit	hockey
play	san	play
games	win	canada
hockey	team	games
season	season	toronto
players	runs	nhl
ca	league	cup
win	game	players
league	won	division
baseball	lost	season
nhl	games	gary

Table 1: Comparison of topic words in similar topics with standard LDA (first column) and informed LDA (last two columns).

the words in a sports-related topic using standard LDA. It clearly finds words related to both hockey and baseball, with no other topics containing any significant amount of hockey or baseball content.

In contrast, the last two columns show the informed topics for baseball and hockey using informed LDA. In addition to the sport names there are additional terms that are discriminative, including *hit*, *runs*, and *wins* (a pitching statistic) for baseball, and *canada*, *nhl*, and *cup* for hockey. Informed LDA also did not have any other topics containing significant amount of hockey or baseball content. This kind of evaluation is of limited use, but it does verify that the algorithm is able to find closely aligned topics.

4.2   Experimental Configuration

The data sets used for evaluation are the 20 Newsgroups data set[3] and the 2008 i2b2 Challenge data set described above. The 20 Newsgroups data set contains around 11,000 training documents, partitioned into 20 topics, which are used as labels for the documents. These include labels such as *alt.atheism* for atheism-related conversations, *sci.crypt* for cryptography-related discussions, and so forth. While each document may have multiple "topics" in the strict semantic sense, it will have one topic *label* – in other words, a classifier must choose a single category from 20 possibilities.

The 2008 i2b2 Challenge data consists of clin-

---

[3]This data set can be downloaded here: http://qwone.com/ jason/20Newsgroups/

ical discharge summaries from patients at an obesity clinic. This data contains 730 notes in the training set, with each note being labeled for 16 disease categories, with both textual and intuitive labels.[4] We use the more challenging intuitive label set, which did not require explicit confirmation of a diagnosis in the text. We discard two labels, hypertriglyceridemia and venous insufficiency, after preliminary work on the training set indicated that those two labels could not be learned satisfactorily even with fully supervised approach. The likely cause of the difficulty is that these two categories contained the fewest number of positive examples, an important issue but one we will have to reserve for future work. In contrast to the 20 Newsgroups data, in the i2b2 data the labels are not mutually exclusive, so we frame the task as 14 binary classification problems.

We used the Weka machine learning toolkit (Hall et al., 2009) during development, and evaluated many different classifiers on both datasets, including Adaboost, support vector machines, logistic regression, and naive bayes. We use the Adaboost algorithm (Freund and Schapire, 1996) with decision stumps as the weak learner for the i2b2 data. For the 20 Newsgroups data we used a support vector machine with linear kernel for the classification experiment and switched to Naive Bayes for the active learning experiments for speed reasons. Besides being relatively accurate, using boosting with decision trees has the beneficial property that the models it builds have some degree of transparency, which clinical researchers appreciate.

For the first experiment we evaluate the effectiveness of informed LDA on generating labels that can train a classifier. We compare first to a random labeling baseline (labeled *RandL*), that generates a random labeling, trains a classifier with those labels, and then uses it to classify the training set. This is not intended to be a competitive baseline, as much as it is a check to set a lower bound on what kind of performance we would get if informed LDA labeling had no signal whatsoever. We also compare to a standard random classifier (*RandC*) which is based on a recall of 0.5 and a precision of the category's prevalence. This baseline is important for the binary classifier to make sure our classifier is learning more

than just how to do random guessing based on our evenly split labels. In the main experimental condition (*Bronze*), we use informed LDA to generate a bronze standard label set for the training data as described in Section 3.3, train a classifier with those labels and evaluate it on those same examples from the training set. The upper bound we compare against is a 5-fold cross-validation of the training set using gold labels.

The next experiment examines the usefulness of these unsupervised classifiers in an augmented active learning scheme described in Section 3.5. We use the two baselines of passive learning and standard active learning. The passive learning baseline is equivalent to just plotting a learning curve for a machine learning problem with random ordering of the instances. The active learning baseline uses an initial seed set of 25 examples from within the pool set. We use uncertainty sampling to select the next example, which uses the example which has the smallest difference in probability estimates between the two most likely classes.

The condition we are testing is labeled *Bronze*. This condition does not use a seed set, but starts with a classifier trained on the entire bronze-labeled pool set. Learning proceeds by finding examples in the pool set that the current iteration of the classifier is uncertain about and uncovering the gold label (i.e. simulating annotation). This means that, in the active learning curve, the x-axis, which traditionally indicates the size of the training data used to train the classifier, now indicates the number of gold instances in the training data (the remaining instances still have bronze labels).

We give gold and bronze instances different weights to reflect varying quality of the labels. This weight is used in calculating the cost function during training – a higher weight on gold labels means the classifier will try harder to get gold-labeled instances correct. Here we use a weight of 0.1 for bronze-labeled instances and a weight of 1.0 for gold-labeled instances.

4.3 Results

Table 2 shows the results of the 14 binary classifiers on the i2b2 data. The random labeling gives rise to a classifier that never obtains an F1 score better than 0.11. The bronze labeling, performs much better than the RandL classifier, with a low performance of 0.26 (for depression) and a high performance of 0.83 (for diabetes). The bronze-

---

[4]In actuality, not every note is labeled for every category, but most are.

Category	RandL	RandC	Bronze	CV
Asthma	0.02	0.20	0.47	0.91
CAD	0.05	0.54	0.66	0.91
CHF	0.07	0.50	0.75	0.86
Depression	0.07	0.29	0.26	0.77
Diabetes	0.06	0.58	0.83	0.95
Gallstones	0.02	0.22	0.33	0.83
GERD	0.04	0.33	0.39	0.77
Gout	0.05	0.21	0.42	0.88
HC	0.08	0.51	0.56	0.83
HTN	0.06	0.62	0.67	0.96
OA	0.10	0.26	0.27	0.66
Obesity	0.07	0.46	0.56	0.97
OSA	0.02	0.22	0.28	0.91
PVD	0.11	0.25	0.37	0.76

Table 2: F1 scores for traditional supervised classifier (CV) vs. unsupervised classifier trained using informed LDA (Bronze), classifiers trained with random labels (RandL), and a classifier that makes random guesses (RandC). (CAD=Coronary Artery Disease, CHF=Congestive Heart Failure, GERD=Gastroesophageal Reflux Disease, HC=Hypercholesterolemia, HTN=Hypertension, OA=Osteoarthritis, OSA=Obstructive Sleep Apnea, PVD=Peripheral Vascular Disease)

	RL	RC	Bronze	CV
Accuracy	0.05	0.05	0.64	0.85

Table 3: Multi-way classifier accuracy on the 20 Newsgroups dataset using random labels (RL), a random classifier (RC), bronze labels obtained from informed LDA (Bronze) and a supervised cross-validation (CV).

trained classifier also outperforms the RandC random classifier in 13 out of 14 categories, by an average of approximately 12 points F1 score. Cross-validation using the gold standard can be very accurate, ranging from 0.66 to 0.96.

There are a few interesting things to point out from these results. First, our analysis of the errors shows that the classifiers trained by the bronze labeling did not systematically favor either precision or recall. A linear regression with the Gold score as the independent variable and the Bronze score as the dependent variable shows that the Gold score is a statistically significant predictor of the Bronze score ($p = 0.01$), but with so few data

Disease	Active Learning		
	Passive	Active	Bronze
Asthma	469.6	486.1	**500.1**
CAD	455.5	462.7	**469.6**
CHF	415.3	422.3	**435.1**
Depression	371.7	**414.5**	410.4
Diabetes	491.7	503.2	**510.8**
Gallstones	400.7	450.8	**457.0**
GERD	317.3	350.0	**360.2**
Gout	477.4	506.1	**519.1**
HC	372.4	389.4	**398.9**
Hypertension	460.1	**476.2**	465.1
Osteoarthritis	305.6	328.7	**349.9**
Obesity	490.4	501.9	502.0
OSA	463.6	487.7	**495.8**
PVD	363.0	**390.9**	372.2
Average Curve	451.0	473.8	**479.8**

Table 4: Performance of augmented active learning on 14 categories from the 2008 i2b2 Challenge data .

	Passive	Active	Bronze
ALC	7203	7469	**7678**

Table 5: Performance of augmented active learning on 20-way classifier for 20 Newsgroups data. Unit is Area Under the Learning curve (ALC).

points the exact nature of this effect is not clear.

Table 3 shows the results of the three classifiers on the Newsgroups data. For this multi-category experiment we use accuracy as the metric instead of F1 score. Here the accuracy of the RandL and RandC are both quite low, at 0.05. Bronze labeling can train a classifier that attains an accuracy of 0.64. The Gold labeling gives us an approximate ceiling performance of 0.85.

Table 4 shows the area under the active learning curve (ALC) for 14 categories in the i2b2 data under three conditions. Both the active learning and the bronze-augmented active learning outperform passive learning on all 14 categories. In 11 of the 14 categories the bronze-augmented version is superior to traditional active learning. We also averaged the curves together and computed the average learning curve, for which the bronze-augmented algorithm is again optimal.

Figure 1 shows the average learning curve across i2b2 category labels. The x-axis has been truncated at 100 instances to clarify the region

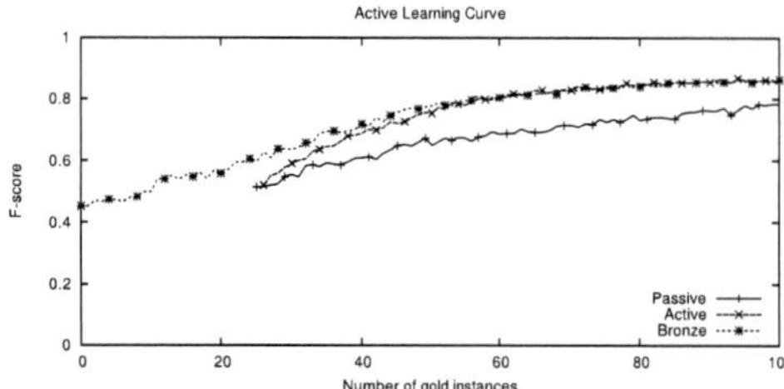

Figure 1: Average active learning curve across 14 disease categories.

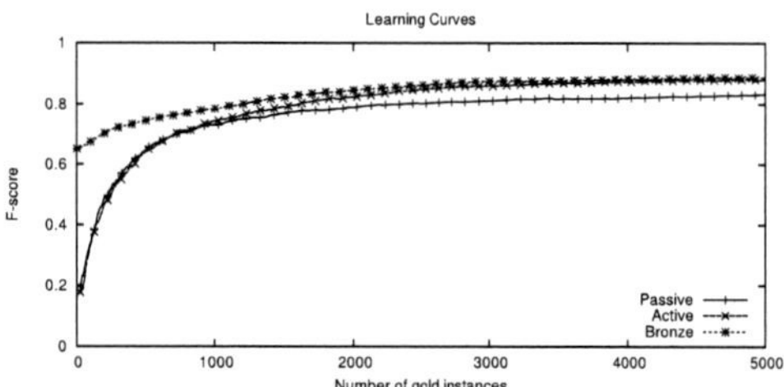

Figure 2: Active learning plot for 20 Newsgroups data.

where there is a clear distinction. Qualitatively, the distinction between passive and active is quite clear – this much is not surprising, given previous success in active learning. While the bronze curve shows an advantage up to maybe 30 instances, it quickly converges with the active curve.

Table 5 shows the Area Under the Learning curve (ALC) results of active learning on the 20 Newsgroups data. Active learning again beats passive learning, and the augmented version using bronze labels performs best. The learning curves for all three conditions are in Figure 2, truncated to 5000 examples to highlight the area showing the most difference. Here the bronze label-based version of active learning seems to have a clearer advantage than in the i2b2 corpus.

4.4 Discussion and Future Work

One aspect that deserves further mention is that of class prevalence and skew. The decision to assign bronze labels on the i2b2 corpus with an even class distribution was vastly superior to any thresholding that was attempted. However, we should note that in the i2b2 data we used, the prevalence is relatively high for most categories. Diabetes, for example, is present in 70% of the patients here, while gout is present in 13%.

Evaluating unsupervised methods on super-vised tasks is tricky. Our experiments here focused on the training set of each corpus rather than following the default train/test splits. Our primary concerns here were evaluating whether this method had any promise at all, and that it was applicable to more than one corpus. One could argue that future work should develop and tune the methods on the training data and then evaluate them on the test set. However, the very nature of this method breaks the traditional training/test model because tuning on the training data is already cheating relative to how the method would actually be applied on unlabeled data.

We are not sure that this problem has any perfect solutions, but we suggest that evaluating on as many different corpora as possible will be the best validation for this method. In this work, we tried to do that by starting on i2b2 data and then moving to the 20 Newsgroups data. Doing this helped us understand how informed priors need to be modified based on the size of the corpus.

One sticking point to portability with this method is the choice of classifier. We could have chosen a single classifier to stick with across corpora but then if one is particularly weak for a given corpus (e.g., SVM performed poorly on i2b2), it is less clear how much credit to assign the bronze labels for the performance. One possible solution to this issue is to require a much smaller sample of gold-labeled validation set if validated performance is strictly necessary.

One final point is that the classifier trained on bronze training labels probably would not generalize to a new corpus very well. This is not much of a problem, because the idea of the method is that for a new corpus one should generate new bronze labels using informed LDA on that data set. This does raise the question of what the difference would be between two classifiers trained on different corpora but with the same topic label, and whether there is some way of extracting additional information from comparing the decisions of these different classifiers on new data.

5 Conclusion

This work has shown that informed topic models seeded with topic information from Wikipedia can be used to train classifiers that perform much better than random. These classifiers are given no gold standard information and yet obtain results that may be useful in some applications. We show that in active learning this method can improve learning rate for many categories. This method may be beneficial in domains where a large number of classifiers are required and state of the art performance is not necessary.

Acknowledgements

Research reported in this publication was supported by National Institute for General Medical Sciences (NIGMS) and National Library of Medicine (NLM) of the National Institutes of Health under award number R01GM114355 (HealthNLP) and U54LM008748 (i2b2). The content is solely the responsibility of the authors and does not necessarily represent the official views of the National Institutes of Health.

References

David M. Blei, Andrew Y. Ng, and Michael I. Jordan. 2003. Latent dirichlet allocation. *Journal of Machine Learning Research*, 3:993–1022.

Yoav Freund and Robert E Schapire. 1996. Experiments with a new boosting algorithm. In *ICML*, volume 96, pages 148–156.

Thomas L. Griffiths and Mark Steyvers. 2004. Finding scientific topics. *PNAS*, 101(suppl. 1):5228–5235.

Mark Hall, Eibe Frank, Geoffrey Holmes, Bernhard Pfahringer, Peter Reutemann, and Ian H. Witten. 2009. The WEKA data mining software: An update. *SIGKDD Explorations*, 11.

Jagadeesh Jagarlamudi, Hal Daumé III, and Raghavendra Udupa. 2012. Incorporating lexical priors into topic models. In *Proceedings of the 13th Conference of the European Chapter of the Association for Computational Linguistics*, pages 204–213. Association for Computational Linguistics.

Thorsten Joachims. 1999. Making large scale svm learning practical. In B. Schlkopf, C. Burges, and A. Smola, editors, *Advances in Kernel Methods - Support Vector Learning*. Universität Dortmund.

Wei Li and Andrew McCallum. 2006. Pachinko allocation: Dag-structured mixture models of topic correlations. In *Proceedings of the 23rd international conference on Machine learning*, pages 577–584. ACM.

Andrew McCallum. 2002. Mallet: A machine learning for language toolkit.

Daniel Ramage, David Hall, Ramesh Nallapati, and Christopher D. Manning. 2009. Labeled LDA: a supervised topic model for credit attribution in multi-labeled corpora. In *Proceedings of the 2009 Conference on Empirical Methods in Natural Language Processing: Volume 1 - Volume 1*, EMNLP '09, pages 248–256, Stroudsburg, PA, USA. Association for Computational Linguistics.

Jesse Read, Bernhard Pfahringer, Geoff Holmes, and Eibe Frank. 2011. Classifier chains for multi-label classification. *Machine Learning*, 85(3):333–359.

Dietrich Rebholz-Schuhmann, Antonio José Jimeno Yepes, Erik M Van Mulligen, Ning Kang, Jan Kors, David Milward, Peter Corbett, Ekaterina Buyko, Elena Beisswanger, and Udo Hahn. 2010. CALBC silver standard corpus. *Journal of bioinformatics and computational biology*, 8(01):163–179.

Michal Rosen-Zvi, Thomas Griffiths, Mark Steyvers, and Padhraic Smyth. 2004. The author-topic model for authors and documents. In *Proceedings of the 20th conference on Uncertainty in artificial intelligence*, pages 487–494. AUAI Press.

Fabrizio Sebastiani. 2002. Machine learning in automated text categorization. *ACM computing surveys (CSUR)*, 34(1):1–47.

Burr Settles. 2010. Active learning literature survey. Technical report, University of Wisconsin–Madison.

Grigorios Tsoumakas and Ioannis Katakis. 2007. Multi-label classification: An overview. *International Journal of Data Warehousing and Mining (IJDWM)*, 3(3):1–13.

Hanna M Wallach, David M Mimno, and Andrew McCallum. 2009. Rethinking LDA: Why priors matter. In *NIPS*, volume 22, pages 1973–1981.

Vocabulary Development To Support Information Extraction of Substance Abuse from Psychiatry Notes

Sumithra Velupillai[1,2]**, Danielle Mowery**[3]**, Mike Conway**[3]**, John Hurdle**[3] **and Brent Kious**[4]
[1] School of Computer Science and Communication, KTH, Stockholm
`sumithra@kth.se`
[2] IoPPN, King's College London
[3] Department of Biomedical Informatics, University of Utah
[4] Department of Psychiatry, University of Utah
`firstname.lastname@utah.edu`

Abstract

Extracting information from mental health records can be useful for large-scale clinical studies (e.g., to predict medication adherence or to understand medication effects) in this clinical specialty largely underserved by the Natural Language Processing (NLP) community. Vocabularies that contain medical terms for specific clinical use-cases, such as signs, symptoms, histories, social risk factors, are valuable resources for the development of NLP systems that aid clinicians in extracting information from text. Substance abuse is an important variable for many clinical use-cases, but, to our knowledge, there are no publicly available vocabularies that cover these types of terms. In this study, we apply and combine three methods for generating vocabularies related to substance abuse. We propose a simple and systematic method to generate highly relevant vocabularies and evaluate these vocabularies with respect to size and content, as well as coverage and relevance when applied to authentic psychiatric notes.

1 Introduction

Information about a mental health patient's clinical condition is documented routinely in mental health records, mostly in the form of free-text. Extracting information from these documents can be useful for large-scale clinical studies to develop new treatment alternatives, to understand disease progression and medication effects, etc. Vocabularies that contain relevant terms for specific clinical use-cases are useful resources for the development of Natural Language Processing (NLP) systems that aid clinicians in extracting information from text.

In this study, we focus on the problem of automated vocabulary generation, specifically, to automate the generation of relevant synonyms and related terms, focusing on *substance abuse*, an area not well-studied. Specifically, we aim to:

1. compare, assess, and combine three different automated vocabulary generation methods
2. determine vocabulary coverage and relevance in substance abuse sections from authentic psychiatric clinical notes, and
3. generate a publicly available vocabulary with substance abuse terms

Our goal is to develop efficient vocabulary generation methods that can be used in larger NLP pipelines for new clinical use-cases, where domain experts with minimal-to-no NLP background can develop tailored solutions for new problems.

1.1 Treatment Management for Acute Anxiety

Patients with depression and anxiety disorders admitted for hospital care commonly receive medications for the management of acute anxiety on an as-needed basis (Curtis and Capp, 2003; Stein-Parbury et al., 2008). These may include benzodiazepines, antihistamines, antipsychotic medications, and others. Although these treatments can reduce a patient's acute distress level, they often have adverse effects. Apart from class-specific side-effects (e.g., oversedation related to benzodiazepines), as-needed anxiolytics may also impair response to psychotherapy and impede long-term recovery (Curran, 1986; Curran and Birch, 1991; Westra et al., 2004; Mystkowski et al., 2003; Otto et al., 2005).

In an effort to better understand the effect of as-needed anxiolytic medications on a patient's ability to manage their anxiety during and after psy-

Proceedings of the 15th Workshop on Biomedical Natural Language Processing, pages 92–101,
Berlin, Germany, August 12, 2016. ©2016 Association for Computational Linguistics

chiatric hospitalization, one of the authors (BK) has undertaken a large-scale retrospective study. The study aims to determine whether anxiolytic use correlates with poorer outcomes for psychiatric inpatients being treated for depression and anxiety, such as prolonged hospitalization or increased risk of readmission. This study involves a large cohort (n=about 3000) of patients admitted to several psychiatric hospitals in the same university system. Because the effects of anxiolytic use on the outcomes of interest are likely modulated by a number of other variables, such as her history of substance use disorders, the study requires the coding of almost 30 variables for each patient, all of which must be abstracted from free-text clinical notes.

1.2 Treatment Variables from Clinical Texts

NLP approaches could accelerate the coding process for this data set, while also providing the foundation for future studies with similar aims. Although research in clinical NLP has matured over the last decades, and there are several publicly available clinical text processing pipelines and modules e.g., cTAKES (Savova et al., 2010), MedLee (Friedman et al., 1994), and pyConText (Chapman et al., 2011), adapting and refining these resources to fit the information needs for specific use-cases is not straightforward.

Furthermore, although there have been a few efforts in the NLP community to address mental health-related use-cases, e.g., understanding a patient's suicidal ideations from suicide notes (Pestian et al., 2010) and detecting signals of post-traumatic stress disorder (PTSD), depression, bipolar disorder, and seasonal affective disorder (SAD) from tweets (Coppersmith et al., 2014), NLP for mental health is still in its early stages.

In this study, we focus on substance abuse as it relates to patients suffering from depression and anxiety disorders. As a first step toward encoding substance abuse variables from clinical text, our domain expert (BK) had manually listed a number of terms thought to be relevant to substance abuse. However, these select keywords may not identify all relevant reports due to the variable use of synonyms, abbreviations, acronyms, and misspellings in clinical texts. To assist our domain expert in identifying all relevant patient reports, one initial solution involves automatically extending the initial set of keywords with relevant synonyms and

related terms, also known as vocabulary expansion, and marking identified terms for further review. The domain expert (or an NLP module) must then review the report and infer the labels assigned to each substance abuse variable, e.g., from the context of the identified mention "etoh" in the sentence "history of ETOH abuse," assign a label *current, past, both* or *none* for the variable *Alcohol*.

2 Related Research

The creation of useful domain-specific vocabularies requires a balance between identifying enough terms for adequate coverage (vocabulary expansion) while pruning terms with limited or no utility (vocabulary reduction).

2.1 Vocabulary Expansion

In the biomedicine domain, common vocabulary expansion methods include dictionary-based (e.g., using terminologies and edit-distances), rule-based (e.g., leveraging orthographic/morphological/lexico-syntactic patterns and grammars), machine learning/statistical-based (e.g., applying feature-engineering and transitional states to identify term boundaries), and hybrid approaches (e.g., integrating combinations of the former approaches) (Krauthammer and Nenadic, 2004).

In the clinical domain, vocabulary expansion efforts have included several of these approaches. The Unified Medical Language System, UMLS (Lindberg et al., 1993), has been an influential and important resource for vocabulary development in this domain. Grabar et al. (2009) use the UMLS and other available terminologies to generate synonyms through a compositional analysis along with syntactic dependency information, resulting in high precision (Grabar et al., 2009). Parts of the UMLS have also been enhanced by synonym substitution methods using WordNet (Fellbaum, 1998) and a set of constraints on the number of generated synonyms, resulting in a 10% increase of valid terms related to GI endoscopic examinations in the Minimal Standard Terminology (Huang et al., 2010) Zeng et al. (2012) demonstrated that synonym expansion from the UMLS, topic modeling with Latent Dirichlet Allocation and predicate-based query expansions achieved higher average recalls and average F-measures when compared with the baseline keyword query for retrieving relevant texts from the United States Veteran Affair's Corporate Data Warehouse (Zeng et al., 2012). Henrikson et al. applied a semi-automatic and language-agnostic method for identifying synonyms of SNOMED CT preferred terms using a distributional similarity technique and a large clinical corpus (Henriksson et al., 2013).

2.2 Vocabulary Reduction

However, many terms from controlled vocabularies like the UMLS and SNOMED CT are not found in biomedical or clinical texts. Hettne et al. (2010) conducted experiments for the building of a medical lexicon using the UMLS Metathesaurus (Hettne et al., 2010). Specifically, they applied term suppression and term rewriting techniques to filter out or discard terms which are considered irrelevant or unlikely to occur in biomedical texts. As a result, a more representative lexicon was produced for medical concept recognition. Wu et al. (2012) conducted a large-scale corpus analysis that leveraged the UMLS Metathesaurus term characteristics to determine which terms generalized across multiple data sources including Mayo Clinic clinical notes and i2b2/VA 2010 NLP Challenge notes, resulting in a set of filtering rules that reduced significantly the size of the original Metathesaurus lexicon (Wu et al., 2012).

Similar to these studies, we aim to assess the utility of terms leveraged from a controlled vocabulary plus a large corpus of notes, and apply various automatic learning approaches to identify relevant terms specifically aimed at an under served clinical domain. Although some NLP research has addressed the annotation and automatic recognition of variables for substance abuse and its subtopics including *Tobacco*, *Alcohol*, and *Drug Abuse* (Yetisgen et al., 2016; South et al., 2015; Uzuner et al., 2008), to our knowledge, no one has investigated the generation of substance abuse lexicons using our particular term recognition methods for utility in the psychiatric domain.

3 Methods and Materials

In this preliminary study (IRB 68896), we aimed to develop a useful methodology for expanding and reducing domain-specific vocabularies to the most relevant related terms to improve manual patient record review. Our methodology includes three approaches for vocabulary expansion:

one ontology-based and two corpus-based (one rule-based using linguistic information and one context-based using neural networks). As a baseline, we used a vocabulary of seed terms defined by a domain expert (BK). Our method also includes an evaluation of characteristics to inform vocabulary reduction: 1) size and content of the generated vocabularies, and 2) coverage and relevance as it relates to authentic data[1]. For the latter, we used a set of psychiatric clinical notes, described below.

3.1 Baseline vocabulary

A set of predefined terms related to substance abuse was used as the baseline vocabulary. These terms were manually generated by a domain expert (psychiatrist, BK) for the purpose of identifying relevant terms from psychiatry notes in relation to specific variables, e.g., term=*opioids*, category=*Opiates*, variable label={*none*, *current*, *past*, *both*}. In total, this substance abuse vocabulary contains 91 terms in 8 categories including e.g., *Alcohol*, *Cocaine* and *Current Smoking Status*.[2] We reference this approach as the *Baseline*.

3.2 Ontology-based vocabulary expansion

To identify relevant synonyms in the UMLS, we searched for each term in the Baseline vocabulary using Knowledge Author (KA) (Scuba et al., 2014). This approach is referenced as *UMLS*.

3.3 Corpus-based vocabulary expansion

Although ontology-based vocabulary expansion approaches can generate many relevant terms, most terms may not be used in practice in clinical texts. A corpus-based approach can be used to identify potentially missed terms and validate the use of ontology-generated synonyms. We used the free-text notes from the entire MIMIC II database (Saeed et al., 2011), which contains clinical documentation for >30000 patients, for two corpus-based vocabulary expansion approaches using: 1) linguistic resources in combination with transformation rules and corpus-based frequency information, and 2) contextual information from a neural network model. Only alphanumeric tokens were

---

[1] Further details and information about this work, including evaluation script and supplementary material, is available here: `http://toolfinder.chpc.utah.edu/content/vocabulary-expansion-and-reduction-algorithms-vera`.

[2] A subset of a larger vocabulary defined for other variables, e.g. *Education*, *Suicidal Ideation* and *Homelessness*.

used, and all words were converted to lower-case. We included both 1- and 2-token words (uni-and bi-grams) in our methods.

Linguistic and rule-based approach generates lexical variants by querying each seed term (e.g., "alcohol abuse") in WordNet after which four steps are applied on each generated WordNet synonym: 1) term reordering ("abuse alcohol"), 2) inflection generation ("alcohol abused"), 3) abbreviation generation ("aa") and 4) typographic error generation ("alchol abuse"). Each generated term variant was then checked against the MIMIC II corpus and candidate terms occurring >15 times were kept (Conway and Chapman, 2012). We reference this approach as *WNLing*.

Neural network approach leverages a word2vec (Mikolov et al., 2013) neural network bigram model for the generation of context-based related terms. We built a model using a window parameter of 5, discarded words occurring < 15 times, and set the vector dimensionality to 400. Each term in the baseline vocabulary was then queried to find the most similar uni- or bigrams with a similarity score $>= 0.5$.[3] This approach is called *word2vec*.

3.4 Evaluation data set

We randomly sampled 100 psychiatric clinical notes (from a total of approx. 2500) from the University Hospital, University of Utah, Salt Lake City, collected for the purpose of extracting information related to *as-needed anxiolytic use*. From each note, sections more likely to contain information about substance abuse (e.g., *PSYCHIATRIC HISTORY AND PHYSICAL* and *PSYCHIATRIC H&P*) were extracted for matching terms from each vocabulary.

3.5 Evaluation

We performed a quantitative evaluation from two perspectives: 1) vocabulary size and content, to understand characteristics of the generated vocabularies, and 2) vocabulary coverage and relevance, to understand their applicability on authentic data.

3.5.1 Vocabulary size and content

Each new vocabulary was generated from the list of terms in the Baseline vocabulary[4]. We calcu-

lated the number of terms in each generated vocabulary, the number of added terms as compared to the other vocabularies, as well as the total number of all terms (set union) and the total number of shared terms (set intersection) between the generated resources. Note that the vocabularies may contain unigrams that are parts of larger n-grams (multi-word tokens e.g., "alcoholics" as a part of "alcoholics anonymous"). Each unique term was counted separately.

3.5.2 Vocabulary coverage and relevance

Each term in each generated resource was matched against the evaluation data to calculate number of terms found and frequencies of occurrence. A simple string matching procedure was employed in each substance abuse section using regular expressions, where a match was counted if a term[5] was found between a word boundary ("$\backslash b$").

As this evaluation data set is not manually annotated for substance abuse-related terms, we instead calculated *approximations* for both precision (positive predictive value) and recall (sensitivity) by comparing the terms generated from each approach to terms generated by all four approaches.

To calculate these versions of precision and recall for each approach, *relevant and correct* terms (true positives, TP) were defined as the set union of the pairwise intersection sets between all four approaches, i.e. all terms that were found by at least two approaches. *Missed* terms (false negatives, FN) were defined as the terms *not* generated by a specific approach but generated by one (or more) combination of other approaches. *Spurious* terms (false positives, FP) were defined as the number of terms found by a specific approach, but not any other approaches.

Precision and *recall* were then calculated from these results for each approach (*precision=TP/(TP+FP)* and *recall=TP/(TP+FN)*), respectively. Note that this approximation ought to be analyzed with caution - it only gives results in relation to the terms that the vocabularies generate (there is no knowledge about potentially relevant terms outside of these vocabularies). Since the vocabulary approaches are rather different, we believe that this evaluation does give a hint towards what could be expected to at least be relevant terms, and illustrates the relationship between

---

[3]We used the gensim package (Řehůřek and Sojka, 2010) to build this model.

[4]Note that some terms in the Baseline vocabulary were not found in the generated models, Table 7 in Supplement: `http://toolfinder.chpc.utah.edu/sites/ default/files/psychiatry_substance_use_`

`related_terms_supplement.pdf`.

[5]excluding English stopwords from the nltk (http://www.nltk.org/) package.

the employed approaches. This evaluation also permits us to learn common terms learned by multiple approaches and contemplate which combinations should be presented back to the domain expert for expanding the initial query.

4 Results

We report characteristics of the vocabularies in terms of size and content, and we report on vocabulary coverage and relevance when applied to the evaluation data.

Vocabulary	size
Baseline	91
UMLS	863
WNLing	1253
word2vec	1758

Table 1: Size of each vocabulary (number of unique terms).

4.1 Vocabulary size and content

The number of terms in each vocabulary (Baseline, UMLS, WNLing and word2vec) is reported in Table 1. Excluding the Baseline, the word2vec model generated the most terms (n=1758), while UMLS generated the fewest (n=863). In total, 3661 unique terms were generated (Baseline $\cup$ UMLS $\cup$ WNLing $\cup$ word2vec).

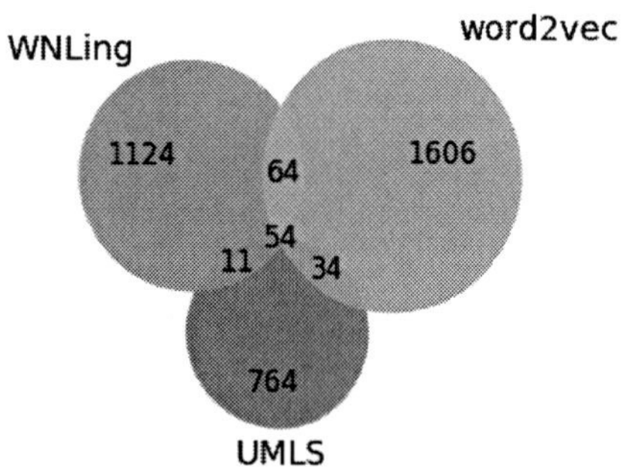

Figure 1: Venn diagram: number of terms in the generated vocabularies from the three approaches: UMLS, WordNet with linguistic heuristics (WNLing), and word2vec.

One term from the Baseline vocabulary was not found by any of the other approaches (*substance use history*). For the corpus-based approaches, there were also some terms in the Baseline vocabulary that were not present in the models generated from the MIMIC II corpus[6].

Forty-three terms were shared between all four vocabularies (Baseline $\cap$ UMLS $\cap$ WNLing $\cap$ word2vec)[7]. and eleven additional terms (a total of 54) were shared between the three approaches (UMLS $\cap$ WNLing $\cap$ word2vec): *addictions, alcoholic beverage, alcoholic drink, amphetamines, beer, benzodiazepines, drug abuse, ethanol, ethyl alcohol, glass,* and *substances.*

Figure 1 shows a Venn diagram with the results from the three vocabulary expansion approaches (UMLS, WNLing, word2vec). In total, 163 terms were shared between at least two approaches (182 in total when including the Baseline vocabulary). Among these 163 terms, added terms as compared to the Baseline vocabulary include misspelling variants (*morpine, coccaine*), inflections (*smokers, addictions*) in addition to new, potentially relevant terms such as *narcotic, etoh, codeine.* The proportion of shared terms for each pairwise vocabulary combination roughly reflects the sizes of the vocabularies, e.g. word2vec $\cap$ WNLing (n=64) $>$ UMLS $\cap$ WNLing (n=11).

4.2 Vocabulary coverage

Vocabulary	u	tot	min	max	avg
Baseline	37	416	1	11	4
UMLS	49	536	2	11	5
WNLing	85	828	2	18	8
word2vec	104	786	1	21	7

Table 2: Number of terms found in the evaluation data. u = number of unique terms found irrespective of frequency, tot = total number of term occurrences found, min, max, avg = minimum, maximum and average number of terms per section.

The number of matched terms (unique and total) in the 100 random substance abuse sections from each vocabulary is shown in Table 2. The sections contain in total 4036 words[8] (min=4, max=192, avg=40). We observed an average of 4 (Baseline) to 8 (WNLing) substance abuse terms (min: 1;

---

[6]Table 7 in Supplement.
[7]Table 1 in Supplement.
[8]Counted using a simple whitespace tokenizer

max: 21) in each substance abuse section using the different vocabularies, Table 2.

The proportion of observed unique terms found in the substance abuse sections varied from about 5-7% for UMLS, WNLing, and word2vec compared to about 41% for the Baseline. As the size of the vocabularies increased, so did the total number of term occurrences (about 7.5 to 11 fold).

A comparison of the coverage between the generated vocabularies is depicted in the Venn diagram in Figure 2. Overall, the number of matched terms is higher for the larger vocabularies (WNLing, word2vec) and the proportion of shared terms is also higher. Twenty-eight terms are shared between all three new vocabularies, and 52 terms are shared between at least two vocabularies (16+28+2+6).

To evaluate the approximated precision and recall, we use the union of each pairwise intersection of *all* vocabularies (including the Baseline), which resulted in 57 unique terms[9].

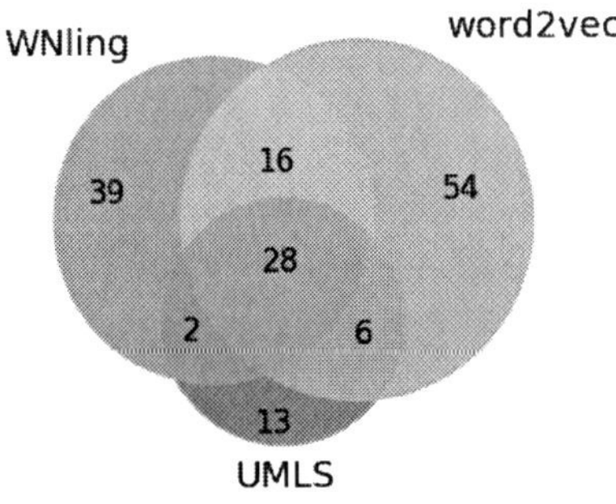

Figure 2: Venn diagram: number of unique terms from the generated vocabularies found in the evaluation data: UMLS, WordNet with linguistic heuristics (WNling), and word2vec.

As expected, the Baseline approach resulted in the highest approximated macro-and micro-precision (0.97/0.998), Table 3. In contrast, the vocabulary-based approaches resulted in the highest macro-and micro-recall (0.91/0.98: word2vec). The fact that the micro-results are higher for the two corpus-based approaches indicates that these two approaches generate term vari-

_____
[9]Table 2 in Supplement.

ants that are also more frequent in the evaluation data set.

4.3 Vocabulary relevance

To analyze relevance, the 20 most frequent terms found by each approach is presented in Table 4, along with information about which vocabulary the term was found in. Nine of these terms were found in all vocabularies. Two terms were not clearly relevant to substance abuse (*last, years*). The three most frequent terms, that were found in all 100 substance abuse sections, are uni- and bi-gram variants of the same term (*substance use, substance* and *use*).

Vocabulary	Macro		Micro	
	P	R	P	R
Baseline	0.97	0.63	0.998	0.67
UMLS	0.77	0.67	0.78	0.68
WNLing	0.55	0.82	0.69	0.93
word2vec	0.5	0.91	0.77	0.98

Table 3: Results: approximated precision and recall, macro (per unique term) and micro (per term occurrence). The number of relevant (and correct) terms is defined as the set union of all pairwise intersections.

The Baseline vocabulary included e.g. the term *packs* but not its singular inflection *pack*, which turned out to be more frequent (*pack* freq=18 as opposed to *packs* freq=6[10]. and found by the two corpus-based approaches WNLing and word2vec. Each approach also resulted in a number of potentially relevant terms that were not found in any of the other approaches, e.g. *amphetamine abuse* (UMLS), *withdrawal* (WNLing), *demerol* (word2vec)[11].

5 Discussion and Conclusion

We present a simple and systematic approach for automated vocabulary generation (expansion and reduction) in the domain of *substance abuse*, applied and evaluated on a set of substance abuse sections from authentic psychiatric notes. Three vocabularies were generated from a set of seed terms using publicly available resources (ontologies, software, and corpora) and combined to: 1) generate a substance abuse vocabulary of highly relevant terms and 2) characterize and analyze

_____
[10]Table 2 in Supplement.
[11]Tables 3–6 in Supplement.

term	Baseline	UMLS	WNLing	word2vec	freq
substance use		x			100
substance	x	x	x	x	100
use				x	100
alcohol	x	x	x	x	62
history				x	41
drug			x	x	40
abuse	x		x	x	40
tobacco	x	x	x	x	35
marijuana	x	x	x	x	32
smokes	x	x	x	x	27
drug use		x		x	26
drug abuse		x	x	x	23
illicit drug				x	22
last			x		21
cocaine	x	x	x	x	21
cigarettes	x	x	x	x	19
pack			x	x	18
years				x	15
heroin				x	15
smoking	x	x	x	x	14

Table 4: 20 Most frequent terms from the union set of all vocabularies that were found in the evaluation data. Presence of term in each respective vocabulary is marked with "x". Note that unigrams could be a substring of an n-gram in each vocabulary (e.g. *substance* and *substance use* in the UMLS vocabulary).

coverage and relevance in an authentic psychiatric dataset.

Through our definition of an approximated precision and recall, we observed that the baseline and ontology-based approaches resulted in the highest approximated precisions, suggesting these methods are useful for identifying the most relevant related terms. This finding is not surprising because the list was vetted by a domain expert and core to the set of terms for all four approaches. In contrast, the vocabulary-based approaches resulted in the highest recalls suggesting these methods are useful for identifying potentially new related terms.

The denominator for calculating these results was based solely on a combination of the four generated vocabularies, which only illustrates relations between approaches. Interestingly, the ontology-based approach (UMLS) resulted in moderate performance for both precision and recall. We hypothesize that this result occurs because although the UMLS provides a notable number of unique terms, these terms do not frequently occur in clinical text due to term characteristics (e.g., inclusion of semantic type and special characters) and concept granularity (use of chemical nomenclature for specific drugs). In future work, we will apply methods to filter based on these characteristics similar to (Hettne et al., 2010; Wu et al., 2012; Demner-Fushman et al., 2010) to address these and other challenges with knowledge authoring leveraging noisy resources.

The baseline vocabulary was biased to more specific terms of substance abuse usage including terms for substances (*alcohol, marijuana, tobacco, cocaine,* and *cigarettes*). Both the ontology- and corpus-based approaches identified more general terms for substance abuse and drug usage as well as terms related to linguistic/semantic attribute information (e.g. *drinking heavily, rarely drinks, quit smoking*). In future work, we will develop methods for learning these patterns to infer these attribute information. These methods will then be placed into a larger infrastructure called *the Information Extraction-Visualization pipeline (IE-Viz)* to aid domain experts with no-to-minimal NLP experience in developing NLP systems for domain-specific use cases.

A preliminary manual analysis of the resulting list of relevant terms (true positives) revealed that a clear majority of the resulting terms were related to substance abuse - only one term was obviously problematic (*years*). The false positives, on the other hand, were in many cases actually relevant and correct terms (e.g. *ecstasy* from the word2vec model), although the WNLing model also produced a number of irrelevant terms (e.g. *charges*). To assess our approximated coverage and relevance metrics, we will conduct a manual assessment of the performance of this approach with respect to true coverage, i.e. analyze which terms were missed, as well as correctness for found terms to determine how well our approximated precision and recall corresponds to the actual precision and recall of terms from this evaluation data set. Moreover, we will assess the relation between terms and categories.

We aim to extend our vocabulary expansion and reduction methods. Most importantly, we have only performed one iteration, using domain expert curated terms, to create the final list of terms. This list could be extended by performing a number of iterations on the resulting list, thereby generating a richer and more comprehensive set of terms. Moreover, we plan to utilize additional publicly available resources, e.g. relevant Wikipedia pages. Once these methods have been integrated into our NLP pipeline, we will extend our experiments to the other psychiatric variables from our data set, e.g., social risk factors of *Homelessness*, *Education level*, *Abuse as a Child*, *Suicide attempts/Self Harm*, and new clinical use-cases such as the detection of bleeding events associated with anticoagulant medication usage by patients with high-risk of stroke.

5.1 Limitations

Our preliminary study evaluation has several limitations including an evaluation using a small data set and calculation of term matches without consideration of term overlap (unigrams/multi-word token counts). We aim to extend our evaluation data set and calculate the effect of term matching criteria in follow up work.

Some of our vocabulary expansion methods have limitations and might be improved. Specifically, word2vec and similar approaches generate related terms that could be a relatedness of se-

mantic types other than synonyms (antonyms, hyponyms, etc.), which is well-known. However, we believe co-occurrence of these terms may correlate with variable terms and perhaps subsequent labels, e.g., alcohol occurring with smoking, which may help with extraction efforts in our NLP pipeline downstream. WNLing abbreviation methods can generate many false positives. Although we reduced some false positives with a stopword check, we could leverage medical acronym and abbreviation dictionaries such as the *Medilexicon*[12] and the *STANDS4 network*[13] to further reduce false positives. Moreover, we believe that combining these types of approaches can be a useful way of limiting the impact of each method's disadvantages.

Finally, our thresholds were chosen rather arbitrarily; therefore, we will experiment with determining the effect of similarity scores and word count thresholds, as well as the use of larger *n*-grams.

5.2 Contribution

To our knowledge, this study is the first systematic study of terms related to substance abuse generated from publicly available resources and the combinations of these approaches, and then evaluated on authentic psychiatric notes. The generated vocabularies can be used to automate parts of the variable encoding process for the ongoing study on treatment management of hospital admitted patients with depression and anxiety disorders, as well as other clinical use-cases where substance abuse information is of importance. This work represents a first step in a larger framework to empower domain experts, in this case psychiatrists, to develop queries and apply NLP methods to identify and extract substance abuse and other variables from large clinical data sets to support mental health research.

Acknowledgments

We would like to thank the anonymous reviewers for valuable comments. This work is partly funded by the Department of Veteran Affairs (CRE 12-312), the National Library of Medicine (R00LM011393), the Patient-Centered Outcomes Research Initiatives (CDRN-1306-04912), the Swedish Research Council (2015-00359), and the

---

[12]http://www.medilexicon.com/medicalabbreviations.php
[13]http://www.abbreviations.com/about.php

Marie Skłodowska Curie Actions, Cofund, Project INCA 600398.

References

Brian E. Chapman, Sean Lee, Hyunseok Peter Kang, and Wendy Webber Chapman. 2011. Document-level classification of CT pulmonary angiography reports based on an extension of the context algorithm. *Journal of Biomedical Informatics*, 44(5):728–737.

Mike Conway and Wendy W. Chapman. 2012. Discovering Lexical Instantiations of Clinical Concepts using Web Services, WordNet and Corpus Resources. In *AMIA 2012 Proceedings*, page 1604, Chicago, USA, November. American Medical Informatics Association.

Glen Coppersmith, Mark Dredze, and Craig Harman. 2014. Quantifying Mental Health Signals in Twitter. In *Association for Computational Linguistics Workshop of Computational Linguistics and Clinical Psychology*.

Helen V. Curran and Brian Birch. 1991. Differentiating the sedative, psychomotor and amnesic effects of benzodiazepines: a study with midazolam and the benzodiazepine antagonist, flumazenil. *Psychopharmacology (Berl)*, 103(4):519–23.

Helen V. Curran. 1986. Tranquillising memories: a review of the effects of benzodiazepines on human memory. *Biol Psychol*, 23(2):179–213, Oct.

Janette Curtis and Kim Capp. 2003. Administration of 'as needed' psychotropic medication: a retrospective study. *Int J Ment Health Nurs*, 12(3):229–34, Sep.

Dina Demner-Fushman, James G Mork, Sonya E Shooshan, and Alan R Aronson. 2010. UMLS content views appropriate for NLP processing of the biomedical literature vs. clinical text. *J Biomed Inform*, 43(4):587–94, Aug.

Christiane Fellbaum. 1998. *WordNet: An Electronic Lexical Database*. Language, Speech and Communication. Mit Press.

Carol Friedman, Philip O. Alderson, John H. Austin, James J. Cimino, and Stephen B. Johnson. 1994. A general natural-language text processor for clinical radiology. *Journal of the American Medical Informatics Association : JAMIA*, 1(2):161–174, March.

N. Grabar, PC. Varoutas, P. Rizand, A. Livartowski, and T. Hamon. 2009. Automatic acquisition of synonym resources and assessment of their impact on the enhanced search in EHRs. *Methods Inf Med.*, 48(2).

Aron Henriksson, Mike Conway, Martin Duneld, and Wendy Webber Chapman. 2013. Identifying synonymy between SNOMED clinical terms of varying length using distributional analysis of electronic health records. In *AMIA 2013, American Medical Informatics Association Annual Symposium, Washington, DC, USA, November 16-20, 2013*.

Kristina M Hettne, Erik M van Mulligen, Martijn J Schuemie, Bob Ja Schijvenaars, and Jan A Kors. 2010. Rewriting and suppressing UMLS terms for improved biomedical term identification. *J Biomed Semantics*, 1(1):5.

Kuo-Chuan Huang, James Geller, Michael Halper, Yehoshua Perl, and Junchuan Xua. 2010. Using WordNet Synonym Substitution to Enhance UMLS Source Integration. *Artif Intell Med.*, 46(2).

Michael Krauthammer and Goran Nenadic. 2004. Term identification in the biomedical literature. *J Biomed Inform*, 37(6):512–26, Dec.

DA. Lindberg, BL. Humphreys, and McCray AT. 1993. The Unified Medical Language System. *Methods Inf Med.*, 32(4):281–91.

Tomas Mikolov, Ilya Sutskever, Kai Chen, Greg Corrado, and Jeffrey Dean. 2013. Distributed Representations of Words and Phrases and their Compositionality. In *Proceedings of NIPS*. NIPS.

Jayson L Mystkowski, Susan Mineka, Laura L Vernon, and Richard E Zinbarg. 2003. Changes in caffeine states enhance return of fear in spider phobia. *J Consult Clin Psychol*, 71(2):243–50, Apr.

Michael W Otto, Steven E Bruce, and Thilo Deckersbach. 2005. Benzodiazepine use, cognitive impairment, and cognitive-behavioral therapy for anxiety disorders: issues in the treatment of a patient in need. *J Clin Psychiatry*, 66 Suppl 2:34–8.

John Pestian, Henry Nasrallah, Pawel Matykiewicz, Aurora Bennett, and Antoon Leenaars. 2010. Suicide note classification using natural language processing: A content analysis. *Biomedical Informatics Insights*, (3):1928.

Radim Řehůřek and Petr Sojka. 2010. Software Framework for Topic Modelling with Large Corpora. In *Proceedings of the LREC 2010 Workshop on New Challenges for NLP Frameworks*, pages 45–50, Valletta, Malta, May. ELRA. http://is.muni.cz/publication/884893/en.

Mohammed Saeed, Mauricio Villarroel, Andrew T Reisner, Gari Clifford, Li-Wei Lehman, George Moody, Thomas Heldt, Tin H Kyaw, Benjamin Moody, and Roger G Mark. 2011. Multiparameter intelligent monitoring in intensive care ii: a public-access intensive care unit database. *Crit Care Med*, 39(5):952–60, May.

Guergana K. Savova, James J. Masanz, Philip V. Ogren, Jiaping. Zheng, Sunghwan Sohn, Karin C. Kipper-Schuler, and Christopher G. Chute. 2010. Mayo clinical text analysis and knowledge extraction system (ctakes): architecture, component evaluation and applications. *Journal of the American Medical Informatics Association*, 17(5):507–513.

William Scuba, Melissa Tharp, Yang Tseytlin Eugene, Liu, Frank A. Drews, and Wendy Chapman. 2014. Knowledge Author: Creating Domain Content for NLP Information Extraction. In *6th International Symposium on Semantic Mining in Biomedicine (SMBM)*.

Brett R. South, Danielle Mowery, Melissa Tharp, Margorie Carter, Adi Gundlapalli, Marzieh Vali, Mike Conway, Salomeh Keyhani, and Wendy W. Chapman. 2015. Extracting social history and functional status from veteran affairs clinical documents. In *AMIA Joint Summits on Translational Science*.

Jane Stein-Parbury, Kim Reid, Narelle Smith, Diane Mouhanna, and Fiona Lamont. 2008. Use of pro re nata medications in acute inpatient care. *Aust N Z J Psychiatry*, 42(4):283–92, Apr.

Özlem Uzuner, Ira Goldstein, Yuan Luo, and Isaac S. Kohane. 2008. Viewpoint paper: Identifying patient smoking status from medical discharge records. *Journal of American Medical Informatics Association*, 15(1):14–24.

Henny A. Westra, Sherry H. Stewart, Michael Teehan, Karen Johl, David J. A. Dozois, and Todd Hill. 2004. Benzodiazepine Use Associated with Decreased Memory for Psychoeducation Material in Cognitive Behavioral Therapy for Panic Disorder. *Cognitive Therapy and Research*, 28(2):193–208, April.

Stephen Tze-Inn Wu, Hongfang Liu, Dingcheng Li, Cui Tao, Mark A. Musen, Christopher G. Chute, and Nigam H. Shah. 2012. Unified medical language system term occurrences in clinical notes: a large-scale corpus analysis. *Journal of Americal Medical Informatics Association*, 19(e1).

Meliha Yetisgen, Elena Pellicer, David R. Crosslin, and Lucy Vanderwende. 2016. Automatic identification of lifestyle and environmental factors from social history in clinical text. In *AMIA 2016 Joint Summits on Translational Science*.

Qing T. Zeng, Doug Redd, Thomas Rindflesch, and Jonathan Nebeker. 2012. Synonym, topic model and predicate-based query expansion for retrieving clinical documents. volume 2012, pages 1050–1059. American Medical Informatics Association.

Syntactic analyses and named entity recognition for PubMed and PubMed Central — up-to-the-minute

Kai Hakala[1,2*]**, Suwisa Kaewphan**[1,2,3*]**, Tapio Salakoski**[1,3] **and Filip Ginter**[1]
1. Dept. of Information Technology, University of Turku, Finland
2. The University of Turku Graduate School (UTUGS), University of Turku, Finland
3. Turku Centre for Computer Science (TUCS), Finland
`kahaka@utu.fi, sukaew@utu.fi,`
`tapio.salakoski@utu.fi, ginter@cs.utu.fi`

Abstract

Although advanced text mining methods specifically adapted to the biomedical domain are continuously being developed, their applications on large scale have been scarce. One of the main reasons for this is the lack of computational resources and workforce required for processing large text corpora.

In this paper we present a publicly available resource distributing preprocessed biomedical literature including sentence splitting, tokenization, part-of-speech tagging, syntactic parses and named entity recognition. The aim of this work is to support the future development of large-scale text mining resources by eliminating the time consuming but necessary preprocessing steps.

This resource covers the whole of PubMed and PubMed Central Open Access section, currently containing 26M abstracts and 1.4M full articles, constituting over 388M analyzed sentences. The resource is based on a fully automated pipeline, guaranteeing that the distributed data is always up-to-date. The resource is available at `https://turkunlp.`
`github.io/pubmed_parses/`.

1 Introduction

Due to the rapid growth of biomedical literature, the maintenance of manually curated databases, usually updated following new discoveries published in articles, has become unfeasible. This has led to a significant interest in developing automated text mining methods specifically for the biomedical domain.

Various community efforts, mainly in the form of shared tasks, have resulted in steady improvement in biomedical text mining methods (Kim et al., 2009; Segura Bedmar et al., 2013). For instance the GENIA shared tasks focusing on extracting biological events, such as gene regulations, have consistently gathered wide interest and have led to the development of several text mining tools (Miwa et al., 2012; Björne and Salakoski, 2013). These methods have been also successfully applied on a large scale and several biomedical text mining databases are publicly available (Van Landeghem et al., 2013a; Franceschini et al., 2013; Müller et al., 2004). Although these resources exist, their number does not reflect the vast amount of fundamental research invested in the underlying methods, mainly due to the nontrivial amount of manual labor and computational resources required to process large quantities of textual data. Another issue arising from the challenging text preprocessing is the lack of maintenance of the existing databases which in effect nullifies the purpose of text mining as these resources tend to be almost as much out-of-date as their manually curated counterparts. According to MEDLINE statistics[1] 806,326 new articles were indexed during 2015 and thus a text mining resource will miss on average 67 thousand articles each month it hasn't been updated.

In this paper we present a resource aiming to support the development and maintenance of large-scale biomedical text mining. The resource includes all PubMed abstracts as well as full articles from the open access section of PubMed Central (PMCOA), with the fundamental language technology building blocks, such as part-of-speech (POS) tagging and syntactic parses, readily available. In addition, recognition of several bio-

---

*These authors contributed equally.

[1]`https://www.nlm.nih.gov/bsd/bsd_key.`
`html`

Proceedings of the 15th Workshop on Biomedical Natural Language Processing, pages 102–107,
Berlin, Germany, August 12, 2016. ©2016 Association for Computational Linguistics

logically relevant named entities, such as proteins and chemicals is included. Hence we hope that this resource eliminates the need of the tedious preprocessing involved in utilizing the PubMed data and allows swifter development of new information extraction databases.

The resource is constructed with an automated pipeline which provides weekly updates with the latest articles indexed in PubMed and PubMed Central, ensuring the timeliness of the distributed data. All the data is downloadable in an easily handleable XML format, also used by the widely adapted event extraction system TEES (Björne and Salakoski, 2015). A detailed description of this format is available on the website.

2 Data

We use all publicly available literature from PubMed and PubMed Central Open Access subset, which cover most of the relevant literature and are commonly used as the prime source of data in biomedical text mining knowledge bases.

PubMed provides titles and abstracts in XML format in a collection of baseline release and subsequent updates. The former is available at the end of each year whereas the latter is updated daily. As this project was started during 2015, we have first processed the baseline release from the end of 2014 and this data has then been extended with the new publications from the end of 2015 baseline release. The rest of the data up to date has been collected from the daily updates.

The full articles in PMC Open Access subset (PMCOA) are retrieved via the PMC FTP service. Multiple types of data format are provided in PM-COA, including NXML and TXT formats which are suitable for text processing. We use the provided NXML format as it is compatible with our processing pipeline. This service does not provide distinct incremental updates, but a list of all indexed articles updated weekly.

3 Processing Pipeline

In this section, we discuss our processing pipeline as shown in Figure 1. Firstly, both PubMed and PMCOA documents are downloaded from NCBI FTP services. For the periodical updates of our resource this is done weekly — the same interval the official PMCOA dataset is updated. From the PubMed incremental updates we only include newly added documents and ignore other updates.

As the PMCOA does not provide incremental updates, we use the index file and compare it to the previous file list to select new articles for processing.

Even though the PubMed and PMCOA documents are provided in slightly different XML formats, they can be processed in similar fashion. As a result, the rest of the pipeline discussed in this section is applied to both document types.

Both PubMed XML articles and PMCOA NXML full texts are preprocessed using publicly available tools[2] (Pyysalo et al., 2013). These tools convert XML documents to plain text and change character encoding from UTF-8 to ASCII as many of the legacy language processing tools are incapable of handling non-ASCII characters. Additionally, all excess meta data is removed, leaving titles, abstracts and full-text contents for further processing. These documents are subsequently split into sentences using GENIA sentence splitter (Sætre et al., 2007) as most linguistic analyses are done on the sentence level. GENIA sentence splitter is trained on biomedical text (GENIA corpus) and has state-of-the-art performance on this domain.

The whole data is parsed with the BLLIP constituent parser (Charniak and Johnson, 2005), using a model adapted for the biomedical domain (McClosky, 2010), as provided in the TEES processing pipeline. The distributed tokenization and POS tagging are also produced with the parser pipeline. We chose to use this tool as the performance of the TEES software has been previously evaluated on a large-scale together with this parsing pipeline (Van Landeghem et al., 2013b) and it should be a reliable choice for biomedical relation extraction. Since dependency parsing has become the prevalent approach in modeling syntactic relations, we also provide conversions to the collapsed Stanford dependency scheme (De Marneffe et al., 2006).

The pipeline is run in parallel on a cluster computer with the input data divided into smaller batches. The size of these batches is altered along the pipeline to adapt to the varying computational requirements of the different tools.

3.1 Named Entity Recognition

Named entity recognition (NER) is one of the fundamental tasks in BioNLP as most of the cru-

---

[2]https://github.com/spyysalo/nxml2txt

Entity type	Our system	State-of-the-art system	References
	Precision/Recall/F-score	Precision/Recall/F-score	
Cell line	89.88 / 84.36 / 87.03	91.67 / 85.47 / 88.46	(Kaewphan et al., 2016)
Chemical	85.27 / 82.92 / 84.08	89.09 / 85.75 / 87.39	(Leaman et al., 2015)
Disease*	86.32 / 80.83 / 83.49	82.80 / 81.90 / 80.90	(Leaman et al., 2013)
GGP**	74.27 / 72.99 / 73.62	90.22 / 84.82 / 87.17	(Campos et al., 2013)
Organism	77.15 / 80.15 / 78.63	83.90 / 72.60 / 77.80	(Pafilis et al., 2013)

Table 1: Evaluation of the named entity recognition for each entity type on the test sets, measured with strict entity level metrics. Reported results for corresponding state-of-the-art approaches are shown for comparison.

* The evaluation of the best performing system for disease mentions is the combination of named entity recognition and normalization.

** The official BioCreative II evaluation for our GGP model results in 84.67, 84.54 and 84.60 for precision, recall and F-score respectively. These numbers are comparable to the listed state-of-the-art method.

cial biological information is expressed as relations among entities such as genes and proteins. To support further development on this dataset, we provide named entity tagging for five entity types, namely diseases, genes and gene products (GGPs), organisms, chemicals, and cell line names. Although several tools with state-of-the-art performance are available for these entity types (Leaman et al., 2015; Leaman and Gonzalez, 2008), we have decided to use a single tool, NERsuite[3], for all types. NERsuite is based on conditional random field classifiers as implemented in the CRFsuite software (Okazaki, 2007). Having a single tool for this processing step instead of using the various state-of-the-art tools is critical for the maintainability of the processing pipeline. NERsuite was selected as several biological models are readily available for this software (Kaewphan et al., 2016; Pyysalo and Ananiadou, 2014) and as it supports label weighting (Minkov et al., 2006) unlike many other NER tools.

For cell line names we use a publicly available state-of-the-art model (Kaewphan et al., 2016), whereas for the other entity types we train our own models with manually annotated data from GENETAG (Tanabe et al., 2005), CHEMDNER (Krallinger et al., 2015), SPECIES (Pafilis et al., 2013) and NCBI disease (Doğan et al., 2014) corpora for GGPs, chemicals, organisms and diseases, respectively. All these corpora are comprised of biomedical articles and should thus reflect well the text types seen in PubMed.

All used corpora provide the data divided to training, development and test sets in advance, the

SPECIES corpus being an exception. For this corpus we do our own data division with random sampling on document level, for each taxonomy category separately. For each entity type, the C2 value, as well as the label weights are selected to optimize the F-score on the development set. For the training of the final models used in the resource, we use the whole corpora, i.e. the combination of training, development and test sets.

Detailed performance evaluations for all entity types are shown in Table 1. We evaluate NERsuite in terms of precision, recall and F-score against the test data using "strict matching" criteria, i.e. only consider the tagged entities correct if they are perfectly matched with the gold standard data. These results may not be directly comparable to the results reported in other studies as relaxed evaluation methods are sometimes used. However, we can conclude that our system is on par with the methods published elsewhere and the limitation of using a single tool does not have a significant negative impact on the overall performance.

4 Data Statistics

During the time of writing this paper the dataset included 25,512,320 abstracts from PubMed and 1,350,119 full articles from PMCOA, resulting in 155,356,970 and 232,838,618 sentences respectively. These numbers are not identical to the ones reported by NCBI for couple of reasons. Firstly, at the moment, we do not process the deletion updates nor do we remove the old versions of PMCOA articles if they are revised, i.e. our dataset may include articles, which have been retracted and an article may be included multiple times if

---

[3] http://nersuite.nlplab.org/

Entity type	Occurrences	Most common entity spans
Cell line	6,967,903	HeLa, MCF-7, A549, HepG2, MDA-MB-231
Chemical	153,285,486	glucose, N, oxygen, Ca2+, calcium
Disease	105,416,758	tumor, cancer, HIV, breast cancer, tumors
GGP	190,543,270	insulin, GFP, p53, TNF-alpha, IL-6
Organism	69,962,111	human, mice, mouse, HIV, humans

Table 2: Occurrence counts and the most frequent entity spans for all entity types in the whole data set.

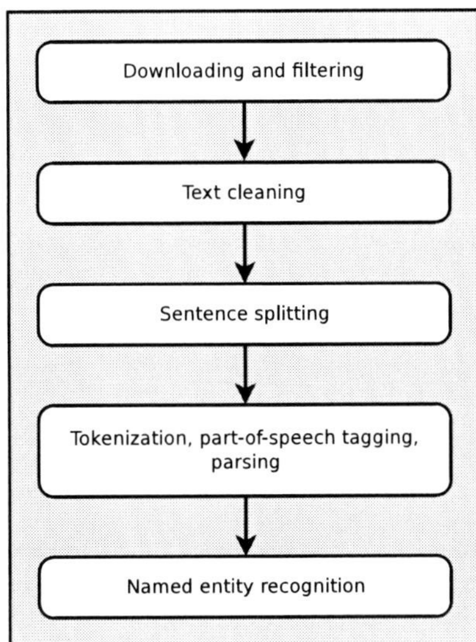

Figure 1: The main processing steps of the pipeline. First, the articles are downloaded from the source and filtered to prevent reprocessing old documents. The documents are then converted to plain text format. This text data is split to independent sentences, tokenized and tagged with POS labels and syntactic dependencies. In addition, named entity recognition for several entity types is carried out.

the content has been modified. We plan to take the deletions into account in near future. Secondly, the external tools in our pipeline may occasionally fail, in which case some of the articles are not processed. Since the pipeline processes the input data in batches, a critical error may lead to a whole batch not being processed. We are currently improving the pipeline to automatically reprocess the failed batches with the problematic articles excluded to minimize the loss of data.

Running the parsing pipeline, including tokenization, POS tagging and conversion to the collapsed Stanford scheme, is the most time consuming part of the whole pipeline. Execution of this step has taken 84,552 CPU hours (9.6 CPU years) for the currently available data.

Unfortunately we do not have exact processing time statistics for named entity recognition and thus estimate its computational requirements by extrapolating from a smaller test run. Based on this experiment NER has demanded 4,100 CPU hours thus far. The text preprocessing and sentence splitting steps are negligible and thus the overall processing time required is approximately 10 CPU years.

In total, our processing pipeline has detected 526,175,528 named entities. GGPs are the most common entities, covering 36.2% of all entity mentions, whereas the cell lines are the most infrequent, forming only 1.3% of the data. The entity type specific statistics along with the most common entity spans are listed in Table 2.

5 Future Work

Our future efforts will focus on expanding the coverage of supported entity types to mutations and anatomical entities (Wei et al., 2013; Pyysalo and Ananiadou, 2014), deepening the captured information of biological processes and bringing text mining one step closer to extracting a realistic view of biological knowledge.

As many of the NER training corpora include only abstracts and are limited to specific domains, the generalizability of the trained NER models to full articles and to the wide spectrum of topics covered in PubMed is not clear. Thus we wish to assess how well these models perform on large-scale datasets and analyze how their performance could be improved on out-of-domain documents.

We plan to also include entity normalization for all supported types, but as we wish to minimize the number of individual tools in the processing pipeline, we are developing a generic approach suitable for most entity types.

6 Conclusions

We have introduced a new resource which provides the basic linguistic analyses, essential in the development of text mining knowledge bases, for the whole of PubMed and PubMed Central Open Access section, thus drastically reducing the amount of required preprocessing efforts.

In addition, we provide named entity tagging for several biologically relevant entity types and show that the models we have used are comparable to the state-of-the-art approaches, although our focus has been on retaining the processing pipeline as simple as possible for easier maintenance.

The resource is periodically updated with an automated pipeline, and currently includes over 26M documents fully parsed with 526M named entity mentions detected. The data is available for download in XML format.

Acknowledgments

Computational resources were provided by CSC - IT Center For Science Ltd., Espoo, Finland. This work was supported by ATT Tieto käyttöön grant.

References

Jari Björne and Tapio Salakoski. 2013. TEES 2.1: Automated annotation scheme learning in the BioNLP 2013 Shared Task. In *Proceedings of the BioNLP Shared Task 2013 Workshop*, pages 16–25.

Jari Björne and Tapio Salakoski. 2015. TEES 2.2: biomedical event extraction for diverse corpora. *BMC Bioinformatics*, 16(16):1–20.

David Campos, Sérgio Matos, and José Luís Oliveira. 2013. Gimli: open source and high-performance biomedical name recognition. *BMC bioinformatics*, 14(1):54.

Eugene Charniak and Mark Johnson. 2005. Coarse-to-fine N-best parsing and MaxEnt discriminative reranking. In *Proceedings of the 43rd Annual Meeting on Association for Computational Linguistics*, ACL'05, pages 173–180, Stroudsburg, PA, USA. Association for Computational Linguistics.

Marie-Catherine De Marneffe, Bill MacCartney, and Christopher D Manning. 2006. Generating typed dependency parses from phrase structure parses. In *Proceedings of LREC*, volume 6, pages 449–454.

Rezarta Islamaj Doğan, Robert Leaman, and Zhiyong Lu. 2014. NCBI disease corpus: A resource for disease name recognition and concept normalization. *Journal of Biomedical Informatics*, 47:1 – 10.

Andrea Franceschini, Damian Szklarczyk, Sune Frankild, Michael Kuhn, Milan Simonovic, Alexander Roth, Jianyi Lin, Pablo Minguez, Peer Bork, Christian von Mering, and Lars J. Jensen. 2013. STRING v9.1: protein-protein interaction networks, with increased coverage and integration. *Nucleic acids research*, 41(D1):D808–D815.

Suwisa Kaewphan, Sofie Van Landeghem, Tomoko Ohta, Yves Van de Peer, Filip Ginter, and Sampo Pyysalo. 2016. Cell line name recognition in support of the identification of synthetic lethality in cancer from text. *Bioinformatics*, 32(2):276–282.

Jin-Dong Kim, Tomoko Ohta, Sampo Pyysalo, Yoshinobu Kano, and Jun'ichi Tsujii. 2009. Overview of BioNLP'09 Shared Task on event extraction. In *Proceedings of the Workshop on Current Trends in Biomedical Natural Language Processing: Shared Task*, pages 1–9. Association for Computational Linguistics.

Martin Krallinger, Obdulia Rabal, Florian Leitner, Miguel Vazquez, David Salgado, Zhiyong Lu, Robert Leaman, Yanan Lu, Donghong Ji, Daniel M. Lowe, Roger A. Sayle, Riza Theresa Batista-Navarro, Rafal Rak, Torsten Huber, Tim Rocktäschel, Sérgio Matos, David Campos, Buzhou Tang, Hua Xu, Tsendsuren Munkhdalai, Keun Ho Ryu, SV Ramanan, Senthil Nathan, Slavko Žitnik, Marko Bajec, Lutz Weber, Matthias Irmer, Saber A. Akhondi, Jan A. Kors, Shuo Xu, Xin An, Utpal Kumar Sikdar, Asif Ekbal, Masaharu Yoshioka, Thaer M. Dieb, Miji Choi, Karin Verspoor, Madian Khabsa, C. Lee Giles, Hongfang Liu, Komandur Elayavilli Ravikumar, Andre Lamurias, Francisco M. Couto, Hong-Jie Dai, Richard Tzong-Han Tsai, Caglar Ata, Tolga Can, Anabel Usié, Rui Alves, Isabel Segura-Bedmar, Paloma Martínez, Julen Oyarzabal, and Alfonso Valencia. 2015. The CHEMDNER corpus of chemicals and drugs and its annotation principles. *Journal of Cheminformatics*, 7(1):1–17.

Robert Leaman and Graciela Gonzalez. 2008. BANNER: an executable survey of advances in biomedical named entity recognition. In *Pacific Symposium on Biocomputing*, volume 13, pages 652–663. Citeseer.

Robert Leaman, Rezarta Islamaj Doğan, and Zhiyong Lu. 2013. DNorm: disease name normalization with pairwise learning to rank. *Bioinformatics*, 29(22):2909–2917.

Robert Leaman, Chih-Hsuan Wei, and Zhiyong Lu. 2015. tmChem: a high performance approach for chemical named entity recognition and normalization. *Journal of Cheminformatics*, 7(1):1–10.

David McClosky. 2010. *Any domain parsing: automatic domain adaptation for natural language parsing*. Ph.D. thesis, Providence, RI, USA.

Einat Minkov, Richard Wang, Anthony Tomasic, and William Cohen. 2006. NER systems that suit user's preferences: adjusting the recall-precision trade-off for entity extraction. In *Proceedings of the Human Language Technology Conference of the NAACL, Companion Volume: Short Papers*, pages 93–96, New York City, USA, June. Association for Computational Linguistics.

Makoto Miwa, Paul Thompson, John McNaught, Douglas B Kell, and Sophia Ananiadou. 2012. Extracting semantically enriched events from biomedical literature. *BMC bioinformatics*, 13(1):1.

Hans-Michael Müller, Eimear E Kenny, and Paul W Sternberg. 2004. Textpresso: An ontology-based information retrieval and extraction system for biological literature. *PLoS Biol*, 2(11), 09.

Naoaki Okazaki. 2007. CRFsuite: a fast implementation of Conditional Random Fields (CRFs).

Evangelos Pafilis, Sune P. Frankild, Lucia Fanini, Sarah Faulwetter, Christina Pavloudi, Aikaterini Vasileiadou, Christos Arvanitidis, and Lars Juhl Jensen. 2013. The SPECIES and ORGANISMS resources for fast and accurate identification of taxonomic names in text. *PLoS ONE*, 8(6):1–6, 06.

Sampo Pyysalo and Sophia Ananiadou. 2014. Anatomical entity mention recognition at literature scale. *Bioinformatics*, 30(6):868–875.

Sampo Pyysalo, Filip Ginter, Hans Moen, Tapio Salakoski, and Sophia Ananiadou. 2013. Distributional semantic resources for biomedical text mining. In *Proceedings of the 5th International Symposium on Languages in Biology and Medicine*, pages 39–44.

Rune Sætre, Kazuhiro Yoshida, Akane Yakushiji, Yusuke Miyao, Yuichiro Matsubayashi, and Tomoko Ohta. 2007. AKANE system: protein-protein interaction pairs in BioCreAtIvE2 challenge, PPI-IPS subtask. In *Proceedings of the BioCreative II*, pages 209–212.

Isabel Segura Bedmar, Paloma Martínez, and María Herrero Zazo. 2013. SemEval-2013 task 9: Extraction of drug-drug interactions from biomedical texts (DDIextraction 2013). In *In Proceedings of the 7th International Workshop on Semantic Evaluation (SemEval*. Association for Computational Linguistics.

Lorraine Tanabe, Natalie Xie, Lynne H Thom, Wayne Matten, and W John Wilbur. 2005. GENETAG: a tagged corpus for gene/protein named entity recognition. *BMC bioinformatics*, 6(1):1.

Sofie Van Landeghem, Jari Björne, Chih-Hsuan Wei, Kai Hakala, Sampo Pyysalo, Sophia Ananiadou, Hung-Yu Kao, Zhiyong Lu, Tapio Salakoski, Yves Van de Peer, and Filip Ginter. 2013a. Large-scale event extraction from literature with multi-level gene normalization. *PLoS ONE*, 8(4):1–12, 04.

Sofie Van Landeghem, Suwisa Kaewphan, Filip Ginter, and Yves Van de Peer. 2013b. Evaluating large-scale text mining applications beyond the traditional numeric performance measures. In *Proceedings of the 2013 Workshop on Biomedical Natural Language Processing*, pages 63–71, Sofia, Bulgaria, August. Association for Computational Linguistics.

Chih-Hsuan Wei, Bethany R Harris, Hung-Yu Kao, and Zhiyong Lu. 2013. tmVar: a text mining approach for extracting sequence variants in biomedical literature. *Bioinformatics*, 29(11):1433–1439.

Improving Temporal Relation Extraction with Training Instance Augmentation

Chen Lin and **Timothy Miller**
Boston Children's Hospital Informatics Program
and Harvard Medical School
{first.last}@childrens.harvard.edu

Dmitriy Dligach
Loyola University Chicago
ddligach@luc.edu

Steven Bethard
University of Alabama at Birmingham
bethard@uab.edu

Guergana Savova
Boston Children's Hospital Informatics Program
and Harvard Medical School
guergana.savova@childrens.harvard.edu

Abstract

Temporal relation extraction is important for understanding the ordering of events in narrative text. We describe a method for increasing the number of high-quality training instances available to a temporal relation extraction task, with an adaptation to different annotation styles in the clinical domain by taking advantage of the Unified Medical Language System (UMLS). This method notably improves clinical temporal relation extraction, works beyond featurizing or duplicating the same information, can generalize between-argument signals in a more effective and robust fashion. We also report a new state-of-the-art result, which is a two point improvement over the best Clinical TempEval 2016 system.

1 Introduction

Temporal relation extraction is important for understanding ordering of events from a narrative text. Recent years have seen annotated corpora created for temporal information extraction, from newspaper text (Pustejovsky et al., 2003; Verhagen et al., 2007; Verhagen et al., 2010; UzZaman et al., 2013), to clinical narratives (Savova et al., 2009; Sun et al., 2013; Styler et al., 2014), all with the aim of developing systems for building event timelines from textual descriptions of events. Such narrative timelines are important for information extraction tasks such as question answering (Kahn et al., 1990), clinical outcomes prediction (Schmidt et al., 2005; Lin et al., 2014), and the identification of temporal patterns (Zhou and Hripcsak, 2007) among many.

In a typical supervised approach to the temporal relation extraction task, argument pairs consist of pairs of events or temporal expressions. Corpora differ in their syntactic annotation of such expres-

sions. For example, the THYME corpus, consisting of oncology, pathology and radiology notes, annotated only event headwords (Styler et al., 2014), while the i2b2 corpus, consisting of discharge summaries, annotated entire noun phrases as events (Sun et al., 2013). As a result, it is necessary to account for these differences when implementing a generalizable relation extraction system.

However, the annotations of the temporal relations between the events remain unaffected by the choice of headwords or phrases for the event annotation. For example, in a relation between the temporal expression *yesterday* and the event *severe lower abdominal pain*, if the argument had been the head word *pain* it still would have been an instance of the same temporal relation. Thus, we can automatically create additional training examples by varying the extent of headword expansion. For example, the relation between *yesterday* and *severe lower abdominal pain* can automatically generate four valid relations of the same type where the second arguments are *pain*, *abdominal pain*, and *lower abdominal pain*.

In this paper, we describe an automatic method that generates more temporal training instances by semantically expanding gold medical events based on a clinical ontology, the Unified Medical Language System (UMLS) (Lindberg et al., 1993). It bridges the gap between different syntactic annotations of events in clinical corpora. We show that this method is superior to representing the same information as additional features, that it differs from plain upsampling, and that the primary mechanism of improvement is in the better representation of between-argument features. Our method can be viewed as a new form of data augmentation, akin to the generation of image variants for vision recognition (Krizhevsky et al., 2012) or the generation of word substitutions for information extraction (Kolomiyets et al., 2011).

Proceedings of the 15th Workshop on Biomedical Natural Language Processing, pages 108–113,
Berlin, Germany, August 12, 2016. ©2016 Association for Computational Linguistics

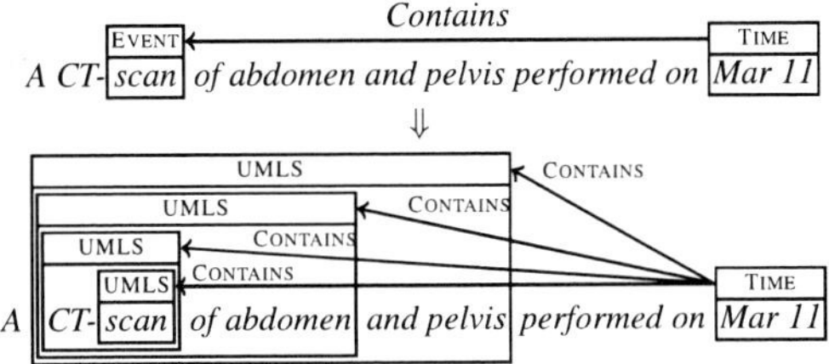

Figure 1: Example expansion of the event "scan"

2 Method

First, the text was scanned for any medical concepts from the UMLS Metathesaurus (http://www.nlm.nih.gov/research/umls/), a collection of concepts from different biomedical terminologies. Apache cTAKES (http://ctakes.apache.org) was used to extract such UMLS concepts. Next, we use these UMLS concepts and gold standard events to expand relation arguments. For a gold standard event e annotated by the headword, we define $\text{EXPAND}(e)$ as the set of UMLS entities whose spans cover e. If e is involved in a temporal relation r, we assume u ($u \in \text{EXPAND}(e)$) is involved in the same relation and therefore we generate a new temporal relation that is identical to r but with the event e replaced by a UMLS entity u. Figure 1 shows an example of expanding the gold event "scan" to its covering UMLS entities and generating related relations.

We differentiate temporal relations into event-time and event-event, and expand relations as detailed in Algorithm 1 and Algorithm 2, respectively. For event-event, we ensure the event spans do not overlap after expansion. Our event-time model classifies all relations – CONTAINS, BEFORE, OVER-LAP, BEGINS-ON, ENDS-ON and NONE, while our event-event model classifies only CONTAINS and NONE relations due to the very low inter-annotator agreement for the other relation types in our evaluation corpus (Styler et al., 2014). For both models, NONE is used to indicate that there is no relation between a pair of arguments.

Algorithm 1 Expansion for event-time relations

1: Given a gold-standard annotated event-time relation $r(e,t)$, where e is an event, t is a temporal expression, $r \in \{\text{CONTAINS}, \text{BEFORE}, \dots, \text{NONE}\}$
2: **for** UMLS entity $u \in \text{EXPAND}(e)$ **do**
3:      Create relation $r'(u, t)$, $r' \leftarrow r$
4:      Add r' to training data
5: **end for**

Algorithm 2 Expansion for event-event relations

1: Given a gold-standard annotated event-event relation $r(e_a,e_b)$, where e_a, e_b are events, $r \in \{\text{CONTAINS}, \text{NONE}\}$
2: **for** UMLS entity $u_a \in \text{EXPAND}(e_a)$ **do**
3:      **if not** overlaps(span(u_a), span(e_b)) **then**
4:          Create relation $r'(u_a, e_b)$, $r' \leftarrow r$
5:          Add r' to training data
6:      **end if**
7: **end for**
8: **for** UMLS entity $u_b \in \text{EXPAND}(e_b)$ **do**
9:      **if not** overlaps(span(e_a, u_b)) **then**
10:          Create relation $r'(e_a, u_b)$, $r' \leftarrow r$
11:          Add r' to training data
12:     **end if**
13: **end for**

---

3 Experiments

3.1 Dataset

We tested our event expansion technique on a publicly available clinical corpus: the colon cancer set of the THYME corpus (Styler et al., 2014) used in SemEval 2015 Task 6 (Bethard et al., 2015) and SemEval 2016 Task 12: Clinical TempEval (Bethard et al., 2016). It contains 600 documents (400 oncology notes and 200 pathology notes) of 200 colon cancer patients. The gold standard annotations contain events (including both medical and general events, all annotated by head words), temporal expressions (e.g. *tomorrow*, *postoperative*, and *March-11-2009*), and temporal relations. We used the same training/development/test split as Clinical TempEval. The development set was used for testing research questions and building final models. Once the models were deemed finalized, they were rebuilt on the combined training and development sets and tested on the test set.

3.2 Models

We built two within-sentence temporal-relation classification models, one for event-time relations and one for event-event relations. We paired every gold event with every gold time expression within the same sentence to form candidate instances for the event-time classifier. We paired all gold events within a sentence to form candidates for the event-event classifier. For training, gold relations were also expanded by calculating the closure sets of all possible relations in a clinical document.

We use the LIBLINEAR (Fan et al., 2008) L2-regularized L2-loss dual SVM as the learning algorithm for both models. Our features for event-time and event-event models are shown in Table 1.

Feature	Description	EE	ET
Tokens	the first and the last word of each concept, all words covered by a concept as a bag, bag-of-words around each concept for a window of [-3, 3], bag-of-words between two concepts, and the number of words between two concepts	✓	✓
Part-of-speech tags	the POS tags of each concept as a bag	✓	
Event attributes	all event-related attributes such as polarity, modality, and type	✓	✓
UMLS feature	UMLS semantic types as features	✓	
Dependency path	the dependency path between two concepts and the number of dependency nodes in-between	✓	✓
Overlapped head	if two concepts share the same head word	✓	
Temporal attributes	the class type of a time expression, e.g. Date, Time, Duration, etc.		✓
Special words	Any words from the time lexicon developed by NRCC (Cherry et al., 2013a) that the concepts or the context in-between contain		✓
Nearest flag	if the event-time pair in question is the closest among all pairs in the same sentence		✓
Conjunction feature	if there is any conjunction word between the arguments		✓

Table 1: Features used for event-event (EE) and event-time (ET) classifiers

3.3 Research questions

We investigate the following questions:

1. Can the effect of the UMLS expansion technique be replicated using additional features? One may wonder if adding instances via UMLS expansion is isomorphic to adding more features that capture the UMLS information. To answer this question, we find all covering UMLS entities, but instead of creating new instances, extract token features from these entities and add those to the other features for the instance.

2. Is it better to expand to the longest UMLS entity or to expand to all possible spans? In our Figure 1 example, the longest UMLS entity covering "scan" is "CT-scan of abdomen and pelvis". But we could also create instances for the UMLS entities "scan", "CT-scan" and "CT-scan of abdomen". We also compare against a purely linguistic expansion to the immediate enclosing noun phrase (NP).

3. Is the improvement due to the replication of instances? Our expansion technique creates many similar relations, and in cases where a UMLS entity has the same span as a gold event, the technique creates true duplicate instances. For example, the relation CONTAINS(scan, March 11) is duplicated in Figure 1. Thus we also compare our UMLS-informed expansion of instances to simple duplication of instances[1].

4. Which types of features benefit most from the expansion? There are three groups of token features: within each argument, between the arguments, and the preceding and following three words (context) of an argument. We test the performance one feature group at a time, with and without the event expansion.

We test all research questions by training on the training set and testing on the development set with token-based features for the event-time relations. Note that expansion is applied only to the training set, not to the development or test set.

3.4 Evaluation

For results on the development set, we calculate closure-enhanced precision, recall and F1-score (UzZaman and Allen, 2011) on just the within-sentence relations (since that's what our models are able to predict). Precision is the percentage of system-generated relations that can be verified in the transitive closure of the gold standard relations. Recall is the percentage of gold standard relations that can be found in the transitive closure of the system-generated relations. The final F1-score is the harmonic mean of the transitive-closure-processed precision and recall.

For results on the test set, we used the official Clinical TempEval evaluation scripts so that our results are directly comparable with the outcomes of Clinical TempEval 2016 (Bethard et al., 2016). These scripts use similar definitions of closure-enhanced precision, recall and F1-score, but evaluate only CONTAINS relations in oncology notes.

4 Results on the development set

Question 1 is answered by the first two rows of Table 2: adding token features representing expanded UMLS entities does not achieve the same perfor-

<hr>

[1] In SVM classification, duplicating training instances can affect the cost penalty by altering the number of instances within the margin. It is thus critical to tune cost parameter C for all experiments, which we do on development data.

P	R	F	#instances	Settings
0.587	0.538	0.561	8423	no Expansion
0.466	0.455	0.460	8423	UMLS as features
0.578	0.533	0.555	16846	duplicate instances
0.580	0.534	0.556	25269	triple instances
0.605	0.557	0.580	9506	longest UMLS
0.592	0.592	0.592	10705	expand to NPs
0.654	0.591	0.621	12966	all UMLS

Table 2: Results on the development set. No expansion vs. encoding UMLS as features; duplicating and triplicating training instances; expand to the longest UMLS span, expand to the immediate enclosing NP vs. expand to all UMLS spans.

P	R	F	#instances	Settings
0.359	0.155	0.217	8423	(A) no expansion
0.582	0.206	0.304	12966	(A) with expansion
0.087	0.116	0.099	8423	(B) no expansion
0.600	0.546	0.572	12966	(B) with expansion
0.587	0.254	0.355	8423	(C) no expansion
0.648	0.264	0.375	12966	(C) with expansion

Table 3: Results on the development set. Comparison of improvement for feature groups: (A) words covered by the arguments; (B) words in between the arguments; (C) words around the arguments.

mance as UMLS expansion, and in fact decreases performance. Question 2 is addressed in the last three rows: expanding to all possible UMLS spans works better than expanding only to the longest span or to the immediate enclosing NP. Expanding to NPs achieved the second best result, suggesting that when a domain-specific ontology is unavailable, expansion via syntax might provide a viable alternative. Question 3 is answered by rows 1, 3 and 4: when the cost parameter is properly tuned, doubling or tripling instances (rows 3 and 4) does not improve performance over no expansion (row 1). Question 4 is addressed by Table 3: features extracted between the two arguments achieve the biggest gain from our expansion method.

5 Results on the test set

Once the parameters were fine-tuned, we trained both event-time and event-event models on the combined training and developments sets, and tested them on the test set. All features described in Table 1 are used. The first two rows of Table 4 evaluate both event-event and event-time models, the next two rows evaluate only the event-time model, and the last two rows evaluate only the event-event model. Statistical significance is computed via Wilcoxon signed-rank tests over document-by-document comparisons, as in (Cherry et al., 2013b).

P	R	F	Settings	P-value
0.635	0.549	0.589	(1) no Expansion	0.117
0.669	0.534	0.594	(1) with Expansion	
0.673	0.291	0.407	(2) no Expansion	
0.708	0.287	0.408	(2) with Expansion	
0.594	0.252	0.354	(3) no Expansion	
0.628	0.243	0.351	(3) with Expansion	

Table 4: Results on the test set with all features. (1) Evaluate both Event-Time and Event-Event models; (2) Evaluate Event-Time model only; (3) Evaluate Event-Event model only. See Section 3.4 for explanation for why shaded scores are different from their counterparts in Table 2.

6 Discussion

Our experiments show our method is helpful for the event-time model, and not harmful for the event-event model. We hypothesize that the multiple instances capture the important surrounding context between arguments and allow more generalization over it. For the example in Figure 1, the most important features are "performed on." Our method weeds out less discriminative features by strengthening the important contextual signals that appear across many different entity boundaries. This is supported by the results of Table 3 (B). We suspect that the small improvement seen on the test data may be a result of the additional development examples canceling the benefit of augmented examples. This suggests that this method may be most effective in tasks with limited training instances.

Event-event relations are more complicated, first in that they have lower annotation quality than event-time relations (see Table 5 from (Styler et al., 2014)). And while almost every temporal expression in a sentence is important, not all events in a sentence are, creating many potential "distractor" events (e.g., *showed*) in the context of the clinical domain. We performed some exploratory experiments (not shown), restricting the data to only adjacent medical events in notes with high inter-annotator agreement, and saw significant performance improvements. But further study is needed to generalize this to all event-event relations.

With the presented method, our temporal relation system achieved F1 of 0.594, a two percentage-point improvement over the best Clinical Temp-Eval 2016 system's F1 of 0.573 (Bethard et al., 2016). Our results also suggest that gains may be possible in the general domain by using syntactic constituents for expansion. The method is available open source at the temporal module of Apache cTAKES[2] (Savova et al., 2010).

7 Acknowledgements

Thanks to Sean Finan for technically supporting the experiments. The study was funded by R01 LM 10090 (THYME), R01GM103859 (PGx), and U24CA184407 (DeepPhe). The content is solely the responsibility of the authors and does not necessarily represent the official views of the National Institutes of Health.

References

Steven Bethard, Leon Derczynski, Guergana Savova, James Pustejovsky, and Marc Verhagen. 2015. SemEval-2015 task 6: Clinical TempEval. In *Proceedings of the 9th International Workshop on Semantic Evaluation (SemEval 2015)*, pages 806–814, Denver, Colorado, June. Association for Computational Linguistics.

Steven Bethard, Guergana Savova, Wei-Te Chen, Leon Derczynski, James Pustejovsky, and Marc Verhagen. 2016. Semeval-2016 task 12: Clinical tempeval. In *Proceedings of the 10th International Workshop on Semantic Evaluation (SemEval 2016)*, San Diego, California, June. Association for Computational Linguistics.

Colin Cherry, Xiaodan Zhu, Joel Martin, and Berry de Bruijn. 2013a. À la recherche du temps perdu: extracting temporal relations from medical text in the 2012 i2b2 NLP challenge. *Journal of the American Medical Informatics Association*, 20(5):843–848.

Colin Cherry, Xiaodan Zhu, Joel Martin, and Berry de Bruijn. 2013b. la recherche du temps perdu: extracting temporal relations from medical text in the 2012 i2b2 nlp challenge. *Journal of the American Medical Informatics Association*, 20(5):843–848.

Rong-En Fan, Kai-Wei Chang, Cho-Jui Hsieh, Xiang-Rui Wang, and Chih-Jen Lin. 2008. Liblinear: A library for large linear classification. *The Journal of Machine Learning Research*, 9:1871–1874.

Michael G Kahn, Larry M Fagan, and Samson Tu. 1990. Extensions to the time-oriented database model to support temporal reasoning in medical expert systems. *Methods of Information in Medicine*, 30(1):4–14.

Oleksandr Kolomiyets, Steven Bethard, and Marie-Francine Moens. 2011. Model-portability experiments for textual temporal analysis. In *Proceedings of the 49th Annual Meeting of the Association for Computational Linguistics: Human Language Technologies*, pages 271–276, Portland, Oregon, USA, June. Association for Computational Linguistics.

Alex Krizhevsky, Ilya Sutskever, and Geoffrey E Hinton. 2012. Imagenet classification with deep convolutional neural networks. In *Advances in neural information processing systems*, pages 1097–1105.

Chen Lin, Elizabeth W Karlson, Dmitriy Dligach, Monica P Ramirez, Timothy A Miller, Huan Mo, Natalie S Braggs, Andrew Cagan, Vivian Gainer, Joshua C Denny, and Guergana K Savova. 2014. Automatic identification of methotrexate-induced liver toxicity in patients with rheumatoid arthritis from the electronic medical record. *Journal of the American Medical Informatics Association*.

Donald A Lindberg, Betsy L Humphreys, and Alexa T McCray. 1993. The unified medical language system. *Methods of information in Medicine*, 32(4):281–291.

James Pustejovsky, Patrick Hanks, Roser Sauri, Andrew See, Robert Gaizauskas, Andrea Setzer, Dragomir Radev, Beth Sundheim, David Day, Lisa Ferro, et al. 2003. The timebank corpus. In *Corpus linguistics*, volume 2003, page 40.

Guergana Savova, Steven Bethard, Will Styler, James Martin, Martha Palmer, James Masanz, and Wayne Ward. 2009. Towards temporal relation discovery from the clinical narrative. In *AMIA annual symposium proceedings*, volume 2009, page 568. American Medical Informatics Association.

Guergana K. Savova, James J. Masanz, Philip V. Ogren, Jiaping Zheng, Sunghwan Sohn, Karin C. Kipper-Schuler, and Christopher G. Chute. 2010. Mayo clinical text analysis and knowledge extraction system (cTAKES): architecture, component evaluation and applications. *J Am Med Inform Assoc*, 17(5):507–513.

Reinhold Schmidt, Stefan Ropele, Christian Enzinger, Katja Petrovic, Stephen Smith, Helena Schmidt, Paul M Matthews, and Franz Fazekas. 2005. White matter lesion progression, brain atrophy, and cognitive decline: the austrian stroke prevention study. *Annals of neurology*, 58(4):610–616.

William Styler, Steven Bethard, Sean Finan, Martha Palmer, Sameer Pradhan, Piet de Groen, Brad Erickson, Timothy Miller, Chen Lin, Guergana Savova, and James Pustejovsky. 2014. Temporal annotation in the clinical domain. *Transactions of the Association for Computational Linguistics*, 2:143–154.

Weiyi Sun, Anna Rumshisky, and Ozlem Uzuner. 2013. Evaluating temporal relations in clinical text: 2012 i2b2 challenge. *Journal of the American Medical Informatics Association*, 20(5):806–813.

Naushad UzZaman and James F Allen. 2011. Temporal evaluation. In *Proceedings of the 49th Annual Meeting of the Association for Computational Linguistics: Human Language Technologies: short papers-Volume 2*, pages 351–356. Association for Computational Linguistics.

---

[2] http://ctakes.apache.org

Naushad UzZaman, Hector Llorens, Leon Derczynski, James Allen, Marc Verhagen, and James Pustejovsky. 2013. SemEval-2013 task 1: TempEval-3: Evaluating time expressions, events, and temporal relations. In *Second Joint Conference on Lexical and Computational Semantics (*SEM), Volume 2: Proceedings of the Seventh International Workshop on Semantic Evaluation (SemEval 2013)*, pages 1–9, Atlanta, Georgia, USA, June. Association for Computational Linguistics.

Marc Verhagen, Robert Gaizauskas, Frank Schilder, Mark Hepple, Graham Katz, and James Pustejovsky. 2007. Semeval-2007 task 15: Tempeval temporal relation identification. In *Proceedings of the 4th International Workshop on Semantic Evaluations*, pages 75–80. Association for Computational Linguistics.

Marc Verhagen, Roser Sauri, Tommaso Caselli, and James Pustejovsky. 2010. Semeval-2010 task 13: Tempeval-2. In *Proceedings of the 5th International Workshop on Semantic Evaluation*, pages 57–62. Association for Computational Linguistics.

Li Zhou and George Hripcsak. 2007. Temporal reasoning with medical dataa review with emphasis on medical natural language processing. *Journal of biomedical informatics*, 40(2):183–202.

Using Centroids of Word Embeddings and Word Mover's Distance for Biomedical Document Retrieval in Question Answering

Georgios-Ioannis Brokos[1], **Prodromos Malakasiotis**[1,2] and **Ion Androutsopoulos**[1,2]

[1]Department of Informatics, Athens University of Economics and Business, Greece
Patission 76, GR-104 34 Athens, Greece
`http://nlp.cs.aueb.gr`
[2]Institute for Language and Speech Processing, Research Center 'Athena', Greece
Artemidos 6 & Epidavrou, GR-151 25 Maroussi, Athens, Greece
`http://www.ilsp.gr`

Abstract

We propose a document retrieval method for question answering that represents documents and questions as weighted centroids of word embeddings and reranks the retrieved documents with a relaxation of Word Mover's Distance. Using biomedical questions and documents from BIOASQ, we show that our method is competitive with PUBMED. With a top-k approximation, our method is fast, and easily portable to other domains and languages.

1 Introduction

Biomedical experts (e.g., researchers, clinical doctors) routinely need to search the biomedical literature to support research hypotheses, treat rare syndromes, follow best practices etc. The most widely used biomedical search engine is PUBMED, with more than 24 million biomedical references and abstracts, mostly of journal articles.[1] To improve their performance, biomedical search engines often use large, manually curated ontologies, e.g., to identify biomedical terms and expand queries with related terms.[2] Biomedical experts, however, report that search engines often miss relevant documents and return many irrelevant ones.[3]

There is also growing interest for biomedical question answering (QA) systems (Athenikos and Han, 2010; Bauer and Berleant, 2012; Tsatsaronis et al., 2015), which allow their users to specify their information needs more precisely, as natural language questions rather than Boolean queries,

and aim to produce more concise answers. Document retrieval is particularly important in biomedical QA, since most of the information sought resides in documents and is essential in later stages.

We propose a new document retrieval method. Instead of representing documents and questions as bags of words, we represent them as the centroids of their word embeddings (Mikolov et al., 2013; Pennington et al., 2014) and retrieve the documents whose centroids are closer to the centroid of the question. This allows retrieving relevant documents that may have no common terms with the question without query expansion. Using biomedical questions from the BIOASQ competition (Tsatsaronis et al., 2015), we show that our method combined with a relaxation of the recently proposed Word Mover's Distance (WMD) (Kusner et al., 2015) is competitive with PUBMED. We also show that with a top-k approximation, our method is particularly fast, with no significant decrease in effectiveness. Given that it does not require ontologies, term extractors, or manually labeled training data, our method could be easily ported to other domains (e.g., legal texts) and languages.

2 The proposed method

The word embeddings and document centroids are pre-computed. For each question, its centroid is computed and the documents with the top-k nearest (in terms of cosine similarity) centroids are retrieved (Fig. 1). The retrieved documents are then optionally reranked using a relaxation of WMD.

2.1 Centroids of documents and questions

In the simplest case, the centroid $\vec{t}$ of a text t is the sum of the embeddings of the tokens of t divided by the number of tokens in t. Previous work on hierarchical biomedical document classification (Kosmopoulos et al., 2016) reported im-

---

[1]See `http://www.ncbi.nlm.nih.gov/pubmed`.

[2]PUBMED uses UMLS (`http://www.nlm.nih.gov/research/umls/`). See also the GoPubMed search engine (`http://www.gopubmed.com/`).

[3]Malakasiotis et al. (2014) summarize the findings of interviews that investigated how biomedical experts search.

Proceedings of the 15th Workshop on Biomedical Natural Language Processing, pages 114–118,
Berlin, Germany, August 12, 2016. ©2016 Association for Computational Linguistics

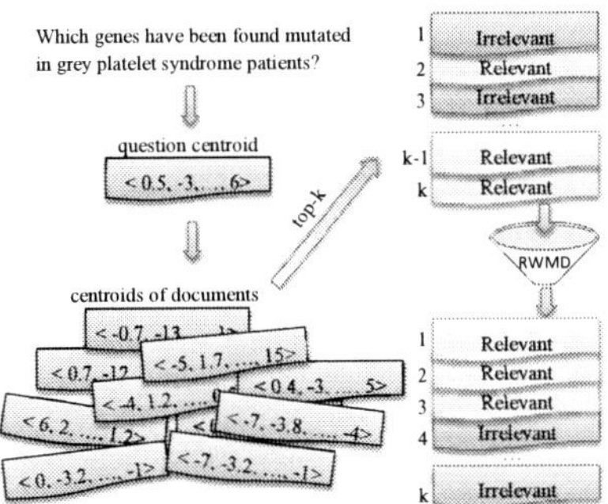

Figure 1: Illustration of the proposed method.

proved performance when the IDF scores of the tokens are also taken into account as follows:

$$
\vec{t} = \frac{\sum_{j=1}^{|V|} \vec{w}_j \cdot \mathrm{TF}(w_j, t) \cdot \mathrm{IDF}(w_j)}{\sum_{j=1}^{|V|} \mathrm{TF}(w_j, t) \cdot \mathrm{IDF}(w_j)}
\tag{1}
$$

where $|V|$ is the vocabulary size (approx. 1.7 million words, ignoring stop words), w_j is the j-th vocabulary word, $\vec{w}_j$ its embedding, $\mathrm{TF}(w_j, t)$ the term frequency of w_j in t, and $\mathrm{IDF}(w_j)$ the inverse document frequency of w_j (Manning et al., 2008). We use the 200-dimensional word embeddings of BIOASQ, obtained by applying WORD2VEC (Mikolov et al., 2013) to approx. 11 million abstracts from PubMed.[4] The IDF scores are computed on the 11 million abstracts.

2.2 Document retrieval and reranking

Given a question with centroid $\vec{q}$, identifying the documents with the k nearest centroids requires computing the distance between $\vec{q}$ and each document centroid, which is impractical for large document collections. Efficient approximate top-k algorithms, however, exist. They divide the vector space into subspaces and use trees to index the instances in each subspace (Arya et al., 1998; Indyk and Motwani, 1998; Andoni and Indyk, 2006; Muja and Lowe, 2009). We show that with an approximate top-k algorithm, document retrieval is very fast, with no significant decrease in performance. The top-k retrieved documents d_i are ranked by decreasing (cosine) similarity of their centroids to $\vec{q}$. We call this method Cent when the

---

[4]The skipgram model of WORD2VEC was used, with hierarchical softmax, 5-word windows, and default other parameters. See http://participants-area.bioasq.org/info/BioASQword2vec/ for further details.

simple (no IDF) centroids are used, and CentIDF when the IDF-weighted centroids (Eq. 1) are used.

The top-k documents are optionally reranked with an approximation of the WMD distance. WMD measures the total distance the word embeddings of two texts (in our case, question and document) have to travel to become identical. In its full form, WMD allows each word embedding to be partially aligned (travel) to multiple word embeddings of the other text, which requires solving a linear program and is too slow for our purposes. Kusner et al. (2015) reported promising results in text classification using WMD as the distance of a k-NN classifier. They also introduced relaxed, much faster WMD versions. In our case, the first relaxation (RWMD-Q) sums the distances the word embeddings $\vec{w}$ of the question q have to travel to the closest word embeddings $\vec{w}'$ of the document d:

$$
\mathrm{RWMD\text{-}Q}(q, d) = \sum_{w \in q} \min_{w' \in d} dist(\vec{w}, \vec{w}')
\tag{2}
$$

Following Kusner et al., we use the Euclidean distance as $dist(\vec{w}, \vec{w}')$. Similarly, the second relaxed form (RWMD-D) sums the distances of the word embeddings of d to the closest embeddings of q. If we set $dist(\vec{w}, \vec{w}') = 1$ if w, w' are identical and 0 otherwise, RWMD-Q counts how many words of q are present in d, and RWMD-D counts the words of d that are present in q. Kusner et al. found the maximum of RWMD-Q and RWMD-D (RWMD-MAX) to be the best relaxation of WMD. In our case, where q is much shorter than d, RWMD-Q works much better, because d contains many irrelevant words that have no close counter-parts in q, and their long distances dominate in RWMD-D and RWMD-MAX.[5] We call CentIDF-RWMD-Q and CentIDF-RWMD-D the CentIDF method with the additional reranking by RWMD-Q or RWMD-D, respectively.

3 Experiments

3.1 Data

We used the 1,307 training questions and the gold relevant PUBMED document ids of the fourth year of BIOASQ (Task 4b).[6] The questions were written by biomedical experts, who also identified the

---

[5]We do not report results with RWMD-MAX reranking, because they are as bad as results with RWMD-D.

[6]The questions and gold document ids are available from http://participants-area.bioasq.org/. The 1,307 questions are all the training and test questions of the previous years of BIOASQ, which were available to the participants of the fourth year. We use all the 1,307 questions for testing, since our method is unsupervised.

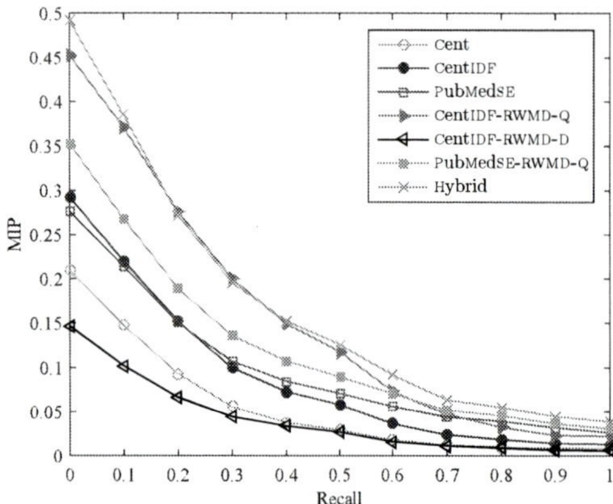

Figure 2: Mean Interpolated Precision at 11 recall points, for k (documents to retrieve) set to 1,000.

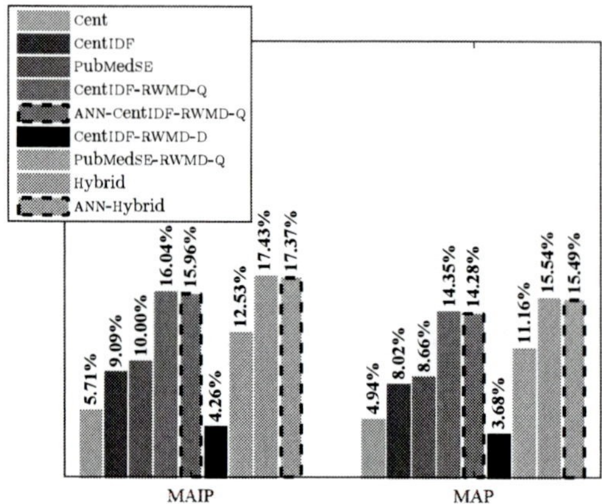

Figure 3: MAIP and MAP scores, for k (documents to retrieve) set to 1,000.

gold relevant documents using PUBMED, and reflect real needs (Tsatsaronis et al., 2015). We pass each question to our methods (after tokenization and stop-word removal) or the PUBMED search engine (hereafter PubMedSE), which performs its own tokenization and query expansion.[7]

The document collection that we search contains approx. 14 million article abstracts and titles from the November 2015 PUBMED dump, which was also used in the fourth year of BIOASQ.[8] Our methods view each document as a concatenation of the title and abstract of an article.[9] The titles and abstracts have an average length of approx. 13 and 143 tokens, respectively. When comparing against PubMedSE, we ignore documents returned by PubMedSE that are not in the dump, but this is very rare and does not affect the results.

3.2 Experimental results

Figures 2–4 show Mean Interpolated Precision (MIP) at 11 recall levels, Mean Average Interpolated Precision (MAIP), Mean Average Precision (MAP), and Normalized Discounted Cumulative Gain (nDCG).[10] Roughly speaking, MAIP is the area under the MIP curve, MAP is the same area without interpolation, and nDCG is an alternative

to MAIP. Unless otherwise stated, the number of retrieved documents is set to $k = 1,000$.

Figure 2 shows that Cent performs much worse than CentIDF. At low recall, CentIDF is as good as PubMedSE, but PubMedSE outperforms CentIDF at high recall. Reranking the top-k documents of CentIDF by RWMD-Q has a significant impact, leading to a system (CentIDF-RWMD-Q) that performs better or as good as PubMedSE up to 0.7 recall. Reranking the top-k documents of PubMedSE by RWMD-Q (PubMedSE-RWMD-Q) also improves the performance of PubMedSE. Reranking the top-k documents of CentIDF by RWMD-D (or RWMD-MAX, not shown) leads to much worse results (CentIDF-RWMD-D), for reasons already explained.[11] Similar conclusions are reached by examining the MAIP, MAP, and nDCG scores.

Keyword-based information retrieval may miss relevant documents that use different terms than the question, even with query expansion. PubMedSE retrieves no documents for 35% (460/1307) of our questions.[12] Further experiments (not reported), however, indicate that PubMedSE has higher precision than CentIDF-RWMD-Q, when PubMedSE returns documents, at the expense of lower recall. Hence, there is scope to combine PubMedSE with our methods. As a first, crude step, we tested a method (Hybrid) that returns the documents of CentIDF-RWMD-Q when PubMedSE retrieves no documents, and those of

---

[7] We use relevance ranking (not recency) in PubMedSE.

[8] The dump is available from `https://www.nlm.nih.gov/databases/license/license.html`. The 14 million articles do not include approx. 10 million articles for which only titles are provided. There are hardly any title-only gold relevant documents, and PubMedSE very rarely returns title-only documents.

[9] It is unclear to us if PUBMED also searches the full texts of the articles, which may put our methods at a disadvantage.

[10] All measures are widely used (Manning et al., 2008). We use binary relevance in nDCG, as in the BIOASQ dataset.

[11] The same holds when the top-k documents of PubMedSE are reranked by RWMD-D or RWMD-MAX (not shown).

[12] The experts that identified the gold relevant documents used simple keyword, Boolean, and advanced PubMedSE queries, whereas we used the English questions as queries.

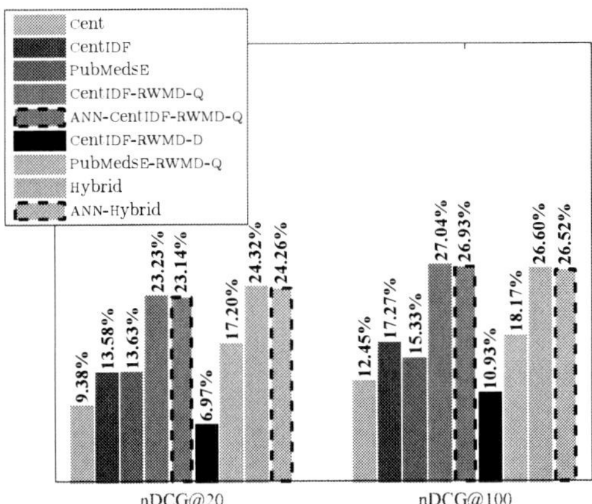

Figure 4: nDCG@k, for $k = 20$ and $k = 100$.

System	Search	Reranking	Total
CentIDF-RWMD-Q	47.41 ($\pm$1.22)	14.45 ($\pm$6.15)	61.86
ANN-CentIDF-RWMD-Q	0.36 ($\pm$0.04)	14.24 ($\pm$6.06)	14.60

Table 1: Average times (in seconds) over all the questions of the dataset ($k = 1000$).

PubMedSE-RWMD-Q otherwise. Hybrid had the best results in our experiments; the only exception was its nDCG@100 score, which was slightly lower than the score of CentIDF-RWMD-Q.

Table 1 shows that an approximate top-k algorithm (ANN) in CentIDF-RWMD-Q (ANN-CentIDF-RWMD-Q) reduces dramatically the time to obtain the top-k documents, with a very small decrease in MAIP, MAP, and nDCG scores (Figures 3 and 4).[13]

We also compared against the other participants of the second year of BIOASQ; the participant results of later years are not yet available.[14] The official BIOASQ score is MAP; MIP, MAIP, and nDCG scores are not provided. Our best method was again Hybrid (avg. MAP over the five batches of the second year 16.18%). It performed overall better than the BIOASQ 'baselines' (best avg. MAP 15.60%) and all eight participants, except for the best one (avg. MAP 28.20%). The best system (Choi and Choi, 2014) used dependency IR models (Metzler and Croft, 2005), combined with UMLS and query expansion heuristics (e.g., adding the titles of the top-k initially retrieved documents to the query). The 'baselines' are actually very competitive; no system beat them in the first year, and only one was better in the second year. They are PubMedSE, but using BIOASQ-specific heuristics (e.g., instructing PubMedSE to ignore types of articles the experts did not consider). Our system is simpler and does not use heuristics; hence, it can be ported more easily to other domains.

4 Other related work

Kosmopoulos et al. (2016) reports that a k-NN classifier that represents articles as IDF-weighted centroids (Eq. 1) of 200-dimensional word embeddings (200 features) is as good at assigning semantic labels (MeSH headings) to biomedical articles as when using millions of bag-of-word features, reducing significantly the training and classification times. To our knowledge, our work is the first attempt to use IDF-weighted centroids of word embeddings in information retrieval, and the first to use WMD to rerank the retrieved documents. More elaborate methods to encode texts as vectors have been proposed (Le and Mikolov, 2014; Kiros et al., 2015; Hill et al., 2016) and they could be used as alternatives to centroids of word embeddings, though the latter are simpler and faster to compute.

The OHSUMED dataset (Hersh et al., 1994) is often used in biomedical information retrieval experiments. It is much smaller (101 queries, approx. 350K documents) than the BIOASQ dataset that we used, but we plan to experiment with OHSUMED in future work for completeness.

5 Conclusions and future work

We proposed a new QA driven document retrieval method that represents documents and questions as IDF-weighted centroids of word embeddings. Combined with a relaxation of the WMD distance, our method is competitive with PUBMED, without ontologies and query expansion. Combined with PUBMED, it performs better than PUBMED on its own. With a top-k approximation, it is fast, and easily portable to other domains and languages.

We plan to consider alternative dense vector encodings of documents and queries, textual entailment (Bowman et al., 2015; Rocktäschel et al., 2016), and full-text documents, where it may be necessary to extend RWMD-Q to take into account the proximity (density) of the words of the (now longer) document the query words are mapped to.

---

[13]We use Annoy (https://github.com/spotify/annoy), 100 trees, 1,000 neighbors, search-$k = 10 \cdot$ |trees| $\cdot$ |neighbors|. Times on a server with 4 Intel Xeon E5620 CPUs (16 cores total), at 2.4 GHz, with 128 GB RAM.

[14]We used the evaluation platform of BIOASQ (http://participants-area.bioasq.org/oracle).

Acknowledgments

The work of the second author was funded by the Athens University of Economics and Business Research Support Program 2014-2015, "Action 2: Support to Post-doctoral Researchers".

References

A. Andoni and P. Indyk. 2006. Near-optimal hashing algorithms for approximate nearest neighbor in high dimensions. In *Proc. of the 47th Annual IEEE Symposium on Foundations of Computer Science*, pages 459–468, Washington, DC.

S. Arya, D. M. Mount, N. S. Netanyahu, R. Silverman, and A. Y. Wu. 1998. An optimal algorithm for approximate nearest neighbor searching fixed dimensions. *Journal of the ACM*, 45(6):891–923.

S. J. Athenikos and H. Han. 2010. Biomedical question answering: A survey. *Computer Methods and Programs in Biomedicine*, 99(1):1–24.

M. A. Bauer and D. Berleant. 2012. Usability survey of biomedical question answering systems. *Human Genomics*, 6(1)(17).

S. R. Bowman, G. Angeli, C. Potts, and C. D. Manning. 2015. A large annotated corpus for learning natural language inference. In *Proc. of the 2015 Conference on Empirical Methods in Natural Language Processing*, Lisbon, Portugal.

S. Choi and J. Choi. 2014. Classification and retrieval of biomedical literatures: SNUMedinfo at CLEF QA track BioASQ 2014. In *Proc. of the QA Lab of the 5th Conference and Labs of the Evaluation Forum*, pages 1283–1295, Valencia, Spain.

W. Hersh, C. Buckley, T. J. Leone, and D. Hickam. 1994. OHSUMED: An interactive retrieval evaluation and new large test collection for research. In *Proc. of the 17th Annual International ACM SIGIR Conference on Research and Development in Information Retrieval*, pages 192–201, Dublin, Ireland.

F. Hill, K. Cho, and A. Korhonen. 2016. Learning distributed representations of sentences from unlabelled data. *arXiv preprint 1602.03483*.

P. Indyk and R. Motwani. 1998. Approximate nearest neighbors: Towards removing the curse of dimensionality. In *Proc. of the 30th Annual ACM Symposium on Theory of Computing*, pages 604–613, Dallas, TX.

R. Kiros, Y. Zhu, R. R. Salakhutdinov, R. Zemel, R. Urtasun, A. Torralba, and S. Fidler. 2015. Skip-thought vectors. In *Advances in Neural Information Processing Systems 28*, pages 3276–3284. Montréal, Canada.

A. Kosmopoulos, I. Androutsopoulos, and G. Paliouras. 2016. Biomedical semantic indexing using dense word vectors in BioASQ. *Journal Of Biomedical Semantics, Supplement On Biomedical Information Retrieval*. To appear.

M. Kusner, Y. Sun, N. Kolkin, and K. Q. Weinberger. 2015. From word embeddings to document distances. In *Proc. of the 32nd International Conference on Machine Learning*, pages 957–966, Lille, France.

Q. Le and T. Mikolov. 2014. Distributed representations of words and phrases. In *Proc. of the 31st International Conference on Machine Learning*, pages 1188–1196, Beijing, China.

P. Malakasiotis, I. Androutsopoulos, A. Bernadou, N. Chatzidiakou, E. Papaki, P. Constantopoulos, I. Pavlopoulos, A. Krithara, Y. Almyrantis, D. Polychronopoulos, A. Kosmopoulos, G. Balikas, I. Partalas, G. Tsatsaronis, and N. Heino. 2014. Challenge Evaluation Report 2 and Roadmap. BioASQ deliverable D5.4.

C.D. Manning, P. Raghavan, and H. Schütze. 2008. *Introduction to Information Retrieval*. Cambridge University Press.

D. Metzler and W.B. Croft. 2005. A Markov Random Field model for term dependencies. In *Proc. of the 28th Annual International ACM SIGIR conference on Research and Development in Information Retrieval*, pages 472–479, Salvador, Brazil.

T. Mikolov, W. Yih, and G. Zweig. 2013. Distributed representations of words and phrases and their compositionality. In *Proc. of the Conference on Neural Information Processing Systems*, Lake Tahoe, NV.

M. Muja and D. G. Lowe. 2009. Fast approximate nearest neighbors with automatic algorithm configuration. In *Proc. of the International Conference on Computer Vision Theory and Applications*, pages 331–340, Lisboa, Portugal.

J. Pennington, R. Socher, and C. D. Manning. 2014. GloVe: Global vectors for word representation. In *Proc. of the Conference on Empirical Methods on Natural Language Processing*, Doha, Qatar.

T. Rocktäschel, E. Grefenstette, K. M. Hermann, T. Kočiský, and P. Blunsom. 2016. Reasoning about entailment with neural attention. In *International Conference on Learning Representations*, San Juan, Puerto Rico.

G. Tsatsaronis, G. Balikas, P. Malakasiotis, I. Partalas, M. Zschunke, M.R. Alvers, D. Weissenborn, A. Krithara, S. Petridis, D. Polychronopoulos, Y. Almirantis, J. Pavlopoulos, N. Baskiotis, P. Gallinari, T. Artieres, A. Ngonga, N. Heino, E. Gaussier, L. Barrio-Alvers, M. Schroeder, I. Androutsopoulos, and G. Paliouras. 2015. An overview of the BioASQ large-scale biomedical semantic indexing and question answering competition. *BMC Bioinformatics*, 16(138).

Measuring the State of the Art of Automated Pathway Curation Using Graph Algorithms - A Case Study of the mTOR Pathway

Michael Spranger
Sony Computer Science
Laboratories Inc.
Tokyo, Japan
michael.spranger
@gmail.com

Sucheendra K. Palaniappan
INRIA,
Campus de Beaulieu,
Rennes, France
sucheendra.palaniappan
@inria.fr

Samik Ghosh
The Systems Biology Institute,
Minato-ku,
Tokyo, Japan
ghosh@sbi.jp

Abstract

This paper evaluates the difference between human pathway curation and current NLP systems. We propose graph analysis methods for quantifying the gap between human curated pathway maps and the output of state-of-the-art automatic NLP systems. Evaluation is performed on the popular mTOR pathway. Based on analyzing where current systems perform well and where they fail, we identify possible avenues for progress.

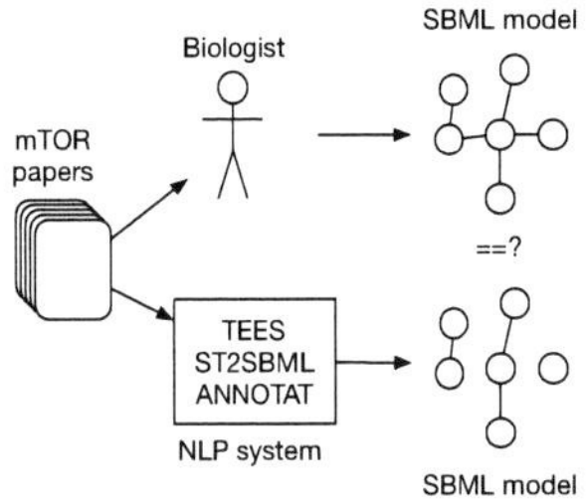

1 Introduction

Biological pathways encode sequences of biological reactions, such as phosphorylation, activations etc, involving various biological species, such as genes, proteins etc., in response to certain stimuli or spontaneous at times (Aldridge et al., 2006; Kitano, 2002). Studying and analyzing pathways is crucial to understanding biological systems and for the development of effective disease treatments and drugs (Creixell et al., 2015; Khatri et al., 2012). There have been numerous efforts to reconstruct detailed process-based and disease level pathway maps such as Parkinson disease map (Fujita et al., 2014), Alzheimers disease Map (Mizuno et al., 2012), mTOR pathway Map (Caron et al., 2010), and the TLR pathway map (Oda and Kitano, 2006)). Traditionally, these maps are constructed and curated by expert pathway curators who manually read numerous biomedical documents, comprehend and assimilate the knowledge in them and construct the pathway.

Manual curation of pathways is rather challenging given the ever increasing barrage of scientific publications. It is basically common place in this community that manual curation is not sufficient (Baumgartner et al., 2007). Consequently, *Automated Pathway Curation* has been an active area of research - particularly in the BioNLP community (Miwa et al., 2012; Valenzuela-Escárcega et al., 2015). It is also the goal of large scale research efforts such as DARPA's Big Mechanism Project (Cohen, 2015).

NLP systems have shown to perform well in BioNLP competitions (Nédellec et al., 2013; Ohta et al., 2013; Ananiadou et al., 2010), but so far we do not have systems that automatically assemble and curate pathways of the scope and complexity of, for example, the mTOR pathway. This paper investigates why this is the case. We measure the state of the art by closing the gap between NLP representations and biological networks, then we apply graph theory and in particular graph matching to quantify how much overlap there is between the NLP output and the information that humans assemble (see also Figure 1). The evaluation is performed on the popular mTOR pathway.

This paper starts by introducing our approach, followed by a description of data sets and evaluation results. We conclude by discussing where current system seem to fail and how to make progress.

Biologist
mTOR papers
SBML model
==?
TEES
ST2SBML
ANNOTAT
NLP system
SBML model

Figure 1: Comparing human pathway curation to NLP extraction.

Proceedings of the 15th Workshop on Biomedical Natural Language Processing, pages 119–127,
Berlin, Germany, August 12, 2016. ©2016 Association for Computational Linguistics

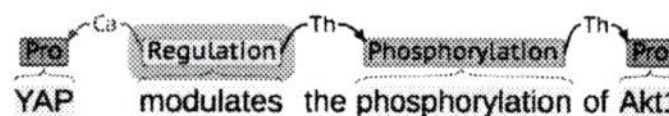

Figure 2: Example sentence with NLP event representations extracted.

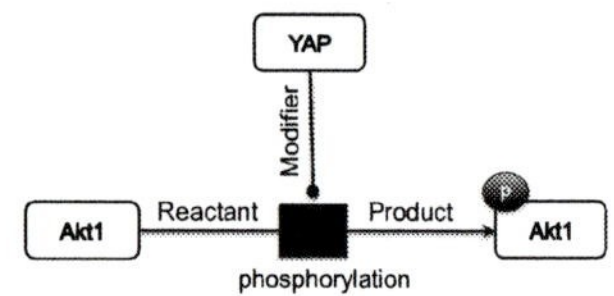

Figure 3: Phosphorylation reaction.

2 Bridging the Gap

In this paper, we close the representational gaps between current NLP systems and human-generated pathways, measure the overlap and analyze possible shortcomings of current systems. Evaluation is performed on the popular, hand-curated mTOR pathway map (Caron et al., 2010). Experts have curated and assembled the information from 522 papers into one large map using CellDesigner (Funahashi et al., 2008) - a software for modeling but also executing mechanistic models of pathways. CellDesigner represents information using a heavily customized XML-based SBML format (Hucka et al., 2003).

mTOR has been published along with a list of the 522 papers used to build the map. This allows us to treat the same papers with state-of-the-art NLP extraction systems. Here we used one of the most successful NLP systems around - the TURKU event extractions system (Björne, 2014, TEES). TEES has won 1st place in BioNLP 2009 ST, 2011 ST and DDI 2011 (Björne et al., 2012). The system integrates various NLP techniques to extract events from text. Processing roughly proceeds as follows 1) A number of external tools detect protein names and parse the sentences. 2) The event detector detects trigger words such as verbs, which is followed by detection of interactions. 3) Complex events are constructed. 4) The system detects modifiers such as negation and speculation.

NLP systems typically operate on something called the standoff format. From a sentence such as in Figure 2, standoff containing *entities* and *events* will be extracted. These in principle correspond to biological species and reactions. We translate the NLP representation into SBML path-

ways and perform additional annotation (Spranger et al., 2015) of species and reactions. For the sentence in Figure 2, the extracted SBML is visualized in Figure 2.

Datasets

We compared 3 different sets of data all related to mTOR pathway.

MTOR-HMN   is a mTOR pathway map manually constructed by human expert pathway curators. (Caron et al., 2010). The pathway is encoded in a dialect of SBML used by CellDesigner (Funahashi et al., 2008). We convert the CellDesigner format into pure SBML and annotate reactions and species further by automatically assigning reaction types and gene/protein identifers (see description below).

MTOR-ANN   consists of 57 abstracts of scientific papers from Pubmed related to the mTOR-pathway map. The data set was *human-annotated* for NLP system training (Ohta et al., 2011, Corpus annotations (c) GENIA Project[1]). This corpus gives an idea of the potential performance of a machine with human-level NLP extraction capabilities. Annotated NLP entities and events were used to create SBML representations and further annotated using various tools (discussed below).

MTOR-NLP   consists of 522 full text papers mentioned in the mTOR pathway map. Paper pdfs were downloaded automatically and translated into raw txt files using CERMINE (Tkaczyk et al., 2015). We managed to extract text from 501 papers. The 501 papers were processed using the Turku Event Extraction System mentioned earlier. From the extracted NLP events we created SBML representations of pathway maps for each text using (Spranger et al., 2015). The SBML was further annotated using various tools (discussed below) and, finally, loaded into a single pathway map.

Notice that MTOR-ANN and MTOR-NLP are different in how they are constructed and consequently what kind of conclusion we can draw from them. MTOR-ANN is a human-annotated dataset which contains much less data than MTOR-NLP. However, because it is human-annotated it allows us to evaluate a human-level performance extraction systems. So we cannot expect that MTOR-ANN is able to reconstruct everything in MTOR-

---

[1]http://nactem.ac.uk/GENIA/current/Other-corpora/mTOR-Pathway-Events/

HMN (recall). However, as we will argue in this paper, we might expect that what is extracted in MTOR-ANN does occur in MTOR-HMN (high precision).

The following table shows number of species, reactions and edges between them for the different datasets.

Dataset	# species	# reactions	# edges
MTOR-HMN	2242	777	2457
MTOR-ANN	2457	857	2343
MTOR-NLP	292049	100130	203042

Annotation

Annotation SBO Reactions in datasets MTOR-HMN, MTOR-ANN and MTOR-NLP were automatically *annotated* using Systems Biology Ontology (SBO) (Le Novère, 2006) and Gene Ontology (GO) terms. SBO provides a class hierarchy of reactions. Reactions can be of a certain type. For instance, NLP systems often identify regulation events. Regulation reactions form a hierarchy. For instance, positive regulation is a subclass of regulation reactions. Phosphorylation reactions are a subclass of conversion reactions.

All reactions in MTOR-HMN, MTOR-ANN, and MTOR-NLP are annotated using SBO/GO (coverage 100%). SBO/GO annotations are computed using different approaches. For MTOR-ANN and MTOR-NLP we used an automated annotation system that is also used to convert NLP event representations to SBML (Spranger et al., 2015). For MTOR-HMN, we used annotations provided by humans extended by automatic annotations. Automatic annotations were deduced by examining the reactants and products of reactions. For example, if a phosphoryl group is added the reaction is annotated using the SBO term for phosphorylation. Notice, in MTOR-HMN each reaction can be annotated with multiple SBO/GO terms. For instance, a single reaction can be annotated as phosphorylation and activation. This is not the case for MTOR-ANN and MTOR-NLP where each reaction corresponds to exactly one SBO/GO term.

Annotation Entrez Gene Species in all three datasets were annotated using the gene/protein named entity recognition and normalization software GNAT (Hakenberg et al., 2011) - a publicly available gene/protein normalization tool. GNAT returns a set of Entrez Gene identifiers (Maglott et al., 2005) for each input string. Species were annotated using all returned Entrez Gene identifiers for a particular species (organism human). We

	MTOR-HMN	MTOR-ANN	MTOR-NLP
activation	72	104	16485
association	210	204	21055
conversion	171	0	0
deacetylation	1	0	0
dephosphorylation	28	14	0
deubiquitination	13	0	0
dissociation	43	55	0
gene expression	4	40	18810
localization	0	16	474
negative regulation	33	99	10723
phosphorylation	85	241	25406
protein catabolism	24	18	1080
regulation	0	0	4832
transcription	78	8	1265
translation	23	1	0
transport	87	53	0
ubiquitination	13	4	0

Table 1: Reaction types extracted and annotated for various data sets. All reactions are annotated with their most specific type. Numbers are non-cumulative. For instance, the 171 conversion operations in MTOR-HMN are only annotated with the general conversion (SBO:182) and not more specific reaction types.

call the set of Entrez Gene identifiers returned by GNAT for each species *Entrez Gene signature*.

	# species	coverage	# Entrez ids
MTOR-HMN	2242	90%	538
MTOR-ANN	2457	87%	317
MTOR-NLP	292049	83%	4194

3 Species

Pathways contain many references to the same protein or gene. We measured the number of unique genes and proteins in each dataset using various ways of identifying (normalizing) genes and proteins in a particular dataset.

	MTOR-HMN	MTOR-ANN	MTOR-NLP
# species	2242	2457	291218
# names	582	359	27928
# appr names	568	316	4517
# Entrez signatures	443	201	6220

The first row repeats the number of species per data set. The second row condenses the species names by removing prefixes such as "phosphorylated" and other adjectives irrelevant for determining the actual biological entity. The third row shows what happens when we reduce the names further by using a Levenshtein-based string distance with a cutoff point of 90. The last row measures how many different unique Entrez Gene id signatures there are. Each species is annotated with a set of Entrez Gene ids. The set of the Entrez Gene identifiers for each species is taken as a signature.

The numbers show the degree of redundancy or reuse of species within each pathway. They also suggest that there are far more species implicated in MTOR-NLP than there are in MTOR-HMN. In other words, human annotators of mTOR have selected 568 species and not the 4517 found by the NLP systems (approx names).

Unique Species Overlap To better understand species identification we can measure the overlap of MTOR-ANN and MTOR-NLP with MTOR-HMN based on the unique species. Here we consider names equal (*nmeq*), names approximately equal (*appeq*), Entrez Gene id signature equal (*enteq*) and Entrez Gene id signature overlap (*entov*). The focus is on unique items.

	precision	recall	f-score
MTOR-HMN/MTOR-ANN			
nmeq	20.89	12.89	15.94
appeq	27.30	15.64	19.88
enteq	45.27	20.54	28.26
entov	83.08	55.53	66.57
MTOR-HMN/MTOR-NLP			
nmeq	0.96	45.88	1.87
appeq	1.59	51.20	3.08
enteq	4.60	64.56	8.58
entov	58.04	99.55	73.33

The rows *nmeq* show precision and recall for unique species names in MTOR-NLP with respect to MTOR-HMN. Precision is low - meaning that only a small percentage of unique species names in MTOR-NLP actually appear in MTOR-HMN. On the other hand, recall is higher. This shows that the few correctly identified species in MTOR-NLP overlap with large parts of MTOR-HMN species. Less than a percent of unique species names in MTOR-NLP cover 46% of species in MTOR-HMN. What is interesting is that MTOR-ANN does not fair too great on precision either. 79% of the unique annotated names do not appear in MTOR-HMN. Especially the annotated version dataset MTOR-ANN, lets us conclude that many species mentioned in papers actually do NOT make it into the pathway or at least not as mentioned in the papers. These analyses point to the fact that researchers building pathways select species. In other words, pathway curation is *not just extraction*, but *active selection* and, in fact, *identification* of species with proteins and genes known to the scientist.

Complex Species MTOR-HMN pathway contains a lot of complex species - i.e. species that contain other species. There are 351 complex species with a total of 1192 total constituents. 16 complexes are part of other complexes. Together this accounts for more than 70% of the species in MTOR-HMN. In other words, this is important information. Both MTOR-NLP and MTOR-ANN do not provide information about complexes explicitly. However, for this paper complexes are essentially treated like any other species.

4 Reactions

We first measured how many unique reaction types there are for each of the datasets.

	# reactions	# SBO/GO terms	# SBO/GO signatures
MTOR-HMN	777	15	29
MTOR-ANN	857	13	13
MTOR-NLP	100130	9	9

MTOR-HMN contains 777 reactions with 12 SBO/GO terms, i.e. reaction types. MTOR-ANN contains 12 and MTOR-NLP slightly less. Each reaction can have multiple SBO/GO terms associated with it. We call this the SBO/GO signature of a reaction. For instance, a particular reaction can be typed as phosphorylation and activation. Its signature are then the SBO/GO terms for these 2 reactions. The table shows that this actually only happens in MTOR-HMN. Human annotators are free to combine various reactions into a single reaction if they see fit. There is no replication of this in the automated data.

Unique Reaction Signature Overlap We then measured how much unique signatures overlap across the different datasets. We checked three different measures: 1) *sboeq* requires that both signatures are the same, 2) *sboov* requires that the intersection of the signatures overlaps - i.e. is not empty - and 3) *sboisa* requires that there is at least one SBO/GO term in each signature that relate in a is_a relationship in the SBO reaction type hierarchy. For instance, if there is a phosphorylation reaction and a conversion reaction, then *sboisa* will match because phosphorylation is a subclass of conversion according to the SBO type hierarchy.

	precision	recall	f-score
MTOR-HMN/MTOR-ANN			
sboeq	69.23	31.03	42.86
sboov	45.51	50.19	47.74
sboisa	92.31	93.10	92.70
MTOR-HMN/MTOR-NLP			
sboeq	55.56	17.24	26.32
sboov	77.78	68.97	73.11
sboisa	88.89	79.31	83.83

MTOR-ANN catches 1/3 of the reaction SBO/GO signatures directly and up to 93% when we allow for overlap sbo_is_a relationship. MTOR-NLP only directly includes 1 out of 5 reaction signatures. However, the overlap is higher when allowing for reaction SBO/GO signatures to overlap and individual SBO terms to be in a is_a relationship.

These results also show that there are reactions in MTOR-NLP and MTOR-ANN that are not part of MTOR-HMN (see also Table 2)

From this preliminary data, we can immediately identify an important difference between human annotation and automated NLP event extraction. Human annotators combine multiple reactions into a single reaction representation to condense information.

5 Networks - Connectedness

Ultimately we are interested in networks of reactions and species. Studying the output of NLP systems it becomes immediately clear that the result of these systems differs from hand-curated data in an important aspect: *connectedness*. To show this we measured isolation of species and networks (reactions cannot be isolated for structural reasons in SBML).

	# isolated networks	# isolated species
MTOR-HMN	4	6
MTOR-ANN	475	632
MTOR-NLP	83,093	110,490

In MTOR-HMN there are 4 separate subgraphs (no connection between them). 3 of them are modeling mistakes by human curators. Basically MTOR-HMN is one connected network. On the other hand, MTOR-ANN and MTOR-NLP consist of numerous unconnected networks. Each of them is quite small as the following data shows.

We measured min, max, mean and median number of species and reactions in each connected component subgraph.

dataset	min	mean	median	max
MTOR-ANN	1	3.00	1.0	24
MTOR-NLP	1	2.02	1.0	215

Results show that subgraphs in MTOR-ANN and MTOR-NLP on average contain between 2 and 3 species and reactions. So very often there will be a single reaction in a subgraph plus some reactant and maybe a product. On the other hand MTOR-HMN consists of essentially one large connected graph. So here is another fundamental difference: human modelers *compose* a single large graph, as opposed to just extracting single reactions.

6 Networks - Overlap

Arguably the most important question is how much overlap there is between disconnected reactions extracted by MTOR-ANN/MTOR-NLP with MTOR-HMN. For this, we measure subgraph isomorphisms of MTOR-ANN and MTOR-NLP subgraphs with the MTOR-HMN graph. We measured *max* overlap and allow multiple hits for each subgraph from MTOR-ANN and MTOR-NLP with parts of MTOR-HMN. We compare different strategies for node (species and reactions) and edge matching.

Species matching We investigated name matches (*nmeq*), approximate name matches (*appeq*), Entrez Gene signature equal (*enteq*) and Entrez Gene signature overlaps (*entov*) and combinations thereof. For example, *appeq/enteq* matches two species if either their names match approximately OR their Entrez Gene signatures are equal. *appeq/entov* matches two species if their names match approximately OR their Entrez Gene signatures overlap. Since there is no information on complexes in MTOR-ANN/MTOR-NLP, we also allowed matches not only on the complex itself but also on its constituents (*wc*). So a link present in MTOR-NLP between some protein and its phosphorylated version, will match if a link is present in a complex that contains that protein in MTOR-HMN.

Reaction matching Reaction matching relies on SBO/GO signatures. We checked with signatures equal (*sboeq*), signatures overlapping overlap (*sboov*) and signatures overlapping with individual SBO terms in is_a relationship (*sboisa*).

Edge matching We only allowed strict edge matching. So if an edge marks a reactant, then it has to be a reactant in MTOR-HMN. Same holds for product and modifier.

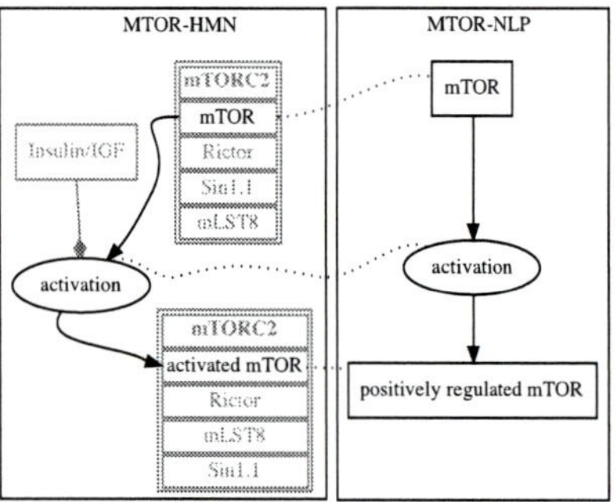

Figure 4: Example of a successful match (*nmeq, sboeq*). Black - matched nodes and edges, grey not mached context. Insulin/IGF is a modifier of this reaction. It is not captured by MTOR-NLP. Modifiers are less frequently detected than reactants and products.

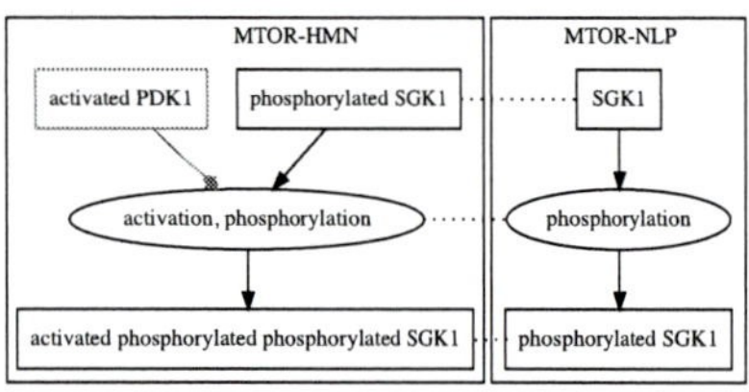

Figure 5: Example of a successful match (*appeq, sbois*) with a reaction that has multiple reaction types.

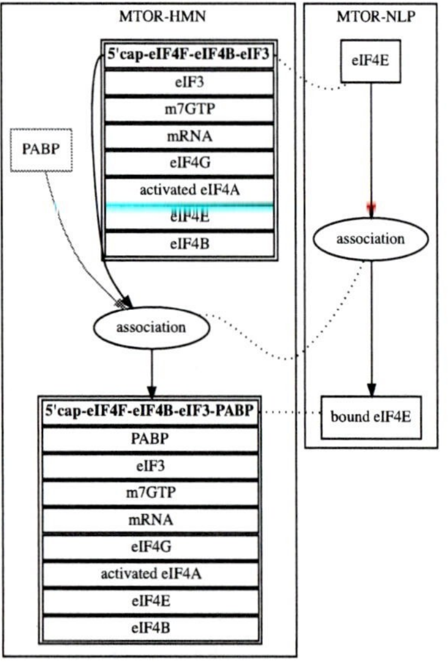

Figure 6: Example of a successful match (*appeq/wc, sboeq*) but ultimately incorrect mapping. It is not eIF4E that gets bound but the whole complex of 5'cap-eIF4F-eIF4B-eIF3 that includes eIF4E.

The final point to note for the results of matching is that we removed isolated nodes (which are always species) from MTOR-ANN and MTOR-NLP, because here we are really interested in graph structure.

Network overlap results Table 2 shows precision and recall for max overlap of different matching strategies (see also Figures 4 to 6). The table shows results for MTOR-ANN and MTOR-NLP successively. In general the first rows (*nmeq, sboeq*) represent very strict matching strategies. The last row (*appeq/entov/wc, sbois*) shows results for the most "relaxed" strategy.

Let us first analyze the performance of MTOR-NLP. The automated NLP system is able to retrieve roughly 9% of all edges given the strictest matching strategy. This means that 1 in 10 edges in the NLP extracted dataset actually appears as is in the human curated data (MTOR-HMN). Also, if we look at the most relaxed matching strategy *appeq/entov/wc, sbois*, we find that roughly 2 of 3 edges and 3 of 4 nodes (species and reactions) in the human curated MTOR have something to do with the NLP extracted data. In particular, the conversion and regulation reactions play a part in the 20 percentage points jump from 45.59 to 65.04 for edges from *appeq/entov/wc, sboov* to *appeq/entov/wc, sboisa* matching. Conversion and regulation are super classes for a whole range of reactions (conversion: phosphorylation etc; regulation: activation, inactivation etc).

Matching strategies that allow for matching complex constituents always have a higher recall and precision performance than their non constituent matching counterparts. For instance, *nmeq, sboeq* matches almost 20 percentage points less edges than *nmeq/wc, sboeq* (MTOR-HMN/MTOR-NLP). This increase in performance of constituent matching points to the fact that human modelers often attribute reactions to the whole complex. For instance, a phosphorylation may be acting on a constituent of a complex but the human modeler chooses to connect the reaction with the whole complex. These matching strategies do account for that and therefore are able to improve the numbers (in some cases) considerably.

Reactions in MTOR-HMN are sometimes incorporating various reaction types. In MTOR-ANN and MTOR-NLP, on the other hand, each reaction only has a single type. Reaction matching strategies *sboov* and *sboisa* account for that by looking at overlaps. This means that reactions in MTOR-ANN and MTOR-NLP will match with a reaction MTOR-HMN if the reaction type signatures intersection is not empty. In reality this means that the reaction in MTOR-ANN or MTOR-NLP has to be an element of the reaction in MTOR-HMN.

Lastly, let us take a look at MTOR-HMN/MTOR-ANN. MTOR-ANN contains much less data than MTOR-NLP but the reason we include it here is because MTOR-ANN consists of human annotated data. It therefore gives an idea about the limits of the annotation data and the limits of human annotation. If all of the problems discussed so far are purely a problem of the NLP system, then MTOR-ANN should do better than MTOR-NLP in terms of precision but not in terms of recall. Recall will be low because the MTOR-ANN consists of less data. However, we would expect high precision numbers. Interestingly, data shows that even for NLP-ANN precision is low. With relaxed matching strategies *appeq/enteq/wc, sboisa* and *appeq/entov/wc, sboisa*, we see some substantial recall 20% (remember NLP-ANN is only abstracts). Nevertheless precision for edges is only 1 in 10 and for nodes about the same.

Caveats There are number of issues that need to be taken into account when analyzing these results. For instance, SBO/GO term annotation for MTOR-HMN is not perfect, as can be seen from the large number of conversion operations. Similarly, Entrez Gene id normalization has its problems, especially when dealing with complex species. Lastly, reaction signature overlap does not count reactions with multiple reaction types as separate. We are currently working on dealing with each of these issues. Some will arguably improve performance, others decrease precision and recall numbers. We are confident though that the general trends in the results will uphold.

7   Discussion

The last section quantitatively demonstrated differences between extraction and curation. Curation involves processes such as annotation, selection and, in particular, composition (of subgraphs into a large graph). The next paragraphs summarize the most important problems.

MTOR-HMN/MTOR-ANN				
	nodes		edges	
	prec	rec	prec	rec
nmeq, sboeq	1.22	1.93	0.94	1.30
nmeq, sboov	1.52	2.72	1.15	1.91
nmeq, sboisa	3.15	4.00	2.43	2.65
nmeq/wc, sboeq	3.48	8.60	2.77	6.76
nmeq/wc, sboov	3.78	9.34	2.99	7.45
nmeq/wc, sboisa	5.59	12.21	4.44	9.73
appeq, sboeq	1.44	2.22	1.11	1.47
appeq, sboov	1.81	3.11	1.37	2.12
appeq, sboisa	3.93	4.50	2.99	2.89
appeq/wc, sboeq	3.85	8.90	3.07	7.04
appeq/wc, sboov	4.22	9.74	3.33	7.77
appeq/wc, sboisa	6.67	12.85	5.25	10.22
appeq/enteq, sboeq	2.74	3.02	2.13	1.95
appeq/enteq, sboov	3.19	3.86	2.43	2.65
appeq/enteq, sboisa	5.81	5.78	4.48	3.74
appeq/enteq/wc, sboeq	9.78	13.69	8.15	10.99
appeq/enteq/wc, sboov	10.48	15.37	8.66	12.54
appeq/enteq/wc, sboisa	14.67	23.88	11.95	19.90
appeq/entov, sboeq	8.85	10.33	7.34	7.41
appeq/entov, sboov	9.41	12.01	7.73	8.79
appeq/entov, sboisa	13.59	19.53	11.01	14.73
appeq/entov/wc, sboeq	9.78	13.69	8.15	10.99
appeq/entov/wc, sboov	10.48	15.37	8.66	12.54
appeq/entov/wc, sboisa	14.67	23.88	11.95	19.90

MTOR-HMN/MTOR-NLP				
	nodes		edges	
	prec	rec	prec	rec
nmeq, sboeq	6.31	13.25	5.84	8.67
nmeq, sboov	7.26	17.40	6.67	11.48
nmeq, sboisa	9.85	27.73	8.88	17.50
nmeq/wc, sboeq	9.83	40.19	9.21	31.14
nmeq/wc, sboov	10.82	44.34	10.08	34.43
nmeq/wc, sboisa	14.48	58.58	13.30	46.68
appeq, sboeq	6.56	14.04	6.07	9.24
appeq, sboov	7.53	18.69	6.92	12.37
appeq, sboisa	10.39	30.35	9.35	19.41
appeq/wc, sboeq	10.27	40.83	9.63	31.62
appeq/wc, sboov	11.28	45.53	10.52	35.33
appeq/wc, sboisa	15.24	60.85	13.98	48.43
appeq/enteq, sboeq	9.33	18.44	8.64	12.21
appeq/enteq, sboov	11.06	23.63	10.16	15.71
appeq/enteq, sboisa	15.94	37.22	14.28	24.50
appeq/enteq/wc, sboeq	21.40	49.73	20.11	40.58
appeq/enteq/wc, sboov	23.59	55.66	22.06	45.95
appeq/enteq/wc, sboisa	32.88	75.33	30.18	65.04
appeq/entov, sboeq	20.18	44.44	18.90	34.88
appeq/entov, sboov	22.35	50.32	20.83	39.97
appeq/entov, sboisa	31.34	69.85	28.65	57.51
appeq/entov/wc, sboeq	21.40	49.73	20.11	40.58
appeq/entov/wc, sboov	23.59	55.66	22.06	45.95
appeq/entov/wc, sboisa	32.88	75.33	30.18	65.04

Table 2: Results of matching MTOR-ANN and MTOR-NLP with MTOR-HMN. Results are always precision/recall.

Species Normalization There has been a lot of work on this topic (Van Landeghem et al., 2013; Wei et al., 2015b; Sohn et al., 2008; Doğan et al., 2014; Hakenberg et al., 2011) provide impressive performance. But there is the problem of how to use the information provided by tools such as GNAT. GNAT, for instance, returns hypotheses of possible identifiers. It is then up to subsequent systems to use this information and reject certain hypotheses based on other information in the text.

Complex formation Identification of complexes is missing from NLP extraction systems. To the best of our knowledge, there is very little work on extraction of complexes and their participants from text (except generally in terms of Named Entity Recognition). However, complexes are extremely important for the mTOR pathway. For a large part the pathway consists of complexes that form and subsequently modify other reactions. Not being able to extract such information is a significant disadvantage for automated systems.

Composition of pathways The NLP system produces pathway maps that consist of scattered reactions without integrating them into one. The human map on the other hand is all about a single network of reactions. Composition is a combinatorial problem constrained by cues in the Natural Language as well as biology. This paper proposed a number of matching strategies. These strategies are not only useful for measuring the state-of-the-art. For instance, matching of species based on Entrez Gene normalization could be useful in pathway composition.

Understanding levels of detail of representation A fundamental problem in pathway curation is that information can be represented on different levels of specificity. For instance, it might be sufficient to capture phosphorylation instead of capturing the exact sites or the number of phosphoryl groups added. Often human modelers make various abstractions and conceptualizations of the same underlying biological process. Final pathway maps are affected by prior knowledge of the curator and this shapes the pathway that a human produces. The problem then becomes how to build machines that can extract knowledge on various levels of abstraction.

It is important to realize that these issues are not just a problem of more data or more precise annotation. Current NLP systems are good at classifying strings and their relations but they have no notion of the underlying processes (in this case the biological processes involved). The learning signal of NLP systems is annotated text and it is not the human-curated biological model. The human as an expert in Systems Biology reading the text will pick out relevant detail and try to build a consistent overall model based on the information in the various texts. The NLP system relies on information detected in the text without any actual notion of what the text actually means, i.e. without building an internal model and integrating it with prior information.

8   Conclusion

To the best of our knowledge, this paper is the first to evaluate automated pathway extraction systems by measuring the difference between automated systems and human curation. We believe this kind of analysis is crucial to make progress towards the ultimate goal of complete automation of pathway curation. The contribution of this paper is twofold: 1) we propose a number of measures that can be used to quantify the state-of-the-art; 2) we identify a number of areas where progress can improve the state-of-the-art measurably.

This paper is part of a larger trend in NLP to move from event extraction to knowledge base creation (Kim et al., 2015) and construction of biologically relevant networks (Rinaldi et al., 2016). It is therefore perfectly aligned with people trying to automatically build mechanistic dynamic pathway models (Cohen, 2015) that could ultimately have a big scientific impact (Kitano, 2016).

References

Bree B Aldridge, John M Burke, Douglas A Lauffenburger, and Peter K Sorger. 2006. Physicochemical modelling of cell signalling pathways. *Nature cell biology*, 8(11):1195–1203.

S. Ananiadou, S. Pyysalo, J. Tsujii, and D. Kell. 2010. Event extraction for systems biology by text mining the literature. *Trends in biotechnology*, 28(7):381–90.

W. Baumgartner, B. Cohen, L. Fox, et al. 2007. Manual curation is not sufficient for annotation of genomic databases. *Bioinformatics*, 23(13):i41–i48.

J. Björne, F. Ginter, and T. Salakoski. 2012. University of turku in the bionlp'11 shared task. *BMC bioinformatics*, 13(11):1.

J. Björne. 2014. *Biomedical Event Extraction with Machine Learning.* Ph.D. thesis, University of Turku.

E. Caron, S. Ghosh, Y. Matsuoka, et al. 2010. A comprehensive map of the mtor signaling network. *Molecular systems biology*, 6(1).

P. Cohen. 2015. Darpa's big mechanism program. *Physical Biology*, 12(4):045008.

P. Creixell, J. Reimand, S. Haider, et al. 2015. Pathway and network analysis of cancer genomes. *Nature methods*, 12(7):615.

Rezarta Islamaj Doğan, Donald C Comeau, Lana Yeganova, and W John Wilbur. 2014. Finding abbreviations in biomedical literature: three bioc-compatible modules and four bioc-formatted corpora. *Database*, 2014:bau044.

K. Fujita, M. Ostaszewski, Y. Matsuoka, et al. 2014. Integrating pathways of parkinson's disease in a molecular interaction map. *Molecular neurobiology*, 49(1):88–102.

A. Funahashi, Y. Matsuoka, A. Jouraku, et al. 2008. Celldesigner 3.5: a versatile modeling tool for biochemical networks. *Proceedings of the IEEE*, 96(8):1254–1265.

Jörg Hakenberg, Martin Gerner, Maximilian Haeussler, Illés Solt, Conrad Plake, Michael Schroeder, Graciela Gonzalez, Goran Nenadic, and Casey M Bergman. 2011. The gnat library for local and remote gene mention normalization. *Bioinformatics*, 27(19):2769–2771.

M. Hucka, A. Finney, H. Sauro, et al. 2003. The systems biology markup language (sbml): a medium for representation and exchange of biochemical network models. *Bioinformatics*, 19(4):524–531.

P. Khatri, M. Sirota, and A. Butte. 2012. Ten years of pathway analysis: current approaches and outstanding challenges. *PLoS Comput Biol*, 8(2).

Jin-Dong Kim, Jung-jae Kim, Xu Han, and Dietrich Rebholz-Schuhmann. 2015. Extending the evaluation of genia event task toward knowledge base construction and comparison to gene regulation ontology task. *BMC bioinformatics*, 16(Suppl 10):S3.

H. Kitano. 2002. Computational systems biology. *Nature*, 420(6912):206–210.

H. Kitano. 2016. Artificial intelligence to win the nobel prize and beyond: Creating the engine for scientific discovery. *AI Magazine*.

N. Le Novère. 2006. Model storage, exchange and integration. *BMC neuroscience*.

Donna Maglott, Jim Ostell, Kim D Pruitt, and Tatiana Tatusova. 2005. Entrez gene: gene-centered information at ncbi. *Nucleic acids research*, 33(suppl 1):D54–D58.

M. Miwa, P. Thompson, J. McNaught, et al. 2012. Extracting semantically enriched events from biomedical literature. *BMC bioinformatics*, 13(1):1.

S. Mizuno, R. Iijima, S. Ogishima, et al. 2012. Alzpathway: a comprehensive map of signaling pathways of alzheimer's disease. *BMC systems biology*, 6(1):52.

C. Nédellec, R. Bossy, J.-D. Kim, et al. 2013. Overview of bionlp shared task 2013. *ACL*, page 1.

K. Oda and H. Kitano. 2006. A comprehensive map of the toll-like receptor signaling network. *Molecular systems biology*, 2(1).

T. Ohta, S. Pyysalo, and J. Tsujii. 2011. From pathways to biomolecular events: opportunities and challenges. In *Proceedings of BioNLP 2011 Workshop*, pages 105–113. ACL.

T. Ohta, S. Pyysalo, R. Rak, et al. 2013. Overview of the pathway curation (pc) task of bionlp shared task 2013. In *Proceedings of the BioNLP Shared Task 2013 Workshop*, pages 67–75. ACL.

F. Rinaldi, T. Effendorf, and S. Madan. 2016. Biocreative 5 track 4: A shared task for the extraction of causal network information in biological expression language. *Database*.

S. Sohn, D. Comeau, W. Kim, and J Wilbur. 2008. Abbreviation definition identification based on automatic precision estimates. *BMC bioinformatics*, 9(1):1.

M. Spranger, S. Palaniappan, and S. Ghosh. 2015. Extracting biological pathway models from nlp event representations. In *Proceedings of the 2015 Workshop on Biomedical Natural Language Processing (BioNLP 2015)*, pages 42—51. ACL.

D. Tkaczyk, P. Szostek, M. Fedoryszak, et al. 2015. Cermine: automatic extraction of structured metadata from scientific literature. *International Journal on Document Analysis and Recognition (IJDAR)*, 18(4):317–335.

M. Valenzuela-Escárcega, G. Hahn-Powell, T. Hicks, and M. Surdeanu. 2015. A domain-independent rule-based framework for event extraction. In *Proceedings of the 53rd Annual Meeting of the Association for Computational Linguistics (ACL-IJCNLP)*, pages 127–132. ACL.

S. Van Landeghem, J. Björne, C.-H. Wei, et al. 2013. Large-scale event extraction from literature with multi-level gene normalization. *PloS one*, 8(4):e55814.

Chih-Hsuan Wei, Robert Leaman, and Zhiyong Lu. 2015. Simconcept: A hybrid approach for simplifying composite named entities in biomedical text. *Biomedical and Health Informatics, IEEE Journal of*, 19(4):1385–1391.

Construction of a Personal Experience Tweet Corpus for Health Surveillance

Keyuan Jiang   **Ricardo A. Calix**   **Matrika Gupta**
Department of Computer Information Technology & Graphics
Purdue University Northwest
{kjiang,rcalix,gupta297}@pnw.edu

Abstract

Studies have shown that Twitter can be used for health surveillance, and personal experience tweets (PETs) are an important source of information for health surveillance. To mine Twitter data requires a relatively balanced corpus and it is challenging to construct such a corpus due to the labor-intensive annotation tasks of large data sets. We developed a bootstrap method of finding PETs with the use of the machine learning-based filter. Through a few iterations, our approach can efficiently improve the balance of two class dataset with a reduced amount of annotation work. To demonstrate the usefulness of our method, a PET corpus related to effects caused by 4 dietary supplements was constructed. In 3 iterations, a corpus of 8,770 tweets was obtained from 108,528 tweets collected, and the imbalance of two classes was significantly reduced from 1:31 to 1:3. In addition, two out of three classifiers used showed improved performance over iterations. It is conceivable that our approach can be applied to various other health surveillance studies that use machine learning-based classifications of imbalanced Twitter data.

1 Introduction

As defined by the Merriam-Webster Dictionary, surveillance is the act of carefully watching someone or something. In the health field, the WHO defines that public health surveillance is the continuous, systematic collection, analysis and interpretation of health-related data needed for the planning, implementation, and evaluation of public health practice. Information directly reported by patients is of significant importance, and having an efficient way of obtaining and analyzing this data is very important. Because of mobile phones and other technologies, patients are inclined to post information on the web. This represents a great opportunity for those concerned with health surveillance if they can only mine the data. As such, the critical issue is where and how to obtain and analyze this health surveillance data.

Nowadays, social media has become a natural platform through which people communicate and share their thoughts, opinions, and experiences. Topics of communication span to a broad range from politics to entertainment to hobbies. Many people are also willing to discuss their personal experiences related to their health problems and treatments on social media. Studies have shown that general purpose social media such as Twitter can be used for surveillance of health-related issues (Dredze, 2012). Examples include: affluenza pandemics (Chew and Eysenbach, 2010; Signorini et al., 2011; Collier et al., 2011; Bilge et al., 2012; Nagel et al., 2013; Gesualdo et al., 2013; Broniatowski et al., 2013; Fung et al., 2013; Nagar et al., 2014), Haitian cholera outbreak (Chunara et al., 2012), Ebola outbreak (Odlum and Yoon, 2015), nonmedical use of a psychostimulant drug (Adderall) (Hanson et al., 2013), drug abuse (Chary et al., 2013), smoking (Sofean and Smith, 2012), suicide risks (Jashinsky et al., 2014), migraine headaches (Nascimento et al., 2014), pharmaceutical product safety (Freifeld et al., 2014; Coloma et al., 2015; Jiang and Zheng, 2013; Sarker et al., 2015), disease outbreaks during festivals (Yom-Tov et al., 2014), detection of Schizophrenia (McManus et al., 2015), foodborne illness (Harris et al., 2014), and even dental pains (Heaivilin et al., 2011).

A common challenge identified in these types of studies is the difficulty in separating the useful

Proceedings of the 15th Workshop on Biomedical Natural Language Processing, pages 128–135,
Berlin, Germany, August 12, 2016. ©2016 Association for Computational Linguistics

or "on-topic" tweets from the majority of the irrelevant tweets. This poses the challenge of finding the tweets that can help to perform the heath surveillance tasks while ignoring the rest.

Twitter is a micro blogging platform on which messages of up to 140 characters can be posted. Despite the shortness of the messages, the size of Twitter user pool may still mean that a lot of information can be posted. As such, for any given topic, there may be a good number of on-topic tweets and a much larger set of off-topic tweets. As a result, one of the key questions to address is how to obtain the relevant data.

In this study, the term personal experience tweet (PET) is used to describe the tweets that are relevant to the analysis. PETs, therefore, are tweets that describe a person's encounters, observations, and important events related to his or her life. In the case of health surveillance, such experience can be related to changes of a person's health, an illness, a disease, or a treatment. In other words, if any of the above affects an individual it signifies a personal experience. For example, if a medicine causes a person to vomit or improves the person's sleeping behavior, then the person is said to have some experience with the medicine. Personal experience tweets (PETs) are an important source of information for health surveillance using Twitter data.

Given the sheer volume of daily posts, Twitter data are known to contain a significant amount of irrelevant off-topic posts (e.g. news, sales promotions, spam, etc.). This can easily result in collections of Twitter data with a significant bias toward the irrelevant posts. For example, in a study of 2 billion tweets collected from May 2009 to October 2010, Bian and colleagues (Bian et al., 2012) found only 489 on-topic tweets for the 5 medicines being studied in clinical trials. As can be seen, from this study discovering on-topic tweets can be a challenge in research problem. Given all the previously stated issues, obtaining relevant data and constructing a relatively balanced corpus can be challenging and a good collection process must be implemented. This paper will discuss the data collection process, the automatic filtering approach, the annotation, and results of the analysis of the corpus. Issues related to class imbalance are also discussed.

Specifically, this study addresses the following research questions: (1) can an automated filtering algorithm help to speed up manual annotation of a PET corpus and (2) can the automated filtering approach help to address the class imbalance issues inherent in Twitter data?

2 Related Work

There have been many studies that validate the use of general purpose social media such as Twitter for surveillance of health related issues. Many of these surveillance activities involve using the information reported by the patients who share their personal health experience on social media. Efforts have been made to construct health-related Twitter corpora (Paul and Dredze, 2012; Collier et al., 2011; Ginn et al., 2014).

Using Mechanical Turk, Dredze's group (Paul and Dredze, 2012) created a corpus of 5,128 tweets classified as related to health or unrelated to health. The results showed only 36.1% of the labeled tweets were health related. It is unclear how the tweets were selected into the corpus.

Collier and colleagues (Collier et al., 2011) created a 5,283 tweets corpus related to influenza from 225,000 tweets collected from March 2010 to April 30th, 2010. These tweets in 5 classes were selected using *hand built patterns* which were unexplained by the authors, and annotated by a single annotator. For each of the 5 classes, the ratio of negative tweets to positive tweets was 2.52, 1.16, 1.95, 7.19 and 2.53 respectively, indicating that there were more negative tweets than positive ones in each class.

In studying adverse drug reactions from Twitter data, Ginn et al. (Ginn et al., 2014) collected 187,450 tweets over 6 months with 74 carefully selected drug names. 71,571 tweets were retained after removing those containing URLs, which were considered as advertisements. Out of 71,571 tweets, 10,822 were randomly chosen with a cap of 300-500 per drug. The 10,822 tweets were manually annotated by three annotators. Among 10,822 tweets, only 1,200 (11%) tweets contain adverse drug reactions (ADRs), showing the imbalance ratio of 1:8. The authors also reported a Kappa inter-annotator agreement metric with a value of 0.69.

3 Methodology

The purpose of health surveillance is to monitor the status of health conditions. To track health information using Twitter data, a data set of Twitter texts is needed. With this dataset, a methodology can be devised to identify the effects in the text. The challenge is in discovering the relevant tweets. Our initial inspection of tweets collected using 4 dietary supplement names showed that many of the tweets were not personal experience tweets relevant to the work. Manual annotation is an expensive process, especially when using large datasets which contain very few on-topic samples. Therefore, an automated filtering tool was needed to address these issues. One of the purposes of this study is to speed up the process of annotation. Many studies have used manual or rule-based approaches for annotation. However, these approaches are time consuming. In this paper, a machine learning-based approach is proposed to try to filter out off-topic tweets.

Inspired by the bootstrap method, we developed an iterative approach of creating Twitter corpus. It starts with a small set of annotated tweets (seed). In each iteration, the annotated tweets (in the training set which is the corpus) are used to retrain classifiers, and the predicted tweets of PET class from the trained classifiers are annotated and added to the training set, in an attempt to obtain a less imbalanced corpus.

In this section, we present our method of finding personal experience tweets and its application in constructing a PET corpus related to the effects caused by 4 dietary supplements. An automated filter was used to try to remove irrelevant samples before the data set was given to annotators. A description of the creation of the PET corpus using this filter is also presented and discussed. The next few sections of the paper describe in more detail the various considerations and methodology used to create the corpus.

3.1 Corpus Construction Procedure

This study was done with the help of two annotators, who were graduate students majoring in biology and computer information technology. They independently labeled the same tweets with personal experience tags if they contained the name of any supplements and stated the experiencing of using the supplements. Below are examples of PET tweets.

Example 1:

1. melatonin gives me some messed up dreams.. or i just have awful dreams and melatonin makes me remember them. either way i dont like it.

Example 2:

2. look into St. John's Wort. Actually helps calm me down at night to sleep. always had the same issues.

First, a small number of tweets were randomly selected as a training set and were annotated manually by annotators. This was a single non-repetitive step to create a seed set. Next, three classifiers were trained using this training set and then used to classify a test set with more tweets, yielding a PET set and a non-PET set. The PET set was then labeled by annotators, and annotated tweets were added to the training set (corpus). Classifiers were retrained with the updated corpus and then used on a new batch of test data. These steps repeated until a relatively balanced corpus was achieved.

Although investigating and annotating only the predicted PET class significantly reduce the effort needed for annotation, it could potentially introduce bias undermining the representation of non-PET (majority) tweets. To compensate this potential bias, we intentionally added a small number of non-PET tweets to the training set in each iteration (Step 06 below).

The above steps are summarized in the following algorithm.

Algorithm **ConstructTweetCorpus()**

Input: A set of tweets $\mathbf{T}$, balance ratio β,
    accuracy δ

Output: A tweet corpus T

01: Randomly choose a small collection of n
    tweets from $\mathbf{T}$ as a training set denoted by T

02: Annotate T

03: Train classifiers with T

04: **Do while** balance ratio of $T < \beta$ and/or
    accuracy of classifiers $< \delta$

05:    Select a collection of l new tweets
      from $\mathbf{T}$ as test set denoted by T_l

06:    Classify T_l using trained
      classifiers, yielding a predicted PET
      set T_y and non-PET set T_n.

07:    Annotate T_y, yielding T_y'

08:    Select m tweets randomly from predicted
      non-PET set T_n and annotate them,
      yielding T_n'

09:    Add T_y' and T_n' to the training set T,
      yielding a new training set:
      $T \leftarrow T + T_y' + T_n'$

10:    Train classifier(s) with T

11: **Loop**

12: **Return** T

where l is greater than m. β is the balance ratio, the ratio between the number of PET and non-PET tweets, δ is the expected accuracy. The value of m is only a fraction of the number of tweets in the newly predicted PET class (Step 06). Both l and m can be constants. The accuracy of a classifier is measured by the ROC Area and /or F-measure.

3.2 Dataset

Using the above algorithm, we constructed a PET corpus related to 4 dietary supplements: Echinacea, Melatonin, St. John's Wort, and Valerian. A total of 108,528 tweets were collected from May 30, 2014 to December 8, 2014, through the use of Twitter REST API. The supplement names were used as keywords to perform Twitter searches. The breakdowns of the collected Twitter data are: 9,210 tweets for Echinacea, 81,915 for Melatonin, 3,176 for St. John's Wort, and 14,227 for Valerian. The collected Twitter data were preprocessed to remove retweets and non-English tweets.

3.3 Features

Two types of features were used by the machine learning-based filter: metadata and textual. Metadata features are features about the tweet itself but not the text. They include user id and Twitter client application. Textual features are the ones extracted directly from the 140 character Twitter text. Most of the tweets collected were unrelated to personal experience. They were usually marketing or promotion tweets or just facts of what a supplement does. According to a study of 106 million tweets with 4262 trending topics, Kwak et al. (2010) found that the majority of the messages were news specific. In another study, Krieck and colleagues found that news information normally repeats official information and has no contribution to the early detection of disease outbreaks (Krieck et al., 2011).

It has been observed that personal pronouns appear frequently in social media posts related to personal experiences (Elgersma and de Rijke, 2008; Jiang and Zheng, 2013). Personal pronouns were considered as a feature to classify personal and impersonal sentences (Li et al., 2010).

Our observation revealed that words or phrases commonly used in one class but not in the opposite class may contribute to the accurate prediction of PET and non-PET tweets. These words or phrases were found in both tweet texts and Twitter user names - unlike the Twitter screen name, a Twitter user name can be a phrase. For example, online stores may use in their names terms such as shop, store, and market. Presence of any of such words can provide classifiers a hint to identify promotional tweets.

A client application is the software application a Twitter author uses to post Twitter messages. Westman and colleagues observed that personal tweets were more often posted from the Twitter website (Westman and Freund, 2010).
The followings are the features used in this study.
1. Occurrences of automatically categorized frequent terms in username in PET class.
2. Occurrences of automatically categorized frequent username in non-PET class
3. Count of URLs in a tweet
4. Count of emotion words in a tweet
5. Count of unique words in a tweet
6. Total word count of a tweet
7. Occurrences of frequent words in PET class
8. Occurrences of frequent words in non-PET class
9. Count of pronouns in a tweet
10. Count of personal pronouns in a tweet
11. Count of first person pronouns in a tweet
12. Count of second person pronouns in a tweet
13. Count of third person pronouns in a tweet
14. Count of singular proper nouns in a tweet
15. Count of automatically categorized frequent terms in PET class
16. Count of automatically categorized frequent terms in non-PET class
17. Occurrences of frequent terms in Twitter user name
18. Client application used to post the tweet
19. Twitter user id

3.4 Classifiers

For filtering the off-topic tweets, three classifiers were used: decision tree (J48), KNN (IB1) and, neural network (Multilayer Perceptron, MLP). Neural networks are known for deriving meaning from complex and imprecise data. Decision trees are simple to understand, interpret and, easily handle feature interaction. KNN is simple and robust for noisy data. For evaluation purposes, both ROC metrics and F-measure were used for the reason that F-measure is not an appropriate measure of performance when the data are imbalanced (Chawla, 2009). Weka (Hall et al., 2009) which contains the implementation of all three algorithms was used in our study. It is well understood that not all classifiers perform the same way. The majority rule was used to determine the outcome of classification. That is, if outputs of two or more classifiers were PET, then the tweet was considered a PET tweet.

4 Results

Using a seed of 3,176 tweets (Run 0), our algorithm had gone three iterations with the test sets shown below. In each iteration (Run 1 through Run 3), the size of training set (corpus) increased as more annotated tweets were added.

Iteration	Training Set	Test Set	# Predicted PET Tweets	# Non-PET Tweets Added
Run 1	3,176	9,210	94	31
Run 2	3,301	14,277	386	128
Run 3	3,815	81,915	3,721	1,235

Table 1: Dataset size over iterations. It shows the number of tweets in the training set, test set, predicted PET set, and added non-PET set in each iteration.

The final annotated data set consisted of 8,770 number of tweets which are available at `https://github.com/medeffects/supplement-corpus/`. Of these, 2,067 were PET tweets and 6,703 non-PET tweets.

4.1 Inter-Annotator Agreement

Inter-annotator agreement metrics are helpful to establish the subjectivity of an annotation scheme. The annotation task was performed by 2 annotators. Two labels were used for the annotation: PET and non-PET. As shown in the table below, the average agreement was 85.4 %. Correcting for expected chance agreement, *kappa* and the other metrics still provide a reasonable score to assess the annotation consistency. The result indicates that the task of finding personal experience tweets does have a level of subjectivity. These values can later be useful to define an expected upper boundary on the PET classification task.

4.2 Corpus Class Balance

As stated earlier, the corpus was built in iterations (or runs). Each iteration used a larger training set that consisted of more examples of PET tweets.

kappa	0.624
alpha	0.624
Average Agreement	0.854
π	0.624
S	0.806

Table 2: Inter-annotator agreement for metrics

As such, it can be noticed that with each iteration more PET tweets were added to the corpus as shown in Table 3, leading to a more balanced distribution of PET and non-PET tweets. This result is beneficial for this study since the goal of it is to find as many personal experience tweets which can later be used to associate effects with dietary supplements for health surveillance.

Iteration	PET	Non-PET	Ratio
Run 0	98	3,078	1:31
Run 1	145	3,156	1:22
Run 2	256	3,559	1:14
Run 3	2,067	6,703	1:3

Table 3: Corpus class balance over iterations

4.3 Classifier Performance

In addition to studying the overall performance of classifiers collectively, we also collected performance data of each individual classifier on predicting PET tweets, and they are shown in the figure below.

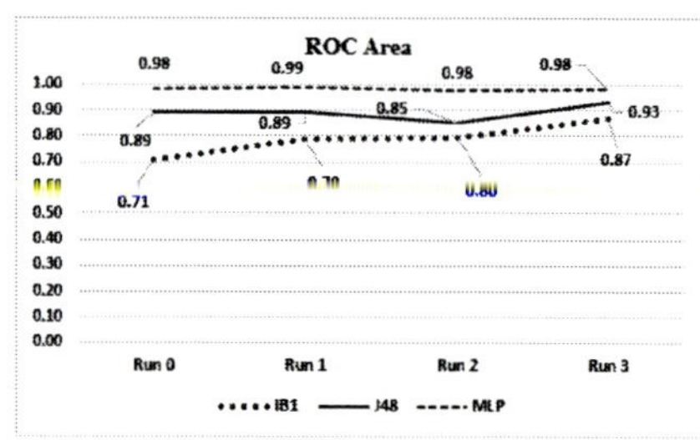

Figure 1: ROC area over iterations

4.4 Feature Ranking

One important aspect in this study is to determine what features helped to automatically detect personal experience tweets. As indicated previously, most of these 19 features by classifiers were extracted from the tweet text using natural language processing techniques. To perform the feature analysis, the Chi-Square ranking method was used. The top ranked features are occurrences of

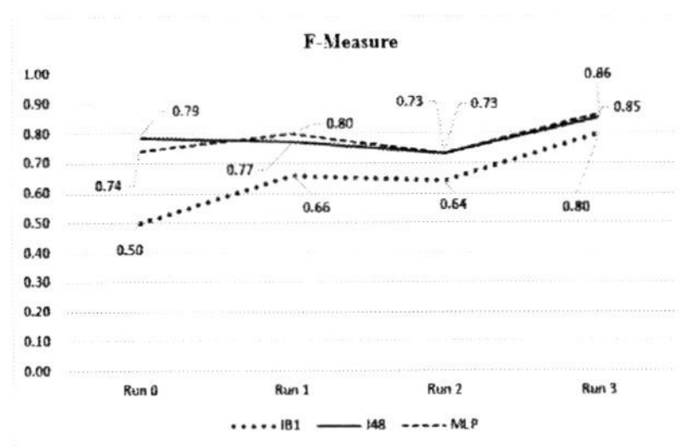

Figure 2: F-Measure over iterations

automatically categorized frequent terms in username in PET class, occurrences of automatically categorized frequent username in non-PET class, occurrences of frequent words in PET class, occurrences of frequent words in non-PET class, pronoun count, personal pronoun count, first person pronoun count, URL count, Twitter client and user id.

4.5 Prediction Precision

The overall performance of the PET classifiers was measured with the training sets. The PET classifiers are the filter used to identify relevant tweets for human annotation. The actual performance of the filter should be measured against the prediction using the test data. Given that only predicted PET tweets were annotated –that is, only true positive and false positive figures were available, prediction precision was measured. Precision is a ratio between actual PET tweets and predicted PET tweets in the same predicted PET set, a performance measurement of classifiers when performing predictions.

Table 4 shows that the precision falls within the range of 0.28 - 0.49. This indicates that for every 100 predicted samples (tweets), between 28 and 49 may be actual PETs.

# PET Tweets			
Iteration	Predicted	Actual	Precision
Run 1	94	46	0.49
Run 2	386	107	0.28
Run 3	3,721	1,597	0.43

Table 4: Prediction precision over iterations

5 Discussions

The amount of work on annotation can be significant when constructing a corpus that requires examination of large sets of data. In this study, if we were to annotate 108,528 tweets, it would take annotators a significant amount of their time to do so. However, using our proposed method, two annotators only needed to annotate 8,770 tweets (= initial seed tweets plus predicted PET tweets and added non-PET tweets in each iteration. Refer to Table 1). If it takes an average of one minute to annotate a single tweet and each annotator spends 8 hours a day on annotation, it will take an annotator 226 days to complete annotation of 108,528 tweets, but 18 days for 8,770 tweets. This represents a significant reduction of annotation time.

By some estimates, the obtained kappa score shown in Table 2 may be considered low which implies that the text is highly subjective and difficult to annotate. This suggests that finding personal experience tweets is highly subjective. Personal experience, in the context of this paper, is text expressed by a person and that is of a very personal nature. The difficulty may lie in the fact that there is not set lexicon to define personal experience. In contrast, emotion text detection, which is also considered subjective, does have its own lexicon (i.e happy words vs. sad words).

As can be seen in Table 3, our approach is also efficient in improving the class balance of the corpus. With only 3 iterations, the ratio of the number of PET tweets to that of non-PET tweets had come down from 1:31 to 1:3, a 10-fold improvement.

The performance of individual classifiers on predicting PET tweets with the training data either remained the same level or improved over iterations. For ROC Area (Figure 1), both IB1 and J48 improved, and MLP remained the same. For F-Measure (Figure 2) which is not an appropriate indicator of performance when data are imbalanced, all three classifiers had improved. In addition, it is noted that the multilayer perceptron (MLP) classifier has the best accuracy in predicting PETs.

Although values of ROC Area and F-Measure are quite promising, when it came to predict the unlabeled data (test set), 3 classifiers could only predict PET tweets with 28% to 49% precision. This implies that if the classifiers are to be used to predict PETs on new sets of unlabeled tweets, only 28% to 49% of tweets in the predicted PET set may be actual PET tweets.

Our result of feature ranking suggests that between metadata and textual features, textual features contribute the most to overall classification accuracy. And the best performing features are the ones related to the frequency of terms used in either tweet text or the user name - that is, the most frequent terms in a class that are infrequent in the opposite class. This approach is sometimes commonly referred to as the Gramulator type approach.

6 Conclusion

We proposed a bootstrap method to construct tweet corpus from noisy Twitter data. Through a few iterations, our approach can help construct quickly a tweet corpus with closely balanced classes, without a significant amount effort on annotation. It is conceivable that our approach can be applied to other health surveillance studies that use machine learning-based classifications of imbalanced social media data.

Acknowledgment

Authors wish to thank anonymous reviewers for their significant effort in critiquing our work and providing constructive comments, Yongbing Tang for data collection for this project, Jiabao Liu for coding for data processing and analysis, and Cecelia Lai for annotating the tweets. This work was supported in part by the National Institutes of Health grant 1R15LM011999-01.

References

Jiang Bian, Umit Topaloglu, and Fan Yu. 2012. Towards large-scale twitter mining for drug-related adverse events. In *Proceedings of the 2012 international workshop on Smart health and wellbeing*, pages 25–32. ACM.

Ugur Bilge, Selen Bozkurt, Basak Oguz Yolcular, and Deniz Ozel. 2012. Can social web helpoff detect influenza related illnesses in turkey? *Stud Health Technol Inform*, 174:100–104.

David A Broniatowski, Michael J Paul, and Mark Dredze. 2013. National and local influenza surveillance through twitter: an analysis of the 2012-2013 influenza epidemic. *PloS one*, 8(12):e83672.

Michael Chary, Nicholas Genes, Andrew McKenzie, and Alex F Manini. 2013. Leveraging social networks for toxicovigilance. *Journal of Medical Toxicology*, 9(2):184–191.

Nitesh V Chawla. 2009. Data mining for imbalanced datasets: An overview. In *Data mining and knowledge discovery handbook*, pages 875–886. Springer.

Cynthia Chew and Gunther Eysenbach. 2010. Pandemics in the age of twitter: content analysis of tweets during the 2009 h1n1 outbreak. *PloS one*, 5(11):e14118.

Rumi Chunara, Jason R Andrews, and John S Brownstein. 2012. Social and news media enable estimation of epidemiological patterns early in the 2010 haitian cholera outbreak. *The American journal of tropical medicine and hygiene*, 86(1):39–45.

Nigel Collier, Nguyen Truong Son, and Ngoc Mai Nguyen. 2011. Omg u got flu? analysis of shared health messages for bio-surveillance. *Journal of Biomedical Semantics*, 2(5):1.

Preciosa M Coloma, Benedikt Becker, Miriam CJM Sturkenboom, Erik M van Mulligen, and Jan A Kors. 2015. Evaluating social media networks in medicines safety surveillance: two case studies. *Drug safety*, 38(10):921–930.

Mark Dredze. 2012. How social media will change public health. *Intelligent Systems, IEEE*, 27(4):81–84.

Erik Elgersma and Maarten de Rijke. 2008. Personal vs non-personal blogs: initial classification experiments. In *Proceedings of the 31st annual international ACM SIGIR conference on Research and development in information retrieval*, pages 723–724. ACM.

Clark C Freifeld, John S Brownstein, Christopher M Menone, Wenjie Bao, Ross Filice, Taha Kass-Hout, and Nabarun Dasgupta. 2014. Digital drug safety surveillance: monitoring pharmaceutical products in twitter. *Drug safety*, 37(5):343–350.

Isaac Chun-Hai Fung, King-Wa Fu, Yuchen Ying, Braydon Schaible, Yi Hao, Chung-Hong Chan, and Zion Tsz-Ho Tse. 2013. Chinese social media reaction to the mers-cov and avian influenza a (h7n9) outbreaks. *Infectious diseases of poverty*, 2(1):1–12.

Francesco Gesualdo, Giovanni Stilo, Michaela V Gonfiantini, Elisabetta Pandolfi, Paola Velardi, Alberto E Tozzi, et al. 2013. Influenza-like illness surveillance on twitter through automated learning of naïve language. *PLoS One*, 8(12):e82489.

Rachel Ginn, Pranoti Pimpalkhute, Azadeh Nikfarjam, Apurv Patki, Karen OConnor, Abeed Sarker, Karen Smith, and Graciela Gonzalez. 2014. Mining twitter for adverse drug reaction mentions: a corpus and classification benchmark. In *Proceedings of the fourth workshop on building and evaluating resources for health and biomedical text processing*. Citeseer.

Mark Hall, Eibe Frank, Geoffrey Holmes, Bernhard Pfahringer, Peter Reutemann, and Ian H Witten. 2009. The weka data mining software: an update. *ACM SIGKDD explorations newsletter*, 11(1):10–18.

Carl L Hanson, Scott H Burton, Christophe Giraud-Carrier, Josh H West, Michael D Barnes, and Bret Hansen. 2013. Tweaking and tweeting: exploring twitter for nonmedical use of a psychostimulant drug (adderall) among college students. *Journal of medical Internet research*, 15(4):e62.

Jenine K Harris, Raed Mansour, Bechara Choucair, Joe Olson, Cory Nissen, Jay Bhatt, et al. 2014. Health department use of social media to identify foodborne illness-chicago, illinois, 2013-2014. *MMWR Morb Mortal Wkly Rep*, 63(32):681–685.

N Heaivilin, B Gerbert, JE Page, and JL Gibbs. 2011. Public health surveillance of dental pain via twitter. *Journal of dental research*, 90(9):1047–1051.

Jared Jashinsky, Scott H Burton, Carl L Hanson, Josh West, Christophe Giraud-Carrier, Michael D Barnes, and Trenton Argyle. 2014. Tracking suicide risk factors through twitter in the us. *Crisis.*

Keyuan Jiang and Yujing Zheng. 2013. Mining twitter data for potential drug effects. In *Advanced data mining and applications*, pages 434–443. Springer.

Manuela Krieck, Johannes Dreesman, Lubomir Otrusina, and Kerstin Denecke. 2011. A new age of public health: Identifying disease outbreaks by analyzing tweets. In *Proceedings of Health Web-Science Workshop, ACM Web Science Conference.*

Haewoon Kwak, Changhyun Lee, Hosung Park, and Sue Moon. 2010. What is twitter, a social network or a news media? In *Proceedings of the 19th international conference on World wide web*, pages 591–600. ACM.

Shoushan Li, Chu-Ren Huang, Guodong Zhou, and Sophia Yat Mei Lee. 2010. Employing personal/impersonal views in supervised and semi-supervised sentiment classification. In *Proceedings of the 48th annual meeting of the association for computational linguistics*, pages 414–423. Association for Computational Linguistics.

Kimberly McManus, Emily K Mallory, Rachel L Goldfeder, Winston A Haynes, and Jonathan D Tatum. 2015. Mining twitter data to improve detection of schizophrenia. *AMIA Summits on Translational Science Proceedings*, 2015:122.

Ruchit Nagar, Qingyu Yuan, Clark C Freifeld, Mauricio Santillana, Aaron Nojima, Rumi Chunara, and John S Brownstein. 2014. A case study of the new york city 2012-2013 influenza season with daily geocoded twitter data from temporal and spatiotemporal perspectives. *Journal of medical Internet research*, 16(10):e236.

Anna C Nagel, Ming-Hsiang Tsou, Brian H Spitzberg, Li An, J Mark Gawron, Dipak K Gupta, Jiue-An Yang, Su Han, K Michael Peddecord, Suzanne Lindsay, et al. 2013. The complex relationship of realspace events and messages in cyberspace: case study of influenza and pertussis using tweets. *Journal of medical Internet research*, 15(10):e237.

Thiago D Nascimento, Marcos F DosSantos, Theodora Danciu, Misty DeBoer, Hendrik van Holsbeeck, Sarah R Lucas, Christine Aiello, Leen Khatib, MaryCatherine A Bender, Jon-Kar Zubieta, et al. 2014. Real-time sharing and expression of migraine headache suffering on twitter: A cross-sectional infodemiology study. *Journal of medical Internet research*, 16(4):e96.

Michelle Odlum and Sunmoo Yoon. 2015. What can we learn about the ebola outbreak from tweets? *American journal of infection control*, 43(6):563–571.

Michael J Paul and Mark Dredze. 2012. A model for mining public health topics from twitter. *Health*, 11:16–6.

Abeed Sarker, Rachel Ginn, Azadeh Nikfarjam, Karen OConnor, Karen Smith, Swetha Jayaraman, Tejaswi Upadhaya, and Graciela Gonzalez. 2015. Utilizing social media data for pharmacovigilance: a review. *Journal of biomedical informatics*, 54:202–212.

Alessio Signorini, Alberto Maria Segre, and Philip M Polgreen. 2011. The use of twitter to track levels of disease activity and public concern in the us during the influenza a h1n1 pandemic. *PloS one*, 6(5):e19467.

Mustafa Sofean and Matthew Smith. 2012. Sentiment analysis on smoking in social networks. *Studies in health technology and informatics*, 192:1118–1118.

Stina Westman and Luanne Freund. 2010. Information interaction in 140 characters or less: genres on twitter. In *Proceedings of the third symposium on Information interaction in context*, pages 323–328. ACM.

Elad Yom-Tov, Diana Borsa, Ingemar J Cox, and Rachel A McKendry. 2014. Detecting disease outbreaks in mass gatherings using internet data. *Journal of medical Internet research*, 16(6):e154.

Modelling the Combination of Generic and Target Domain Embeddings in a Convolutional Neural Network for Sentence Classification

Nut Limsopatham and **Nigel Collier**
Language Technology Lab
Department of Theoretical and Applied Linguistics
University of Cambridge
Cambridge, UK
{nl347,nhc30}@cam.ac.uk

Abstract

Word embeddings have been successfully exploited in systems for NLP tasks, such as parsing and text classification. It is intuitive that word embeddings created from a larger corpus would provide a better coverage of vocabulary. Meanwhile, word embeddings trained on a corpus related to the given task or target domain would more effectively represent the semantics of terms. However, in some emerging domains (e.g. bio-surveillance using social media data), it may be difficult to find a domain corpus that is large enough for creating effective word embeddings. To deal with this problem, we propose novel approaches that use both word embeddings created from generic and target domain corpora. Our experimental results on sentence classification tasks show that our approaches significantly improve the performance of an existing convolutional neural network that achieved state-of-the-art performances on several text classification tasks.

1 Introduction

Word embeddings (i.e. distributed vector representation) represent words using dense, low-dimensional and real-valued vectors, where each dimension represents a latent feature of the word (Turian et al., 2010; Mikolov et al., 2013; Pennington et al., 2014). It has been empirically shown that word embeddings could capture semantic and syntactic similarities between words (Turian et al., 2010; Mikolov et al., 2013; Pennington et al., 2014; Levy and Goldberg, 2014). Importantly, word embeddings have been effectively used for several NLP tasks (Turian et al., 2010; Collobert et al., 2011; Segura-Bedmar et al., 2015; Limsopatham and Collier, 2015a; Limsopatham and Collier, 2015b; Muneeb

et al., 2015). For example, Turian et al. (2010) used word embeddings as input features for several NLP systems, including a traditional chunking system based on conditional random fields (CRFs) (Lafferty et al., 2001). Collobert et al. (2011) used word embeddings as inputs of a multilayer neural network for part-of-speech tagging, chunking, named entity recognition and semantic role labelling. Limsopatham and Collier (2016) leveraged semantics from word embeddings when identifying medical concepts mentioned in social media messages. Kim (2014) showed that using pre-built word embeddings, induced from 100 billion words of Google News using word2vec (Mikolov et al., 2013), as inputs of a simple convolutional neural network (CNN) could achieve state-of-the-art performances on several sentence classification tasks, such as classification of positive and negative reviews of movies (Pang and Lee, 2005) and consumer products, e.g. cameras (Hu and Liu, 2004).

The quality of word embeddings (e.g. the ability to capture semantics of words) highly depends on the corpus from which they are induced (Pennington et al., 2014). For instance, when induced from a generic corpus, such as Google News, the vector representation of 'tissue' would be similar to the vectors of 'paper' and 'toilet'. However, when induced from medical corpora, such as PubMed[1] or BioMed Central[2], the vector of 'tissue' would be more similar to those of 'cell' and 'organ'. Hence, word embeddings induced from the corpus related to the task or target domain are likely to be more useful. Meanwhile, it is intuitive that the more training documents used, the more likely that more vocabulary is covered. Recent studies (e.g. (Faruqui et al., 2015; Xu et al., 2014; Yu and Dredze, 2014)) have attempted to improve the quality of word embeddings by

---

[1] https://www.ncbi.nlm.nih.gov/pubmed/
[2] https://www.biomedcentral.com/

Proceedings of the 15th Workshop on Biomedical Natural Language Processing, pages 136–140,
Berlin, Germany, August 12, 2016. ©2016 Association for Computational Linguistics

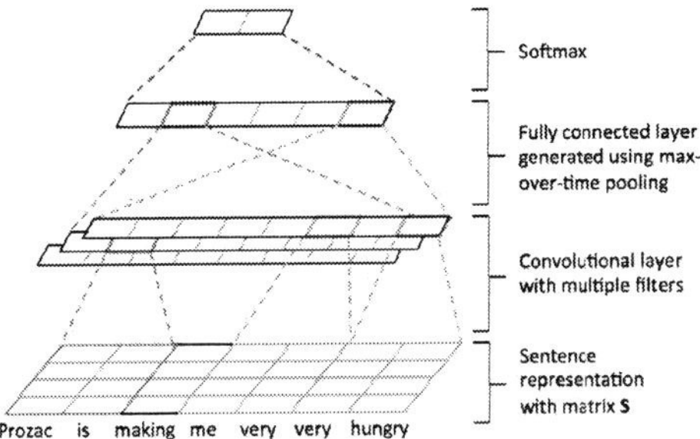

Figure 1: CNN for sentence classification.

enhancing the learning algorithm or injecting an existing knowledge-base, e.g. WordNet (Miller, 1995) or UMLS semantic network[3]. Pennington et al. (2014) incorporated aggregated global word co-occurrence statistics from the corpus when inducing word embeddings. Xu et al. (2014) and Yu and Dredze (2014) exploited semantic knowledge to improve the semantic representation of word embeddings. Nevertheless, in some emerging domains, e.g. detecting adverse drug reactions (ADR) reported in social media, existing knowledge resources or corpora may not be large enough for creating effective embeddings.

In this work, we investigate novel approaches to incorporate both generic and target domain embeddings in CNN for sentence classification. We hypothesise that using both generic and target domain embeddings further improves the performance of CNN, since it can benefit from both the good coverage of vocabulary from the generic embedding, and the effective semantic representation of the target domain embedding. This would enable CNN to perform effectively without requiring new target domain embeddings induced from a large amount of domain documents specifically related to individual tasks. We thoroughly evaluate our proposed approaches using an ADR tweet classification task (Ginn et al., 2014). In addition, to show that our approaches are effective for different target domains, we also evaluate them using a movie review classification task (Pang and Lee, 2005). Our experimental results show that our approaches significantly improve the performance in term of accuracy over an existing strong baseline that uses only either the generic or the target domain embeddings.

2   CNN for Sentence Classification

CNN has been used to model sentences in different NLP tasks, such as sentence classification and

sentence matching (Collobert and Weston, 2008; Kim, 2014; Kalchbrenner et al., 2014; Hu et al., 2014). In this work, we adapt the CNN model of Kim (2014) to exploit both generic and target domain word embeddings, because of its simplicity and effectiveness. The model architecture of Kim (2014) is shown in Figure 1. In particular, for a given input sentence of length n words (padded where necessary), we create a sentence matrix $\mathbf{S} \in \mathbb{R}^{d \times n}$, where each column is the d-dimensional vector (i.e. embedding) $\mathbf{x}_i \in \mathbb{R}^d$ of each word in the sentence:

$$\mathbf{S} = \begin{bmatrix} | & | & | & & | \\ \mathbf{x}_1 & \mathbf{x}_2 & ... & & \mathbf{x}_n \\ | & | & | & & | \end{bmatrix} \qquad (1)$$

The CNN with max pooling architecture (Collobert et al., 2011; Kim, 2014) is then used for modelling the sentence. Specifically, a convolution operation using a filter $\mathbf{w} \in \mathbb{R}^{d \times h}$ is applied to a window of h words to extract a feature c_i from a window of words $\mathbf{x}_{i:i+h-1}$ as follows:

$$c_i = f(\mathbf{w} \cdot \mathbf{x}_{i:i+h-1} + b) \qquad (2)$$

where f is an activation function, such as tanh or rectifier linear unit (ReLU) (Nair and Hinton, 2010), and $b \in \mathbb{R}$ is a bias.

The filter $\mathbf{w}$ is convolved over the sequence of words represented in the sentence matrix $\mathbf{S}$ to create a feature matrix $\mathbf{C}$. In order to capture the most important features, max pooling is applied to take the maximum value of each row in the matrix $\mathbf{C}$:

$$\mathbf{c}_{max} = \begin{bmatrix} max(\mathbf{C}_{1,:}) \\ \vdots \\ max(\mathbf{C}_{d,:}) \end{bmatrix} \qquad (3)$$

This fixed sized vector $\mathbf{c}_{max}$ forms a fully connected layer, before passing to a softmax function for classification. Note that multiple filters (e.g. using different window sizes) can be used to extract features for the fully connected layer.

3   Modelling the Combination of Word Embeddings

We investigate two approaches to model the combination of generic and target domain word embeddings in the described CNN architecture.

3.1   Vector Concatenation

The first approach (namely, *vector concatenation*) is to concatenate vectors from the two embeddings when generating the sentence matrix $\mathbf{S}$ (i.e. at the input layer). In particular, each word vector $\mathbf{x}_i$ in the sentence matrix $\mathbf{S}$ becomes the concatenation

---

[3]https://semanticnetwork.nlm.nih.gov/

of the vectors from both generic and target domain embeddings corresponding to that word. This allows the filter $\mathbf{w}$ to learn the importance of each dimension of both embeddings[4].

3.2 Combining when Forming the Fully Connected Layer

The second approach (namely, *fully connected layer combination*) models the combination of the word embeddings when forming the fully connected layer before applying softmax for classification. Indeed, we apply the convolution operation (i.e. the convolutional layer in Figure 1) on two different sentence matrices, each of which is created using either the generic or the target domain embeddings. Then, the extracted features are concatenated at a single fully connected layer before applying softmax. This enables the model to learn the importance of each feature from both embeddings directly, before allowing the softmax to take into account the extracted features. Intuitively, this approach should be more effective than the first approach, as it allows more parameters to be learned directly based on the effectiveness of the word vectors from each of the embeddings.

4 Experimental Setup

4.1 Test Collection

To evaluate our approaches, we use two different test collections, which represent domain-specific tasks where existing target domain documents for training word embeddings may be limited. First, the adverse drug reaction (ADR) tweet collection (Ginn et al., 2014) contains 5,250 Twitter messages[5] that can be classified as ADR and non-ADR discussions. Second, the movie review collection (Pang and Lee, 2005)[6] consists of 10,662 sentences that can be classified as having a positive or a negative meaning. On average, a sentence contains 20 terms. For both collections, we report the performance based on the accuracy measure (Pang and Lee, 2005; Ginn et al., 2014), and use paired t-test ($p < 0.05$) to measure the significant difference between the performance achieved by the proposed approaches and the baselines.

4.2 Pre-trained Word Embeddings

As a representative of generic word embeddings, we use the publicly available 300-dimension embeddings (vocabulary size of 3M) that were induced from 100 billion words from Google News using word2vec[7], which has been shown to be effective for several tasks (Baroni et al., 2014; Kim, 2014). For target domain embeddings, we use the skip-gram model from word2vec (using default parameters) to create 300-dimension word embeddings from two different publicly available corpora, which are considerably smaller than the Google News. Specifically, the first corpus, representing the target domain corpus of the ADR tweet classification task, contains 854M words from 119k medical articles from BioMed Central. The vocabulary size is 1.3M. For the movie review classification task, we use 24M words of 28k movie reviews from the IMDb archive[8] for inducing the target domain embedding (vocabulary size of 63k). In addition, we use a vector of random values sampled from $[-0.25, 0.25]$ to represent a word that does not exist in any embedding.

4.3 Hyper-parameters and Training Regime

We set the hyper-parameters of CNN in our approaches and the baselines following Kim (2014), whose system achieved state-of-the-art performances on several sentence classification tasks, including the movie review classification task evaluated in this paper. Indeed, we use ReLU as activation functions, and use the filter $\mathbf{w}$ with the window size (h) of 3, 4 and 5, each of which with 100 feature maps. We also apply dropout (dropout rate 0.5) (Srivastava et al., 2014) and L_2 regularisation of the weight vectors at the fully connected layer.

We conduct experiments using 10-fold cross validation. The CNN model is trained over a mini-batch of size 50 by back-propagation. The stochastic gradient decent is performed using Adadelta update rule (Zeiler, 2012) to minimise the negative log-likelihood of correct predictions.

5 Experimental Results

We compare the performance of our approaches, i.e. vector concatenation (Section 3.1) and fully connected layer combination (Section 3.2), with that of the effective CNN model of Kim (2014) (denoted, *simple CNN*). Note that we use the *static*

---

[4]The size of the filter $\mathbf{w} \in \mathbb{R}^{d_* \times h}$ depends on the dimension d_* of the concatenated vectors.

[5]We have a smaller dataset than the original paper because some tweets can no longer be accessed via Twitter API.

[6]https://www.cs.cornell.edu/people/pabo/movie-review-data

[7]https://code.google.com/p/word2vec/

[8]Downloaded from http://www.cs.cornell.edu/people/pabo/movie-review-data/polarity_html.zip.

Approach	Word Embeddings	Accuracy	
		ADR Tweets	Movie Review
Simple CNN (Kim, 2014)	Random	87.97	72.41
Simple CNN (Kim, 2014)	Generic	88.47	80.56*
Simple CNN (Kim, 2014)	Domain	88.75*	80.88*
Vector Concatenation	Generic+Domain	88.85*	81.29*
Vector Concatenation	Random+Random	87.96	72.28
Vector Concatenation	Generic+Generic	88.76	80.69
Vector Concatenation	Domain+Domain	88.88	80.61
Fully Connected Layer Combination	Generic+Domain	**89.74***°•	**81.59***°•
Fully Connected Layer Combination	Random+Random	88.61	72.57
Fully Connected Layer Combination	Generic+Generic	89.47	80.54
Fully Connected Layer Combination	Domain+Domain	89.21	80.81

Table 1: The accuracy performance of the proposed approaches and the *simple CNN* baselines (Kim, 2014). Significant differences ($p < 0.05$, paired t-test) compared to the simple CNN baselines with the *Random*, *Generic* and *Domain* word embeddings, are denoted *, ° and •, respectively.

variant of the CNN model, which does not allow the input embeddings to be updated during training, as we aim to investigate the performance when using original embeddings[9]. In addition to the pre-trained embeddings described in Section 4.2, we use 300-dimension randomly generated word embeddings, as an alternative baseline.

Table 1 reports the accuracy performance of our approaches and the simple CNN baselines on the ADR tweet and movie review classification tasks. We first compare the effectiveness of the simple CNN baselines when applied with different word embeddings. For both tasks, the simple CNN with the target domain word embeddings (accuracy 88.75% and 80.88%) outperforms the simple CNN with either the generic (accuracy 88.47% and 80.56%) or the random (accuracy 87.97% and 72.41%) word embeddings. The performance differences between using the target domain and the random word embeddings are statistically significant ($p < 0.05$) for both tasks. These results show the importance of target domain embedding for the simple CNN on the classification tasks.

Next, we discuss the performance of our two proposed approaches. As shown in Table 1, *Fully Connected Layer Combination (Generic+Domain)* performs better than all of the other approached reported in this paper for both the ADR tweet (accuracy 89.74%) and movie review (accuracy 81.59%) classification tasks. Importantly, it significantly ($p < 0.05$) outperforms the simple CNN baselines that use either the random, generic or target domain word embeddings for both tasks. Meanwhile, *Vector Concatenation (Generic+Domain)* also outperforms all of the simple CNN baselines. These support our hypothesis that exploiting both the generic and target domain word embeddings further improves the performance of CNN for sentence classification.

To further support that our approaches are effective because of exploiting both generic and target domain embeddings rather than because of allowing the model to learn more parameters, we compare our approaches with another set of baselines that use either the generic, target domain, or random embedding twice in both of our proposed approaches. We observe that *Fully Connected Layer Combination (Generic+Domain)* outperforms all of its corresponding baselines, e.g. *Domain+Domain*, for both tasks. The same trends of performance are also observed for the vector concatenation approach, excepting that *Vector Concatenation (Domain+Domain)* marginally outperforms *Vector Concatenation (Generic+Domain)* on the ADR tweet classification task.

6 Conclusions

We have shown the potential of incorporating generic and target domain embeddings in CNN for sentence classification. This provides an alternative method for exploiting generic word embeddings for a given task, where existing domain knowledge or corpora for creating word embeddings are limited, as well as avoiding inducing new word embeddings from a large number of target domain documents for individual tasks. We proposed two approaches that modelled the combination of the two embeddings at the input layer and the fully connected layer of a CNN model. Our experimental results conducted on the ADR tweet and movie review classification tasks showed that both approaches significantly improved the performance over a strong CNN baseline.

---

[9]The performances of both Kim's and our approaches will further improve, if we allow the embeddings to be updated.

Acknowledgements

The authors acknowledge the support of the EP-SRC (grant number EP/M005089/1).

References

Marco Baroni, Georgiana Dinu, and Germán Kruszewski. 2014. Don't count, predict! a systematic comparison of context-counting vs. context-predicting semantic vectors. In *ACL*, pages 238–247.

Ronan Collobert and Jason Weston. 2008. A unified architecture for natural language processing: Deep neural networks with multitask learning. In *ICML*, pages 160–167.

Ronan Collobert, Jason Weston, Léon Bottou, Michael Karlen, Koray Kavukcuoglu, and Pavel Kuksa. 2011. Natural language processing (almost) from scratch. *J. Mach. Learn. Res.*, 12:2493–2537.

Manaal Faruqui, Jesse Dodge, Sujay Kumar Jauhar, Chris Dyer, Eduard H. Hovy, and Noah A. Smith. 2015. Retrofitting word vectors to semantic lexicons. In *NAACL*, pages 1606–1615.

Rachel Ginn, Pranoti Pimpalkhute, Azadeh Nikfarjam, Apurv Patki, Karen OConnor, Abeed Sarker, Karen Smith, and Graciela Gonzalez. 2014. Mining twitter for adverse drug reaction mentions: a corpus and classification benchmark. In *The fourth workshop on building and evaluating resources for health and biomedical text processing*.

Minqing Hu and Bing Liu. 2004. Mining and summarizing customer reviews. In *KDD*, pages 168–177.

Baotian Hu, Zhengdong Lu, Hang Li, and Qingcai Chen. 2014. Convolutional neural network architectures for matching natural language sentences. In *NIPS*, pages 2042–2050.

Nal Kalchbrenner, Edward Grefenstette, and Phil Blunsom. 2014. A convolutional neural network for modelling sentences. In *ACL*, pages 655–665.

Yoon Kim. 2014. Convolutional neural networks for sentence classification. In *EMNLP*, pages 1746–1751.

John D. Lafferty, Andrew McCallum, and Fernando C. N. Pereira. 2001. Conditional Random Fields: Probabilistic models for segmenting and labeling sequence data. In *ICML*, pages 282–289.

Omer Levy and Yoav Goldberg. 2014. Dependency-based word embeddings. In *ACL*, pages 302–308.

Nut Limsopatham and Nigel Collier. 2015a. Adapting phrase-based machine translation to normalise medical terms in social media messages. In *EMNLP*, pages 1675–1680.

Nut Limsopatham and Nigel Collier. 2015b. Towards the semantic interpretation of personal health messages from social media. In *Proceedings of the ACM First International Workshop on Understanding the City with Urban Informatics*, UCUI '15, pages 27–30.

Nut Limsopatham and Nigel Collier. 2016. Normalising medical concepts in social media texts by learning semantic representation. In *ACL*.

Tomas Mikolov, Ilya Sutskever, Kai Chen, Greg S Corrado, and Jeff Dean. 2013. Distributed representations of words and phrases and their compositionality. In *NIPS*, pages 3111–3119.

George A Miller. 1995. Wordnet: a lexical database for english. *Communications of the ACM*, 38(11):39–41.

TH Muneeb, Sunil Kumar Sahu, and Ashish Anand. 2015. Evaluating distributed word representations for capturing semantics of biomedical concepts. In *BioNLP*, pages 158–163.

Vinod Nair and Geoffrey E Hinton. 2010. Rectified linear units improve restricted boltzmann machines. In *ICML*, pages 807–814.

Bo Pang and Lillian Lee. 2005. Seeing stars: Exploiting class relationships for sentiment categorization with respect to rating scales. In *ACL*, pages 115–124.

Jeffrey Pennington, Richard Socher, and Christopher D Manning. 2014. GloVe: Global vectors for word representation. In *EMNLP*, pages 1532–1543.

Isabel Segura-Bedmar, Víctor Suárez-Paniagua, and Paloma Martínez. 2015. Exploring word embedding for drug name recognition. In *LOUHI*, pages 64–72.

Nitish Srivastava, Geoffrey Hinton, Alex Krizhevsky, Ilya Sutskever, and Ruslan Salakhutdinov. 2014. Dropout: A simple way to prevent neural networks from overfitting. *J. Mach. Learn. Res.*, 15(1):1929–1958.

Joseph Turian, Lev Ratinov, and Yoshua Bengio. 2010. Word representations: a simple and general method for semi-supervised learning. In *ACL*, pages 384–394.

Chang Xu, Yalong Bai, Jiang Bian, Bin Gao, Gang Wang, Xiaoguang Liu, and Tie-Yan Liu. 2014. Rc-net: A general framework for incorporating knowledge into word representations. In *CIKM*, pages 1219–1228.

Mo Yu and Mark Dredze. 2014. Improving lexical embeddings with semantic knowledge. In *ACL*, pages 545–550.

Matthew D Zeiler. 2012. ADADELTA: an adaptive learning rate method. *arXiv preprint arXiv:1212.5701*.

PubTermVariants: biomedical term variants and their use for PubMed search

Lana Yeganova, Won Kim, Sun Kim, Rezarta Islamaj Doğan, Wanli Liu,
Donald C Comeau, Zhiyong Lu, W John Wilbur
National Center for Biotechnology Information, NLM, NIH, Bethesda, MD, USA
{yeganova, wonkim, sun.kim, islamaj, liu15,
comeau, luzh, wilbur}@mail.nih.gov

Abstract

Term normalization is frequently used in information retrieval task to reduce variant word forms to a common form. The most general term normalization technique used in practice is stemming, however it has been found to not be completely reliable. Here we present PubTermVariants, a high-quality data-driven resource of term variant pairs that can improve search results in PubMed. For a given pair, we consider two terms to be variants if they stem to the same form, pass the hypergeometric test, and pass the morpho-semantic test. We perform manual evaluation of a subset of PubTermVariants that confirms the high quality of the candidate pairs. We further present experiments that demonstrate their usefulness for PubMed search.

1 Introduction

Information retrieval, and biomedical text processing in general, profoundly depend on sensitive techniques for term normalization. Frequently, the link between a query and a document is not established because they use different forms of a term. These differences may be morphological (related by derivation or inflection) variations of a word, (e.g. autoimmune, autoimmunities, autoimmunity), synonyms (e.g. kidney disease and renal disease), abbreviations, etc. It is to the problem of morphological term variations that we wish to give attention here. Specifically, the goal of this study is to find pairs of string variants that have the same meaning and when used interchangeably benefit PubMed search.

Stemming (Porter 1980) is frequently used for the string normalization task to conflate different forms of a word that have the same meaning. This has been found useful in the task of information retrieval and has been shown to yield small improvements on typical test collections (Hull 1996, Hollink, Kamps et al. 2004, Manning, Raghavan et al. 2009, Moral, de Antonio et al. 2014). Since stemming is not completely reliable, different methods have been applied in an attempt to improve the final results of stemming such as limiting the results to forms found together in a lexicon (Krovetz 1993). This latter approach however is quite limiting and (Xu and Croft 1998) developed a method that uses a mutual information measure of the co-occurrence of two word forms to estimate how related they are and to put them in the same equivalence class if the information is above a threshold. This is done with the aim of improving the equivalence classes of forms with a common stem produced by the original application of the stemmer.

A study of morpho-semantic relationships in Medline (Wilbur and Smith 2013) identifies morphologically related tokens in Medline by using character n-grams as features and then computes the probability that two strings are related based on the context. This approach infers the morphological relatedness of two strings in a way more general than

Proceedings of the 15th Workshop on Biomedical Natural Language Processing, pages 141–145,
Berlin, Germany, August 12, 2016. ©2016 Association for Computational Linguistics

stemming, but based on certain substrings of characters on which they match.

We take what we would describe as a more pragmatic approach. We define two terms to be variants of each other if they can be used interchangeably at the query stage. With the goal to obtain a reliable list of variants, we first find pairs of terms that stem to the same form and have common document in which they appear. We then apply the hypergeometric (HG) test (Larson 1982) to decide whether the observed co-occurrence for a particular pair of terms is above random. For the pairs that pass the HG test, we compute the morpho-semantic similarity score following (Wilbur and Smith 2013) and only retain the pairs that score above the threshold.

When all these conditions hold, we view the two forms as a good candidate term variant pair. We believe this is a less aggressive and clearly safer way to use stemming for query expansion that results in a conservative list of term variants.

In the next section we provide more details on how we generate term variants. We then present the results of manual evaluation of a randomly selected set of candidate term variant pairs. Further we describe two experiments that reveal how PubMed document retrieval is affected when term variants are used. In these experiments we consider both zero-result and nonzero-result queries.

2   Computing Term Variants

Stemming. To begin our processing, we first extracted space-separated tokens that appeared in ten or more PubMed articles. We stemmed every token with the Porter stemmer (Porter 1980) and collected pairs of tokens with the same stem. This process resulted in 201,219 unique pairs of tokens.

The Hypergeometric Test. Here we used the hypergeometric distribution and the p-value test for every pair of words in a group. Let N_s and N_t be the number of documents in Medline that contain terms s and t respectively, let N be the size of Medline, and N_{st} be the number of documents in Medline containing both terms s and t. The random variable Y representing a number of documents containing both terms s and t is a hypergeometric random variable with parameters N_s, N_t and N (Larson 1982) if s is randomly assigned to the N_s documents. The probability distribution of Y is:

$$P(y) = \binom{N_t}{y}\binom{N - N_t}{N_t - y} \Big/ \binom{N}{N_s}.$$

From N_{st} we compute the p-value, i.e. the probability of the observed N_{st} or a higher frequency arising by chance as follows:

$$\text{p-value} = \sum_{y=N_{st}}^{\min(N_s,N_t)} P(y).$$

The p-value reflects the significance of two words co-occurring in N_{st} documents given the frequencies of individual words and the size of the database. A low p-value indicates that the co-occurrence of the two words in not likely to be by chance, but because the words are closely related. By applying the HG test to these pairs at the 0.01 level, we obtain 124,548 word pairs that we will refer to as set StemHG.

The Morpho-Semantic Analysis. Finally, we make use of the study of morpho-semantic relationships in Medline (Wilbur and Smith 2013). For a candidate pair of strings the approach assigns a probability that the strings are semantically related. Using every candidate pair from the above 124,548 pairs, we retain only the pairs for which the probability of being related is 0.9 or higher. This analysis results in a collection of 82,216 term variant pairs that we will refer to as PubTermVariants. There are about 109,725K unique terms in PubTermVariants, since a term may be paired with multiple variants. Because of the HG and the morpho-semantic tests, the relationships between term variants are not transitive.

In the next section we confirm that this collection is of high quality by manual evaluation of a random sample and present experiments designed to demonstrate their usefulness.

3   Experimental Evaluation and Results

We evaluate the quality of PubTermVariants by performing manual evaluation of random pairs sampled from the collection. The manual evaluation reveals the high quality of the variants in the collection. Our further experiments are designed to prove their usefulness for PubMed search.

3.1   Manual Evaluation

Here we report a manual analysis of PubTermVariants with the goal to confirm that two variants are indeed word forms that carry the same meaning and can be safely interchanged in a query.

As mentioned earlier, PubTermVariants is a collection of 82,216 candidate term variant pairs. In addition to PubTermVariants we have 42,332 term pairs in the set StemHG\PubTermVariants that potentially may be enriched in term variants. We believe that the quality of term variants in PubTermVariants is attributable to the effect of independently applying different statistical methods.

We assessed the quality of proposed term variant pairs by manually evaluating 200 random pairs from PubTermVariants, as well as 200 random pairs from StemHG\PubTermVariants. The 400 pairs were shuffled and each pair presented to two annotators. Eight annotators reviewed 100 pairs each, so that each pair was evaluated by two different people. The annotators involved in the manual evaluation all have backgrounds in biomedical information retrieval.

A web-based tool was developed to carry out this evaluation process and this tool was designed to show the term pair, two PubMed abstracts that contained one variant but not the other, and one PubMed abstract that contained both word forms. The annotators were asked to judge whether the two word forms could be used interchangeably. This decision was made by judging the displayed abstracts and deciding whether all of them should be retrieved regardless of which term is being used.

At this round pairs of annotators agreed on 329 of 400 instances considered. All individual evaluations were compared and each pair of annotators met separately to discuss the discrepancies on the remaining 71 pairs. This was later followed by a meeting where all annotators were present, and all remaining cases were discussed. The annotation experiment found that 89% of pairs in PubTermVariants were true variants of the same concept, while only 81.5% of pairs in StemHG\PubTermVariants were true variants, presented in Table 1.

We further examined the quality of term variants as a function of token length, as shown in Figures 1 and 2. We find that tokens of length 3 are typically abbreviations and therefore not good term variant candidates in the absence of context. For example, a pair of terms *ohd* and *ohds* is labeled negative, because, while *ohd/ohds* could be used interchangeably as singular and plural forms of the abbreviation for *"occupational health departments"*, *ohds* may also stand for *"hydroxylase deficiency syndrome"*.

Consequently, of 22 pairs from PubTermVariants that were labeled negative 12 pairs include 3 letter abbreviations. We also observe that the distribution of errors in StemHG\PubTermVariants is more uniform as a function of string length.

3.2 Effect of Term Variants on PubMed Search.

With the goal to understand the usefulness of these term pairs in a real-world setting, we examined the queries in PubMed logs and performed the following analyses:

1. We analyzed zero-result queries and identified real user queries that could have returned results by using PubTermVariants.
2. We analyzed a subset of result-producing queries and identified the difference in the result set had PubTermVariants been used.

	PubTerm-Variants	StemHG\Pub-TermVariants	Total
Positives	178(89%)	163(81.5%)	341
Negatives	22(11%)	37(18.5%)	59
Total	200	200	400

Table 1. Results of manual annotation of 200 random pairs from PubTermVariants and 200 pairs from StemHG\PubTermVariants. 89% of pairs in PubTermVariants and 81.5% of pairs in StemHG\PubTermVariants were found to be true variants.

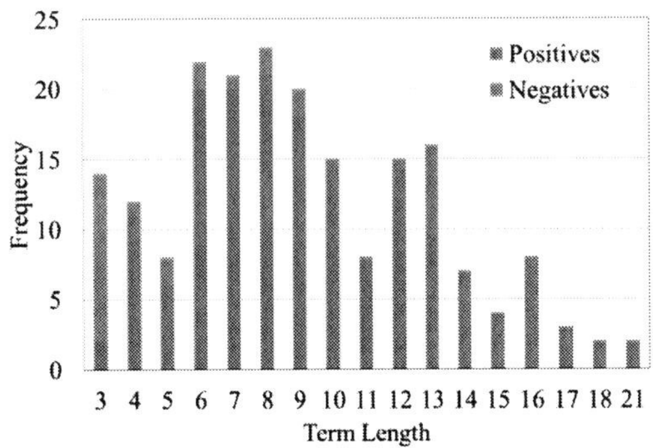

Figure 1. The quality of term variants as a function of token length in PubTermVariants.

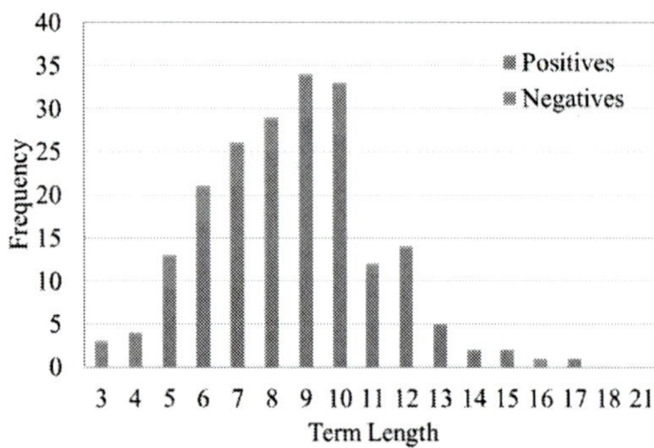

Figure 2. The quality of term variants as a function of token length in StemHG\PubTermVariants.

For these experiments we collected PubMed log data for 2015. We find that about 76% of PubMed queries contain a term from PubTermVariants.

Effect of using PubTermVariants on zero-result queries

Some PubMed queries do not produce a result. We call these queries zero-result queries. We ask whether using PubTermVariants could lead to results being retrieved for these queries.

In order to answer this question we chose a random day of PubMed logs in 2015. We preprocessed the data and kept only multi-term queries that contained alphanumeric characters, dashes and commas. We also removed queries that did not contain a term from PubTermVariants. For each query, we verified that if a term variant is removed, the query consisting of the remaining tokens retrieved a set of PubMed articles. This set contained 55,496 unique queries.

For each of these queries, we replaced a term with its variant from the PubTermVariants list. Since some terms have multiple variants, this process resulted in 110,103 queries which are now used to query PubMed and report results. We call the use of the term variants successful if at least one of the variant queries resulted in a successful search. For example, term *monoubiquitination* is mapped in our collection to terms *monoubiquitin*, *monoubiquitinate*, *monoubiquitinated*, and *monoubiquitinates*. For a given query *hdm2 monoubiquitination*, one substitution to *hdm2 monoubiquitinated* is found to be successful, and so we call that substitution successful. Articles were retrieved for 8.83% (4,902

queries) of the original 55,496 queries. This percentage, however, represents a lower bound of success because for every query only one term was considered for replacement. Queries can have several candidate terms for replacement.

Effect of using PubTermVariants on result-producing queries

Since PubMed is quite successful at producing relevant results for most queries, we wanted to examine the effect of the variants in PubTermVariants on these searches. For this experiment we randomly selected 1,000 user queries that contained a term listed in PubTermVariants and where the number of PubMed results for each of these queries was between 1 and 20. For each of these queries we produced only one variant query so that the original term variant was replaced with one of its paired variants in the PubTermVariants list, randomly selected. The resulting variant queries were used to query PubMed and we retrieved results for 480 of these queries. We compared the results set for the original user queries with their variant queries and found that the average number of results for the original queries was 6.8, however, if we combine the results with the results of the variant queries this number increases to 8.5. Furthermore, 38% variant queries retrieve additional relevant PubMed articles without overwhelming the search results. Similar to the zero result case, this percentage represents a lower bound.

4   Conclusions

We presented a high-quality list of biomedical term variants which we call PubTermVariants. The PubTermVariants resource is generated in a data-driven way by applying two statistical tests to pairs of tokens that stem to the same form. Both, the hypergeometric and the morpho-semantic tests, provide a useful tool for deciding whether terms in the pair are related or not.

PubTermVariants provides a clean and reliable high-quality collection of terms that can be used interchangeably in PubMed queries. The manual examination revealed that 89% of the pairs are true variants, and removing three letter tokens results in higher quality. Our experiments on PubMed log data demonstrated that some zero-result queries that contain a term variant can return results by applying a substitution from PubTermVariants. Our other experiments revealed that when a term variant is applied to create a variant query, in 38% of the cases the result set was enriched with articles which were not present in the initial request, thus increasing recall.

PubTermVariants is available for other applications of biomedical term variants from ftp://ftp.ncbi.nlm.nih.gov/pub/wilbur/PubTermVariants/pairs.txt.gz.

5 Acknowledgements

The authors thank Grigory Balasanov for his help in preparing the web-based tool for the manual annotation task.

Funding: This research was supported by the Intramural Research Program of the NIH, National Library of Medicine.

References

Church, K. and P. Hanks (1989). Word association norms, mutual information, and lexicography. Proceedings of the 27th ACL Meeting: 76-83.

Hollink, V., et al. (2004). Monolingual Document Retrieval for European Languages. Information Retrieval 7(1).

Hull, D. (1996). Stemming Algorithms - A Case Study for Detailed Evaluation. Journal of the American Society for Information Science 47(1): 70-84.

Krovetz, R. (1993). Viewing Morphology as an Inference Process. Proceedings of the 16th annual international ACM SIGIR conference on Research and development in information retrieval.

Larson, H. J. (1982). Introduction to Probability Theory and Statistical Inference. New York, John Wiley & Sons.

Manning, C., et al. (2009). Introduction to Information Retrieval. Cambridge, England, Cambridge University Press.

Mikolov, T., et al. (2013). Distributed representations of words and phrases and their compositionality. Advances in Neural Information Processing Systems.

Moral, C., et al. (2014). A survey of stemming algorithms in information retrieval. Information Research 19(1).

Porter, M. F. (1980). An algorithm for suffix stripping. Program 14(3): 130-137.

Wilbur, W. J. and L. Smith (2013). A Study of the Morpho-Semantic Relationship in Medline. Open Inf Syst J 6.

Xu, J. and W. B. Croft (1998). Corpus-based stemming using cooccurrence of word variants. ACM Transactions on Information Systems (TOIS) 16(1): 61-81.

This before That:
Causal Precedence in the Biomedical Domain

Gus Hahn-Powell   Dane Bell   Marco A. Valenzuela-Escárcega   Mihai Surdeanu

University of Arizona
Tucson, AZ 85721, USA
`hahnpowell@email.arizona.edu`

Abstract

Causal precedence between biochemical interactions is crucial in the biomedical domain, because it transforms collections of individual interactions, e.g., bindings and phosphorylations, into the causal mechanisms needed to inform meaningful search and inference. Here, we analyze *causal* precedence in the biomedical domain as distinct from open-domain, *temporal* precedence. First, we describe a novel, hand-annotated text corpus of causal precedence in the biomedical domain. Second, we use this corpus to investigate a battery of models of precedence, covering rule-based, feature-based, and latent representation models. The highest-performing individual model achieved a micro F1 of 43 points, approaching the best performers on the simpler temporal-only precedence tasks. Feature-based and latent representation models each outperform the rule-based models, but their performance is complementary to one another. We apply a sieve-based architecture to capitalize on this lack of overlap, achieving a micro F1 score of 46 points.

1   Introduction

In the biomedical domain, an enormous amount of information about protein, gene, and drug interactions appears in the form of natural language across millions of academic papers. There is a tremendous ongoing effort (Nédellec et al., 2013; Kim et al., 2012; Kim et al., 2009) to extract individual chemical interactions from these texts, but these interactions are only isolated fragments of larger causal mechanisms such as protein signaling pathways. Nowhere, however, including any

database, is the complete mechanism described in a form that lends itself to causal search or inference. The absence of such a database is not for lack of trying; Pathway Commons (Cerami et al., 2011) aims to address the need, but its authors estimate it currently covers 1% of the literature due to the high cost of annotation[1]. This issue only grows more pressing with the yearly growth in biomedical publishing, which presents an otherwise insurmountable challenge for biomedical researchers to query and interpret.

The Big Mechanism program (Cohen, 2015) aims to construct exactly such large-scale mechanistic information by reading and assembling protein signaling pathways that are relevant for cancer, and exploit them to generate novel explanatory and treatment hypotheses. Although prior work (Chambers et al., 2014; Mirza, 2016) has addressed the challenging area of temporal precedence in the open domain, the biomedical domain presents very different data and, consequently, requires novel techniques. Precedence in mechanistic biology is *causal* rather than *temporal*. Though event temporality is crucial to understanding electronic health records for individual patients (Bethard et al., 2015; Bethard et al., 2016), its contribution to the understanding of biomolecular reactions is less clear as these events and processes may repeat in extremely short cycles, continue without end, or overlap in time. At any level of abstraction, causal precedence encodes mechanistic information and facilitates inference over spotty evidence. For the purpose of this work, *precedence* is defined for two events, A and B, as

A precedes B if and only if the output of A is necessary for the successful execution of B.[2]

---

[1]Personal communication.

[2]See the "precedes" examples in Table 1.

Proceedings of the 15th Workshop on Biomedical Natural Language Processing, pages 146–155,
Berlin, Germany, August 12, 2016. ©2016 Association for Computational Linguistics

Very little annotated data exists for causal precedence, especially efforts focusing on signaling pathways. BioCause (Mihăilă et al., 2013), for instance, is centered on connections between claims and evidence and contains only 51 annotated examples of causal precedence[3]. Our work[4] offers three contributions in aid of automatically extracting causal ordering in biomedical text. First, we provide and describe a dataset of real text examples, manually annotated for causal precedence. Second, we analyze the efficacy of a battery of different models in automatically determining precedence, built on top of the Reach automatic reading system (Valenzuela-Escárcega et al., 2015a; Valenzuela-Escárcega et al., 2015c) and measured against this novel corpus. In particular, we investigate three classes of models: (a) deterministic rule-based models inspired by the precedence sieves proposed by Chambers et al. (2014), (b) feature-based models, and (c) models that rely on latent representations such as long short-term memory (LSTM) networks (Hochreiter and Schmidhuber, 1997). Our analysis indicates that while independently the top-performing model achieves a micro F1 of 43, these models are largely complementary with a combined recall of 58 points. Lastly, we conduct an error analysis of these models to motivate and inform future research.

2 A Corpus for Causal Precedence in the Biomedical Domain

Our corpus annotates several types of relations between mentions of biochemical interactions. Following common terminology promoted by the BioNLP shared tasks, we will interchangeably use "events" to refer to these interactions. To generate candidate events for our planned annotations, we ran the Reach event extraction system (Valenzuela-Escárcega et al., 2015a; Valenzuela-Escárcega et al., 2015c) over the full text[5] of 500 biomedical papers taken from the

Relation	Example
E1 precedes E2	A is phosphorylated by B. Following its phosphorylation, A binds with C.
E2 precedes E1	A is phosphorylated by B. Prior to its phosphorylation, A binds with D.
Equivalent	The phosphorylation of A by B. A is phosphorylated by B.
E1 specifies E2	A is phosphorylated by B at Site 123. A is phosphorylated by B.
E2 specifies E1	A is phosphorylated by B. A is phosphorylated by B at Site 123.
Other	B does not regulate C when C is bound to A.
None	A phosphorylates B. A ubiquitinates C.

Table 1: The seven inter-event relation labels annotated in the corpus. The "precedes" labels are causal. Subsumption is captured with the "specifies" labels.

Open Access subset of PubMed[6]. The events extracted by Reach are biochemical events of two types: simple events such as phosphorylation that modify one or more entities (typically proteins), and nested events (regulations) that have other events as arguments.

To improve the likelihood of finding pairs of events with a relevant link, we filtered event pairs by imposing the following requirements for inclusion in the corpus:

1. *Event pairs must share at least one participant.* This constraint is based on the observation that interactions that share participants are more likely to be connected.

2. *Event pairs must be within 1 sentence of each other.* Similarly, discourse proximity increases the likelihood of two events being related.

3. *Event pairs must not share the same type.* This helps to maximize the diversity of the dataset.

4. *Event pairs must not already be contained in an extracted Regulation event.* For example, we did not annotate the relation between the binding and the phosphorylation events in "The binding of X and Y is inhibited

---

[3]These are marked in the BioCause corpus as `Causality` events with `Cause` and `Effect` arguments. The remaining 800 annotations are claim-evidence relations.

[4]The corpus, tools, and system introduced in this work are publicly available at `https://github.com/myedibleenso/this-before-that`

[5]We chose to ignore the "references", "materials", and "methods" sections, which generally do not contain mechanistic information.

[6]`http://www.ncbi.nlm.nih.gov/pmc/tools/openftlist/`

by X phosphorylation", because it is already captured by most state-of-the-art biomedical event extraction systems.

After applying these constraints, only 1700 event pairs remained. In order to rapidly annotate the event pairs, we developed a browser-based annotation UI that is completely client-side (see Figure 3). Using this tool, we annotated 1000 event pairs for this work; 84 of these were discarded due to severe extraction errors. The annotations include the event spans, event triggers (i.e., the verbal or nominal predicates that indicate the type of interaction such as "binding" or "phosphorylated"), source document, minimal sentential span encompassing both event mentions, and whether or not the event pair involves coreference for either the event trigger or the event participants. For events requiring coreference resolution, we expanded the encompassing span of text to also capture the antecedent. Note that domain-specific coreference resolution is a component of the event extraction system used here (Bell et al., 2016).

When describing the relations between these event pairs, we refer to the event that occurs first in text as Event 1 (E1) and the event that follows as Event 2 (E2). Each (E1, E2) pair was assigned one of seven labels: "E1 precedes E2", "E2 precedes E1", "Equivalent", "E1 specifies E2", "E2 specifies E1", "Other", or "None". Table 1 provides examples for each of these labels. We converged on these labels because they are fundamental to the assembly of causal mechanisms from a collection of events. Collectively, the seven labels address three important assembly tasks: *equivalence*, i.e., understanding that two event mentions discuss the same event, *subsumption*, i.e., the two mentions discuss the same event, but one is more specific than the other, and, most importantly, *causal precedence*, the identification of which is the focus of this work. During the annotation process, we came across examples of other relevant phenomena. We grouped these instances under the label "Other" and leave their analysis for future work.

Though simplified, the examples in Table 3 illustrate that this is a complex task sensitive to linguistic evidence. For example, the direction of the precedence relations in the first two rows in the table changes based on a single word in the context ("prior" vs. "following").

In terms of the distribution of relations, causal

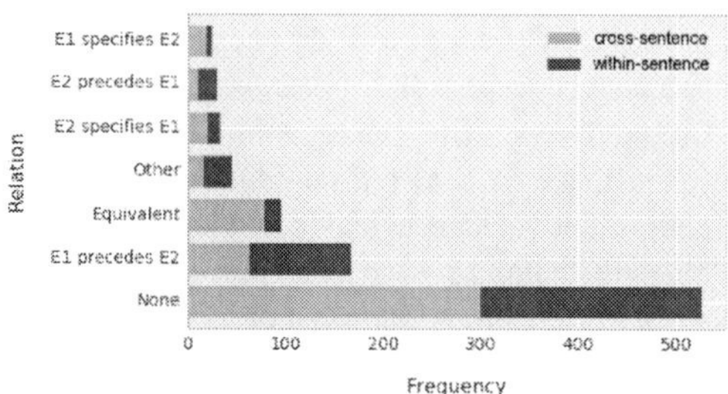

Figure 1: The distribution of assembly relation labels both within and across sentences.

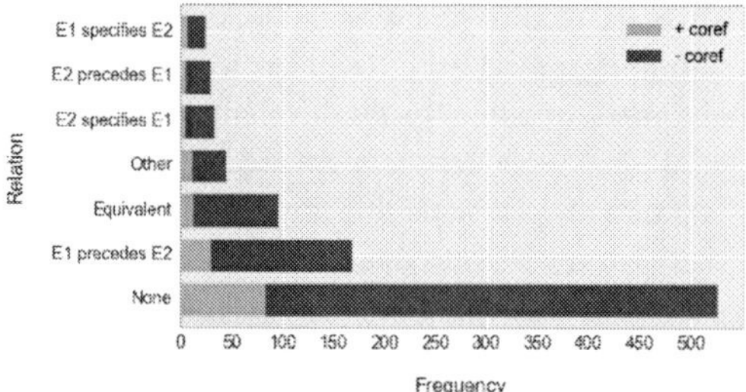

Figure 2: The distribution of event pairs involving coreference across assembly relations.

precedence pairs appear more frequently within the same sentence, while cases of the subsumption ("specifies") and equivalence relations are far more common across sentences (see Figure 1). Coreference is involved in 10–15% of the instances for each relation label (see Figure 2).

The annotation process was performed by two linguists familiar with the biomedical domain. To minimize errors, the annotation task was initially performed together at the same workstation.[7] On a randomly selected sample of 100 event pairs, the two annotators had a Cohen's kappa score (Cohen, 1960) of 0.82, indicating "almost perfect" agreement for the *precedes* labels (Landis and Koch, 1977).

3 Models of Causal Precedence

We have developed both deterministic, interpretable models and automatic, machine-learning models for detecting causal precedence in our dataset. Importantly, the models covered in this work focus solely on causal precedence, which is the most complex relation annotated in the dataset previously introduced. Thus, for all experiments discussed here, we reduce these annotations to three labels: "E1 precedes E2", "E2 precedes E1", and Nil, which covers all the other labels in the

---

[7] Similar to pair programming.

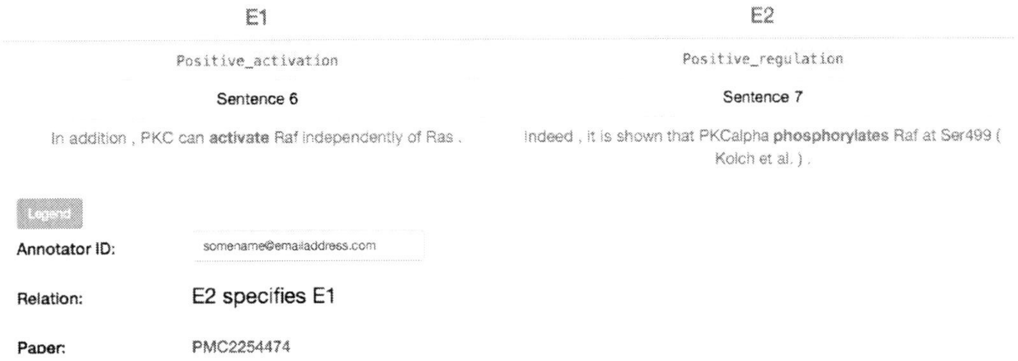

Figure 3: Browser-based tool for annotating assembly relations in text. An annotation instance consists of a pair of event mentions. The annotator assigns a label to each pair of events using the number keys and navigates from annotation to annotation using the arrow keys. E1 refers to the event in the pair that appears first in the text. The event span is formatted to stand out from the surrounding text. The "Paper" field provides the annotator with easy access to the full text of the source document for the current annotation instance. Annotations can be exported to JSON and reloaded via a local storage cache or through file upload.

corpus.

Model	Rules
Intra-sentence	29
Inter-sentence	5
Reichenbach	8

Table 2: Few rules defined each deterministic model of precedence compared with the number of features for the machine learning models.

3.1 Deterministic Models

The deterministic models are defined by a small number of hand-written rules using the Odin event extraction framework (Valenzuela-Escárcega et al., 2015b). The number of rules for each model is shown in Table 2, and sharply contrast with the 92,711 features introduced later (Table 3) that are used by our machine-learning models. In order to avoid overfitting, all of the deterministic models were created without reference to the annotation corpus, using general linguistic expertise and domain knowledge.

Intra-sentence ordering Within sentences, syntactic regularities can be exploited to cover a large variety of grammatical constructions indicating precedence relations. Rules defined over dependency parses (De Marneffe and Manning, 2008) capture precedence in sentences like those in (1) and (2) as well as many others.

(1) [The RBD of PI3KC2B binds HRAS]$_{after}$, when [HRAS is not bound to GTP]$_{before}$

(2) [The ubiquitination of A]$_{before}$ is followed by [the phosphorylation of B]$_{after}$

Other phrases captured include: "precedes", "due to", "leads to", "results in", etc.

Inter-sentence ordering Although syntax operates over single sentences, cross-sentence time expressions can indicate ordering, as shown in Examples (3) and (4). We exploit these regularities as well by checking for sentence-initial word combinations.

(3) [A is phosphorylated by B]$_{before}$. As a downstream effect, [C is ...]$_{after}$

(4) [A is phosphorylated by B]$_{before}$. [C is then ...]$_{after}$

Other phrases captured include: "Later", "In response", "For this", and "Ultimately".

Verbal tense- and aspect-based (Reichenbach) ordering Following Chambers et al. (2014), we use deterministic rules to establish precedence between events that have certain verbal tense and aspect. These rules are derived from linguistic analysis of tense and aspect by (Reichenbach, 1947; Derczynski and Gaizauskas, 2013). Example (5) illustrates a case in which we can accurately infer order just from this information. Because *has been phosphorylated* has past tense and perfective aspect, this model concludes that it precedes *share* (present tense, simple aspect) and thus the binding of histone H2A.

> (5) These [PTIP] proteins also share the ability to bind histone H2A (or H2AX in mammals) that has been phosphorylated....

The logic determining which tense-aspect combinations receive which precedence relations is identical to CAEVO, which is possible because it is open source[8]. However, CAEVO operates over annotations that include gold tense and aspect values, whereas this model additionally detects tense and aspect using Odin rules before applying this logic.

3.2 Feature-based Models

Most instances of causal precedence cannot be captured with deterministic rules, because they lack explicit words, phrases, or syntactic structures that unambiguously mark the relation. Using a combination of the surface, syntactic, and taxonomic features outlined in Table 3, we trained a set of statistical classifiers to detect causal precedence relations between pairs of events in our corpus. For training and testing purposes, we treated any instance not labeled as either "E1 precedes E2" or "E2 precedes E1" as a negative example. We examined the following statistical models: a linear kernel SVM (Chang and Lin, 2011), logistic regression (Fan et al., 2008), and random forest[9] (Surdeanu et al., 2014). For the SVM and logistic regression (LR) models, we also compared the effects of L1 and L2 regularization.

3.3 Latent Representation Models

Due to the complexity of the task and variety of causal precedence instances encountered during the annotation process, it is unclear whether a linear combination of engineered features is sufficient for broad coverage classification. For this reason, we introduce a latent feature representation model using an LSTM (Hochreiter and Schmidhuber, 1997; Bergstra et al., 2010; Chollet, 2015) to capture underlying semantic features by incorporating long-distance contextual information and selectively persisting memory of previous event pairs to aid in classification.

The basic architecture is shown in Figure 5. The input to this model is the provenance of the relation, i.e., the whole text containing the two events and the text in between. Formally, this is represented as a concatenated sequence of 200 dimensional vectors where each vector in the sequence corresponds to a token in the minimal sentential span encompassing the event pair being classified. Intuitively, this LSTM "reads" the text from left to right and outputs a classification label from the set of three when done. We consider two variations of this model: the basic model (LSTM) with the vector weights for each token uninitialized and a second form (LSTM+P) where the vectors are initialized using pre-training. In the pre-training configuration, the vector weights are initialized using word embeddings generated by a word2vec (Mikolov et al., 2013; Řehůřek and Sojka, 2010) model trained on the full text of over 1 million biomedical papers taken from the Open Access subset of PubMed. Because the corpus is only 1000 annotations, it was thought that pre-training could improve prediction of causal precedence and guide the model with distributional semantic representations specific to this domain.

Building on this simple blueprint, we designed a three-pronged "pitchfork" (FLSTM) where the span of E1, the span of E2, and the minimal sentential span encompassing E1 and E2 each serve as a separate input, allowing the model to explicitly address each of them as well as discover how these three inputs relate to one another. This architecture is shown in Figure 6. Each input feeds into its own LSTM and corresponding dropout layer before the three forks are merged via a concatenation of tensors. Like the basic model, one version of the "pitchfork" is trained with vector weights initialized using the pre-trained word embeddings (FLSTM+P).

4 Results

We summarize the performance of all these models on the dataset previously introduced in Table 4. We report results using micro precision, recall, and F1 scores for each model. With fewer than 200 instances of causal precedence occurring in 1000 annotations, training and testing for both the feature-based classifiers and latent feature models was performed using stratified 10-fold cross validation. For the latent feature models, training was parameterized using a maximum of 100

---

[8] https://github.com/nchambers/caevo
[9] Abbreviated as RF

	Feature	Description
Event	Event labels	The taxonomic labels Reach assigned to the event (e.g. *phosphorylation* $\rightarrow$ Phosphorylation, AdditiveEvent, …).
	Event trigger	The predicate signaling an event mention (ex. "phosphorylated", "phosphorylation").
	Event trigger + label	A concatenation of the event's trigger with the event's label.
	token *n*-grams with entity replacement	*n*-grams of the tokens in the mention span, where each entity is replaced with the entity label (ex. "the ABC protein" $\rightarrow$ "the PROTEIN"). If an entity is shared between pairs of events, replace it with the label SHARED.
	token *n*-grams with role replacement	*n*-grams of the tokens in the mention span, where each argument is replaced with the argument role (ex. "A inhibits the phosphorylation of B" $\rightarrow$ "CONTROLLER inhibits the CONTROLLED")
	Syntactic path from trigger to args	Variations of the syntactic dependency path from an event's trigger to each of its arguments (unlexicalized path, path + lemmas, trigger $\rightarrow$ argument role, trigger $\rightarrow$ argument label, etc.).
Event-Event (surface)	Interceding tokens (*n*-grams)	*n*-grams (1-3) of the tokens between E1 and E2.
Event-Event (syntax)	Cross-sentence syntactic paths	A concatenation of the syntactic path from the sentential ROOT to an event's trigger (see the example in Figure 4).
	Trigger-to-trigger syntactic paths (within sentence)	the syntactic path from the trigger of E1 to the trigger of E2
	Shortest syntactic paths	The shortest syntactic path between E1 and E2 (restricted to intra-sentence cases).
	Syntactic distance	The length of each syntactic path (restricted to intra-sentence cases).
Coreference	*Event* features for anaphors	Whether or not an event mention is resolved through coreference. For cases of coreference, generate the *Event* features prefixed with "coref-anaphor" for the text labeled "E1-anaphor" in the following example: (6) [A binds with B]$_{\text{E1-antecedent}}$ (7) [This interaction]$_{\text{E1-anaphor}}$ precedes the [phosphorylation of C]$_{\text{E2}}$
	Resolved arguments	Which arguments, if any, were resolved through coreference. For example: [The mutant$_{\text{THEME}}$ binds with B$_{\text{THEME}}$]$_{\text{E1}}$ $\rightarrow$ `THEME:resolved`

Table 3: An overview of the primary features used in the feature-based classifier, grouped into four classes: *Event* – features extracted from the two participating events, in isolation; *Event-Event (surface)* – features that model the lexical context between the two events; *Event-Event (syntax)* – features that model the syntactic context between the two events; and *Coreference* – features that capture coreference resolution information that impact the participating events.

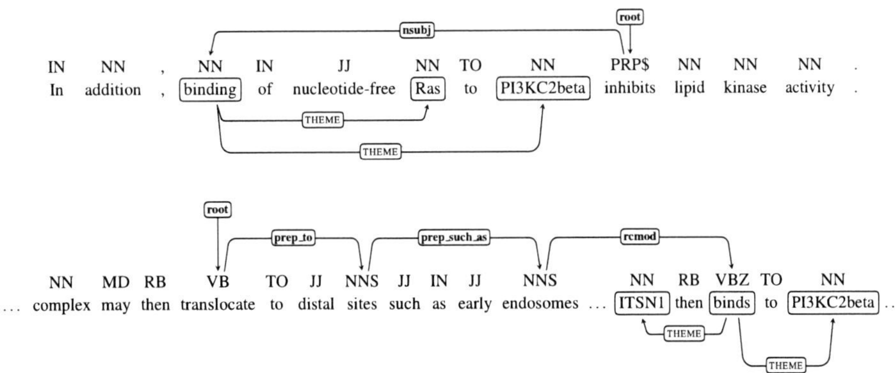

Figure 4: Generation procedure for the cross-sentence syntactic path feature. For each event in a pair, we find the shortest syntactic path originating from the sentential root node leading to a token in the event's trigger. The two syntactic paths are then joined using the + symbol to form a single feature.

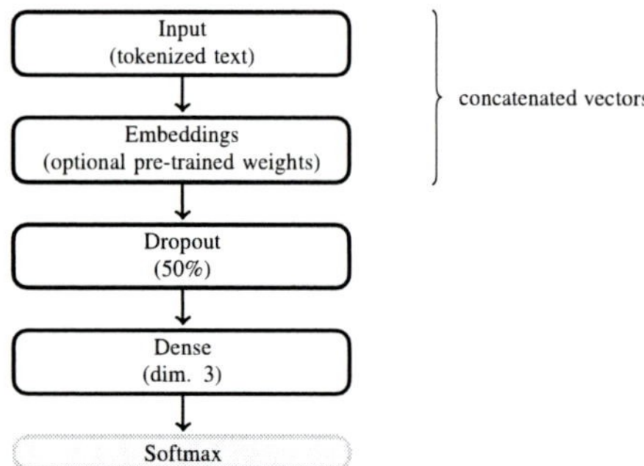

Figure 5: Architecture for the basic latent feature model using the minimal sentential span encompassing events 1 and 2 as input.

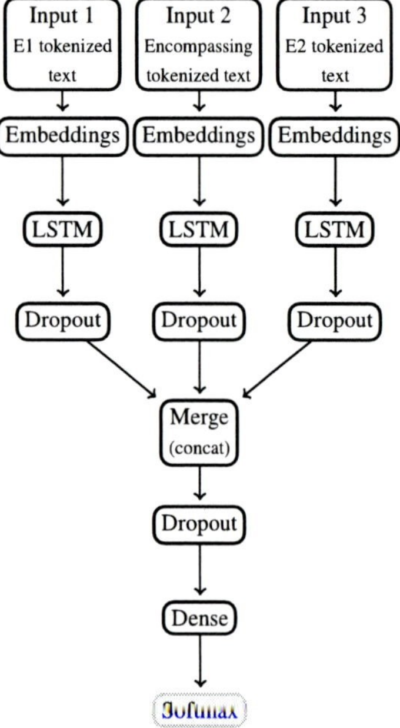

Figure 6: Modified architecture for a latent feature model with three-pronged input: the text of event 1 (left), the minimal sentential span encompassing events 1 and 2 (middle), and the text of event 2 (right).

epochs with support for early stopping through monitoring of validation loss[10]. Weight updates were made on batches of 32 examples and all folds completed in fewer than 50 epochs.

The table also includes a sieve-based ensemble system, which performs significantly better than the best-performing single model. In this architecture, the sieves are applied in descending order

---

[10]The validation set used for each fold came from a different class-balanced fold.

of precision, so that the positive predictions of the higher precision sieves will always be preferred to contradictory predictions made by subsequent, lower-precision sieves. Figure 7 illustrates that as sieves are added, the F1 score remains fairly constant, while recall increases at the cost of precision.

Model	p	r	$f1$
Intra-sentence	0.5	0.01	0.01
Inter-sentence	0.5	0.01	0.01
Reichenbach	0	0	0
LR+L1	0.58	0.32	0.41
LR+L2	0.65	0.26	0.37
SVM+L1	0.54	0.35	**0.43**
SVM+L2	0.54	0.29	0.38
RF	0.62	0.25	0.36
LSTM	0.40	0.25	0.31
LSTM+P	0.39	0.20	0.26
FLSTM	0.43	0.15	0.22
FLSTM+P	0.38	0.22	0.28
Combined	0.38	0.58	**0.46***

Table 4: Results of all proposed causal models, using stratified 10-fold cross-validation. The combined system is a sieve-based architecture that applies the models in decreasing order of their precision. The combined system significantly outperforms the best single model, SVM with L1 regularization, according to a bootstrap resampling test ($p = 0.022$).

Despite some obvious patterns noted in Table 1, the deterministic models perform the worst due in large part to their rarity in the corpus. An analysis of this result is given in Section 5. Overall, our top-performing model was the linear kernel SVM with L1 regularization. In all cases, the feature-based classifiers outperform the latent feature representations, suggesting that in cases such as this where little data is available, feature-based classifiers capitalizing on high-level linguistic features are able to better generalize and thus outperform latent feature models. However, as our discussion in Section 5.1 will show, our combined model demonstrates that the latent and feature-based models are largely complementary.

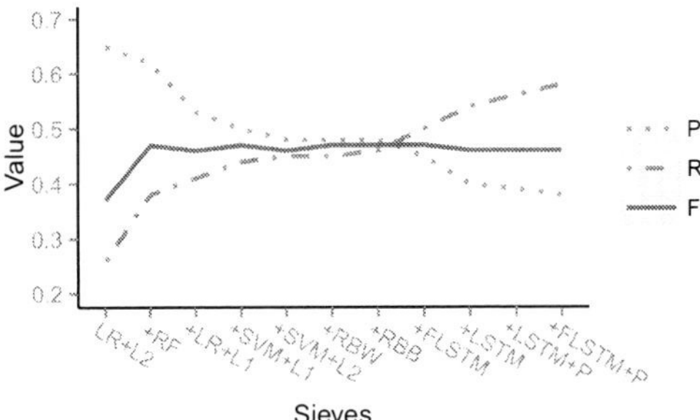

Figure 7: The performance of the sieve-based combined model varies with each model added.

5 Discussion

Overall, results are promising, particularly in light of the conscious choice to omit (causal) regulation reactions from this task, as they are already captured by the Reach reading system.

However, the deterministic models created so far have extremely low recall, such that it is difficult even to determine their precision. An analysis of the Reichenbach model reveals one source of this low coverage. In short, although writers *could* describe causal mechanisms using temporal indicators such as tense and aspect, temporal description is rare enough in this domain not to be represented in our randomly sampled database. Table 5 illustrates the lack of overlap with informative tense-aspect combinations; a single tense is used per passage, and no perfective aspect is used.

E1↓, E2→		past		pres.		fut.	
		simple	perf.	simple	perf.	simple	perf.
past	simple	69	0	38	0	0	0
	perf.	0	0	0	0	0	0
pres.	simple	49	0	134	0	1	0
	perf.	0	0	0	0	0	0
fut.	simple	0	0	0	0	0	0
	perf.	0	0	0	0	0	0

Table 5: Event tense and aspect for events containing verbs in the present study. Highlighted cells are tense-aspect combinations that are informative for establishing temporal precedence, following Chambers et al. (2014). All but one event pair fall outside of these informative combinations, and that exceptional pair was a false positive case.

Similarly, the time expressions required by the deterministic intra- and inter-sentence precedence rules are rare enough to make them ineffective on this sample.

5.1 Model overlap

As Chambers et al. (2014), Mirza (2016), and many other algorithms have shown, models can be applied sequentially in "sieves" to produce higher-quality output. Ideally, each model in a sieve-based system will capture different portions of the data through a mixture of approaches, distinguishing this method from more naive ensembles in which the contributions of a lone component would be washed out. Figure 8 details this observation by showing the coverage difference between the models described here.

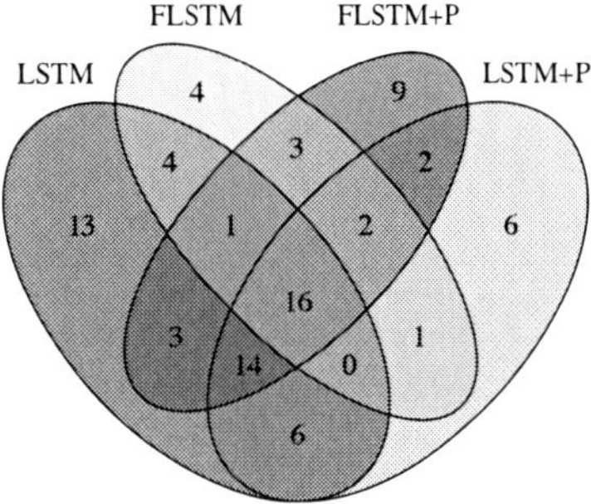

(a) Overlap of true positive predictions made by LSTM models. Though in Table 4 the models appear to perform similarly, the learned representations are largely distinct and complementary in their coverage.

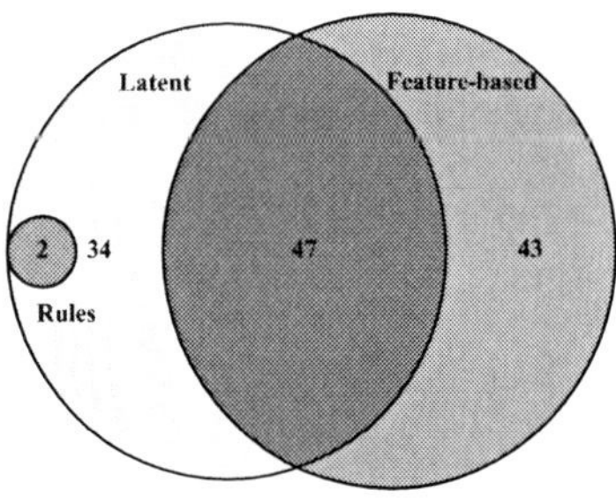

(b) Similarly, the overlap between the feature-based models and the latent models was low overall.

Figure 8: The overlap of true positives among the investigated models was low.

5.2 Error analysis

We performed an analysis of the false positives shared by all feature-based classifiers, in addition to the false negatives shared by all models. Here we limit our discussion to only the most prominent characteristic shared by the majority of false positives.

Discourse information More than half of the false positives share contrastive discourse features, suggesting that a model of discourse could improve classifier discrimination. Example (8) demonstrates such a contrastive structure, which *whereas* introduces a clause (and event) that is contrasted and therefore both temporally and causally distinct from the following clause (and event). The existence of regular cues like *whereas* indicates that a feature to explicitly model these structures is possible.

(8) Whereas [PRAS40 inhibits the mTORC1 activity via raptor]$_{E1}$, DEPTOR was identified to interact directly with mTOR in both [mTORC1 and mTORC2 complexes]$_{E2}$

6 Related Work

Though focused on temporal ordering, Chambers et al. (2014) adopt a sieve-based approach, with high-precision deterministic sieves preceding and constraining lower-precision, higher-recall machine learning sieves. As with our system, the deterministic sieves were linguistically motivated, and had the additional advantage of operating over time expressions (*during*, *Friday*, etc.) as well as events, the former of which are typically lacking in the biomedical domain.

Mirza (2016) implemented a hybrid sieve-based approach for causal relation detection between events that includes a set of causal verb rules and corresponding syntactic dependencies and a feature based classifier. However, both of these works focus on open-domain texts. To our knowledge, we are the first to investigate causal precedence in the biomedical domain.

7 Conclusion

These are the first experiments regarding automatic annotation of causal precedence in the biomedical domain. Although the dearth of temporal expressions and other regular linguistic cues make the task especially difficult in this domain, the initial results are promising, and demonstrate that a sieve-based system of the models tested here improves performance over the top-performing individual component. Both the annotation corpus and the models described here represent large steps toward linking automatic reading to a larger, more informative biological mechanism.

Acknowledgments

This work was funded by the Defense Advanced Research Projects Agency (DARPA) Big Mechanism program under ARO contract W911NF-14-1-0395.

References

Dane Bell, Gus Hahn-Powell, Marco A. Valenzuela-Escárcega, and Mihai Surdeanu. 2016. An investigation of coreference phenomena in the biomedical domain. In *Proceedings of the 10th International Conference on Language Resources and Evaluation. LREC 2016*. Paper available at http://arxiv.org/abs/1603.03758.

James Bergstra, Olivier Breuleux, Frédéric Bastien, Pascal Lamblin, Razvan Pascanu, Guillaume Desjardins, Joseph Turian, David Warde-Farley, and Yoshua Bengio. 2010. Theano: a CPU and GPU math expression compiler. In *Proceedings of the Python for Scientific Computing Conference (SciPy)*, June. Oral Presentation.

Steven Bethard, Leon Derczynski, Guergana Savova, Guergana Savova, James Pustejovsky, and Marc Verhagen. 2015. Semeval-2015 task 6: Clinical tempeval. In *Proceedings of the 9th International Workshop on Semantic Evaluation (SemEval 2015)*, pages 806–814.

Steven Bethard, Guergana Savova, Wei-Te Chen, Leon Derczynski, James Pustejovsky, and Marc Verhagen. 2016. Semeval-2016 task 12: Clinical tempeval. In *Proceedings of the 10th International Workshop on Semantic Evaluation (SemEval 2016), San Diego, California, June. Association for Computational Linguistics*.

Ethan G Cerami, Benjamin E Gross, Emek Demir, Igor Rodchenkov, Özgün Babur, Nadia Anwar, Nikolaus Schultz, Gary D Bader, and Chris Sander. 2011. Pathway commons, a web resource for biological pathway data. *Nucleic acids research*, 39(suppl 1):D685–D690.

Nathanael Chambers, Taylor Cassidy, Bill McDowell, and Steven Bethard. 2014. Dense event ordering with a multi-pass architecture. *Transactions of the Association for Computational Linguistics*, 2:273–284.

Chih-Chung Chang and Chih-Jen Lin. 2011. LIBSVM: A library for support vector machines. *ACM Transactions on Intelligent Systems and Technology*, 2:27:1–27:27. Software available at http://www.csie.ntu.edu.tw/~cjlin/libsvm.

Franois Chollet. 2015. keras. https://github.com/fchollet/keras.

Jacob Cohen. 1960. A coefficient of agreement for nominal scale. *Educational and Psychosocial Measurement*, 20:37–46.

Paul R. Cohen. 2015. DARPA's Big Mechanism program. *Physical Biology*, 12(4):045008.

Marie-Catherine De Marneffe and Christopher D Manning. 2008. Stanford typed dependencies manual. Technical report, Stanford University.

Leon Derczynski and Robert Gaizauskas. 2013. Empirical validation of Reichenbach's tense framework. In *Proceedings of the 10th International Conference on Computational Semantics*, pages 71–82.

Rong-En Fan, Kai-Wei Chang, Cho-Jui Hsieh, Xiang-Rui Wang, and Chih-Jen Lin. 2008. Liblinear: A library for large linear classification. *J. Mach. Learn. Res.*, 9:1871–1874, June.

Sepp Hochreiter and Jürgen Schmidhuber. 1997. Long short-term memory. *Neural computation*, 9(8):1735–1780.

Jin-Dong Kim, Tomoko Ohta, Sampo Pyysalo, Yoshinobu Kano, and Jun'ichi Tsujii. 2009. Overview of bionlp'09 shared task on event extraction. In *Proceedings of the Workshop on Current Trends in Biomedical Natural Language Processing: Shared Task*, pages 1–9. Association for Computational Linguistics.

Jin-Dong Kim, Ngan Nguyen, Yue Wang, Junichi Tsujii, Toshihisa Takagi, and Akinori Yonezawa. 2012. The genia event and protein coreference tasks of the bionlp shared task 2011. *BMC bioinformatics*, 13(11):1.

J Richard Landis and Gary G Koch. 1977. The measurement of observer agreement for categorical data. *biometrics*, pages 159–174.

Claudiu Mihăilă, Tomoko Ohta, Sampo Pyysalo, and Sophia Ananiadou. 2013. Biocause: Annotating and analysing causality in the biomedical domain. *BMC Bioinformatics*, 14(1):1–18.

Tomas Mikolov, Kai Chen, Greg Corrado, and Jeffrey Dean. 2013. Efficient estimation of word representations in vector space. In *Proceedings of the International Conference on Learning Representations (ICLR)*.

Paramita Mirza. 2016. Extracting temporal and causal relations between events.

Claire Nédellec, Robert Bossy, Jin-Dong Kim, Jung-Jae Kim, Tomoko Ohta, Sampo Pyysalo, and Pierre Zweigenbaum. 2013. Overview of bionlp shared task 2013. In *Proceedings of the BioNLP Shared Task 2013 Workshop*, pages 1–7.

Radim Řehůřek and Petr Sojka. 2010. Software Framework for Topic Modelling with Large Corpora. In *Proceedings of the LREC 2010 Workshop on New Challenges for NLP Frameworks*, pages 45–50, Valletta, Malta, May. ELRA. http://is.muni.cz/publication/884893/en.

Hans Reichenbach. 1947. *Elements of Symbolic Logic*. Macmillan, London.

Mihai Surdeanu, Marco Valenzuela-Escárcega, Gus Hahn-Powell, Peter Jansen, Daniel Fried, Dane Bell, and Tom Hicks. 2014. processors. https://github.com/clulab/processors.

Marco Valenzuela-Escárcega, Gus Hahn-Powell, Dane Bell, Tom Hicks, Enrique Noriega, and Mihai Surdeanu. 2015a. Reach. https://github.com/clulab/reach.

Marco A. Valenzuela-Escárcega, Gus Hahn-Powell, and Mihai Surdeanu. 2015b. Description of the odin event extraction framework and rule language.

Marco A. Valenzuela-Escárcega, Gustave Hahn-Powell, Thomas Hicks, and Mihai Surdeanu. 2015c. A domain-independent rule-based framework for event extraction. In *Proceedings of the 53rd Annual Meeting of the Association for Computational Linguistics and the 7th International Joint Conference on Natural Language Processing of the Asian Federation of Natural Language Processing: Software Demonstrations (ACL-IJCNLP)*, pages 127–132. ACL-IJCNLP 2015.

Syntactic methods for negation detection in radiology reports in Spanish

Viviana Cotik[1] and **Vanesa Stricker**[1] and **Jorge Vivaldi**[2] and **Horacio Rodriguez**[3]

[1]Departamento de Computación, FCEyN, UBA, Argentina, {vcotik, vstricker}@dc.uba.ar
[2]Universitat Pompeu Fabra, UPF, Barcelona, Spain, jorge.vivaldi@upf.edu
[3]Universitat Politècnica de Catalunya, UPC, Barcelona, Spain, horacio@lsi.upc.edu

Abstract

Identification of the certainty of events is an important text mining problem. In particular, biomedical texts report medical conditions or findings that might be factual, hedged or negated. Identification of negation and its scope over a term of interest determines whether a finding is reported and is a challenging task. Not much work has been performed for Spanish in this domain.

In this work we introduce different algorithms developed to determine if a term of interest is under the scope of negation in radiology reports written in Spanish. The methods include syntactic techniques based in rules derived from PoS tagging patterns, constituent tree patterns and dependency tree patterns, and an adaption of NegEx, a well known rule-based negation detection algorithm (Chapman et al., 2001a). All methods outperform a simple dictionary lookup algorithm developed as baseline. NegEx and the PoS tagging pattern method obtain the best results with 0.92 F1.

1 Introduction

Text mining and natural language processing (NLP) techniques have been applied to the biomedical domain for a long time. Automatic identification of relevant terms in medical reports is a preliminary step for indexing and for search tools and it is useful for clinical, educational and research purposes.

A clinical condition mentioned in a biomedical text does not necessarily mean that a factual condition is reported, since the term or terms referring to the condition could be under the scope of negation or epistemic modality markers (hedges). For example, in *"no lymphadenopathies were detected"*, *"no ... were detected"* indicates that the medical condition (*"lymphadenopathy"*) is negated.

We refer to language constructions that denote negations as *negations* or *triggers* and to medical conditions and observations made about a particular illness in medical examinations as *findings* or *terms of interest*.

According to (Chapman et al., 2001b), many of the medical conditions described in unstructured texts in medical health records are negated. For this reason, the detection of negations in texts of the biomedical domain is an important task in the field of NLP, called BioNLP. Scope of negation has also received attention in other domains (Wiegand et al., 2010; Potts, 2011; Wor, 2010).

In this work we implement five techniques: 1) a simple approach, used as baseline, that determines if a finding is negated based on the presence of a negation term and a finding in the same sentence. The negation term is detected by dictionarylookup of negation terms; 2) an adaptation of NegEx to Spanish; the use of negation rules that were created based on 3) PoS tagging patterns, 4) constituent tree patterns, and 5) dependency tree patterns. Our goal is to decide which of the implemented methods is the best to automatically detect negations of important findings tagged in radiology reports written in Spanish.

Our methods are applied to Spanish and to a particular domain: radiology. This domain (and particularly our dataset) has the characteristic of having short reports, with usually short sentences, using informal language, containing non-standard abbreviations, and with highly noisy text. As far as we know, of our methods only NegEx has been implemented for Spanish and our implementation obtains better results. Using a Spanish dataset presents some challenges: we had to build a cor-

Proceedings of the 15th Workshop on Biomedical Natural Language Processing, pages 156–165,
Berlin, Germany, August 12, 2016. ©2016 Association for Computational Linguistics

pus and annotate it, syntactic parsing tools are less developed for languages other than English, and translations needed for the development of the work incorporates errors.

Experiments were performed over a dataset prepared from a set of ultrasonography reports written in Spanish, that have been previously tagged automatically with a tool based on RadLex[1], a specific radiology lexicon. A fragment of a tagged ultrasonography report in Spanish and its translation to English can be seen below: *"Pancreas: tamano y ecoestructura normal. Retroperitoneo vascular: sin <finding>alteraciones</finding>. No se detectaron <finding>adenomegalias </finding>. (...)"("Pancreas: normal size and echotexture. Vascular retroperitoneum: without <finding> changes </ finding>. No <finding> lymphadenopathies </finding> were detected.(...) ").*

The rest of the paper is organized as follows. Section 2 presents previous work in the detection of negation terms in the medical domain. In Section 3 we present our main contributions, by explaining the methods and datasets used. Section 4 shows the results of evaluating each of the algorithms with the testing dataset. Finally, Discussions, Conclusion and Future Work are presented.

2 Previous work

The use of information retrieval techniques for automatically indexing narrative medical reports and creating terminological resources has been present at least since mid-late 90s (Aronson et al., 1994; Rindflesch and Aronson, 1994; Sundaram, 1996).

In order to determine if a finding mentioned in a narrative medical report is under the scope of negation, (Chapman et al., 2001a) developed NegEx, a simple algorithm based on regular expressions that obtained very good results for English. Several methods were built upon this simple algorithm. (Wu et al., 2011) developed a word-based radiology report search engine based in a modification of NegEx. (Harkema et al., 2009) developed ConText, based in NegEx, employing a different definition for the scope of triggers and adopting it to different type of medical reports. NegEx has been adapted to Swedish (Skeppstedt, 2011), French (Deléger and Grouin, 2012), Dutch (Afzal et al., 2014), and Spanish for clinical records written in that language (Costumero et al., 2014) and radiology reports (Stricker et al.,

[1]http://www.radlex.org/

2015). The NegEx lexicon has been extended for Swedish, French and German (Chapman et al., 2013).

Syntactic methods have also been used. (Huang and Lowe, 2007) construct manually grammar rules using Part of Speech tagging in order to detect negations in radiology reports. (Uzuner et al., 2009) compare a NegEx extension with a machine learning technique that uses lexical and syntactic information using two corpora of discharge summaries and one of radiology reports. (Mehrabi et al., 2015) use dependency parsing to reduce NegEx False Positives. (Sohn et al., 2012) applies techniques of dependency parsing to detect negations. Therefore he compiles negation rules derived from the dependency paths.

Finally, machine learning techniques are also used for the negation detection task. (Cruz Díaz et al., 2010) compare these techniques to a regular expression-based method. (Morante and Daelemans, 2009) use them in order to establish the scope of negation in biomedical texts. (Rokach et al., 2008) perform automatic negation identification in clinical reports by means of extracting automatically regular expressions and patterns from annotated data and using them to create a learning method.

Several challenges have been performed on this topic. CoNLL 2010 Shared Task: Learning to Detect Hedges and Their Scope in Natural Language Text (Farkas et al., 2010), 2010 i2b2 NLP challenge, focused on the negation and uncertainty identification (Uzuner et al., 2011) and SEM 2012 Shared Task: Resolving the Scope and Focus of Negation(Morante and Blanco, 2012).

3 Methods

In this section we introduce the different methods developed to detect negations in radiology reports written in Spanish. The idea underlying syntactic techniques is to identify patterns of negations, manually compile negation rules, and use them to determine if a finding is under the scope of a negation or not. These methods used rules that were elaborated based on: 1) PoS tag patterns, 2) constituent tree (or shallow parsing) patterns of of the sentences and 3) dependency tree patterns (paths obtained from the dependency parsing of sentences). Rules were evaluated with the testing dataset.

Our methods only take into account the sentence where the term of interest appears in order to determine whether it is negated or not, i.e. it does not use information of other sentences.

3.1 *Dictionary lookup* algorithm

This simple algorithm developed is based on the lookup in the text of a list of negations marked by the expert radiologist as usual negation terms used in radiology reports. Sentences containing tagged findings where a negation appears (in any order) are tagged as Negated, and those with findings and without negations are tagged as Affirmed. This algorithm will be used as baseline.

3.2 The NegEx algorithm

NegEx algorithm for negation detection takes as input medical records with tagged findings and looks for phrases (triggers) that are mostly used to denote negation, for example *"no signs of"*. It checks if the phrase is applied to negate the finding or disease using rules that take into account the distance among the finding and the negation phrase.

The set of triggers provided by the NegEx tool[2] was translated using automatic translation[3] (since translation is an expensive task and we are not experts in the domain) and revised by two non-domain experts. Those triggers that were not correctly translated were eliminated or corrected. Given that English lacks grammatical gender, while Spanish has two (male and female), additional trigger instances were generated due to inflectional properties (for example from *"no"* to *"ningún" "ninguna"*). NegEx triggers are divided into: *pseudo negation phrases*, *negation terms*, *termination terms* and *conjunction terms*. A label is used to classify each trigger in one of these groups. Triggers were classified according to their use.

This implementation differs from others (Costumero et al., 2014; Stricker et al., 2015) (and is part of our contribution) mainly in that:

- tests were performed with two different trigger sets: 1) NegEx translated triggers (described in previous paragraph). A total of 210 translated triggers were obtained. 2) triggers obtained by combining translated triggers, a

set of bi and trigrams[4], and a list of triggers provided by a physician expert in the radiology domain (a total of 350 triggers),

- some end of scope triggers were added,

- coordinated negations, that were not taken into account in the English, nor in the Spanish versions were included as a trigger (*ni - nor-*) and NegEx algorithm was modified to include this term.

3.3 POS tagging patterns

Tags were assigned to each word of the sentence in order to determine the Part of Speech with the use of Freeling analyzer (Carreras et al., 2004). A small set of sentences were used to define negation patterns based on PoS tags. Patterns defined were:

- no +...+ verb + ...+ *<finding>*

- no +...+ *<finding>*

- sin +...+ *<finding>*

- sin +...+ *<finding>* +...+ ni +...+ *<finding>*

- no +...+ *<finding>* +...+ ni +...+ *<finding>*

- no +...+ verb +...+ *<finding>* +...+ ni +...+ *<finding>*

where "..." denotes zero or more words. The algorithm looks for these patterns in PoS tagged sentences. If a pattern occurs, the sentence is labeled as *Negated* indicating that the finding is under the scope of negation. For example: For "no se detectaron adenomegalias" we would have "RN P00CN000 VMIS3P0 FINDING", that satisfies the pattern "no +...+ verb + ...+*<finding>*".

RN represents "no". The words "sin" (without) and "ni" (nor) do not have specific negation tags (they are tagged as preposition and conjunction). That is why we look for these words directly in the text, instead of looking for some specific tag that represents them.

4Bi and trigrams were obtained from the 85600 report dataset (see Data subsection). Those, whose first word was *no*, were selected and the resulting were manually analyzed in order to discard those that did not correspond to triggers. 94 triggers were obtained.

3.4 Constituent tree patterns

Shallow parsing identifies the constituents of a sentence. We use this technique to manually elaborate patterns based on the phrase constituents avoiding the use of word distance to determine negation scope. Following patterns were used (patterns and phrase constituents[5] are shown):

1) no... verb... <finding>

```
     /   neg
S    -   grup-verb
     \   sn→grup-nom-mp→ w-ms→<finding>
```

2) without <finding>

```
S    -   grup-sp-> prep->sin
     \   sn→grup-nom-mp→w-mp →<finding>
```

3) no <finding>

```
S    -   neg
     \   sn→grup-nom-mp→w-mp →<finding>
```

4) no ... verb ... <finding> nor ... <finding>

```
     /   neg
     /   grup-verb
S    —   sn →grup-nom-mp→w-mp →<finding>
     \   coord
     \   sn→grup-nom-mp→w-mp →<finding>
```

5) <finding>: no

```
     /   sn →grup-nom-ms→w-ms →<finding>
S    -   no-c→:
     \   neg
```

Three steps were performed to obtain patterns from the constituent tree: 1) the finding is replaced by "finding" and using Freeling the shallow parsing tree is obtained. 2) the tree structure is represented in an array. 3) the array is used to check whether the sentence satisfies one of the patterns previously discovered. For example, in order to check if a sentence satisfies pattern 1, it is verified if node with label S has as children a node with label *neg*, a node with label *grup-verb* and a node with label *sn* (in this order), and if node with label *sn* has as child a node with label *grup-nom-ms*, which also has as child a node with label *w-ms* and this has as child the node with content *finding*.

3.5 Dependency tree patterns

Dependency parsing allows us to know the syntactic structure of a phrase. The method is based on syntactic context and does not take into account word distance to determine negation scope. Negation patterns are manually created based on syntactic dependency paths in the following way:

1. a small set of sentences containing all known type of negations (no, ni, sin) (no, nor, without) were parsed with a MATE dependency parser (Bohnet et al., 2013)[6]. A parse tree was obtained for each sentence (see Fig. 1),

2. negation terms were located automatically and an algorithm was developed in order to retrieve the path in the dependency tree between the negation term and the *finding* previously tagged,

3. paths were analyzed and a set of patterns that imply negation of findings was manually developed, and

4. patterns obtained in the previous step were tested with the testing dataset.

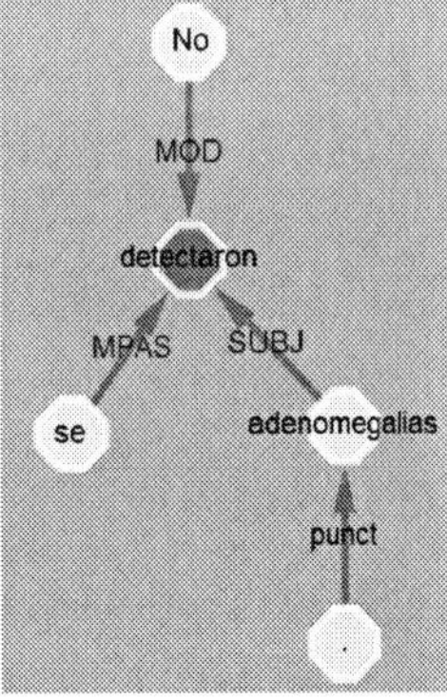

Figure 1: Example of a dependency parser tree for a sentence of the form of Pattern 1 (P1).

Patterns detected were following:

---

[5]*neg* stands for "no", *grup-verb* for "verbal syntagma", *sn* for nominal syntagma. See https://github.com/iknow/FreeLing/blob/master/doc/grammars/esCHUNKtags for further references.

[6]The model was obtained as indicated in (Arias et al., 2014).

- P1: sentences of the form "**no** se detectaron adenomegalias" (The Spanish structure of this particular sentence corresponds to NEG (no) verb finding). The negation has a dependency relation with a word that the finding depends on.

- P2: sentences of the form "retroperitoneo vascular: **sin** alteraciones" (vascular retroperitoneum: without alterations) (anatomical part: NEG (sin) *<finding>*). The finding depends of "sin".

- P3: sentences like "via biliar **no** dilatada" (*bile duct not delated*) (anatomical part NEG *<finding>*, where NEG is "no").

- P4: sentences of the form "**No** se detectaron colecciones **ni** liquido libre" (neither collections nor free liquid has been detected) (NEG(**no**) verb *<finding>* NEG(**ni**) *<finding>*).

Data

Two datasets were used. The *analysis dataset* to infer the patterns of each of the proposed methods used and the *test dataset* to test the methods and compare their results.

Our original dataset is composed of about 85600 reports of ultrasonography studies performed in a public hospital. Reports are written in Spanish in non-structured format. They are brief (approximately five lines each) and they state what was found in the study performed on the patient. Text is noisy, characterized by frequent typos, abbreviations, sentences which are not syntactically well-formed and there is lack of punctuation in some cases.

The process to obtain both datasets was the following: An algorithm was used in order to automatically detect terms of interest (findings in the radiology domain) in the reports (Cotik et al., 2015). Then, a sentence tokenization was performed using NLTK (Loper and Bird, 2002). Only sentences with findings are selected (randomly) to create analysis and testing datasets. Finally, those sentences were annotated as containing negation with scope over the term of interest (Negated) or not (Affirmed). For the creation of the testing dataset a set of sentences were randomly selected and the following steps were performed: 1) we verified manually that sentences were neither the

same (among them) nor very similar, 2) segmentation issues -e.g. different sentences that were not separated by the tokenizer- were corrected, 3) sentences with findings tagged by the algorithm and that were not considered actual findings by the annotators were eliminated and replaced by new ones. The analysis set is composed of 979 sentences and the testing set of 1000 sentences.

Findings detection

There are various inventories that serve as a basis to detect relevant terms in medical reports. Some of them are ICD10[7], a standard diagnostic terminology for epidemiology, health management and clinical purposes; and SNOMED CT[8], a clinical health terminology ontology -all of them included in UMLS (Unified Medical Language System)[9] Metathesaurus-; and RadLex[10], a lexicon centered only on radiology terms. SNOMED CT and ICD-10 are available in Spanish, RadLex is only available in English and in German. Previous implementations vary the type of inventory used to detect terms (UMLS, adaptations of ICD-10 and MeSH[11], among others). The information extraction algorithm we used to detect findings is based on the appearance of RadLex *pathological terms* in the reports. RadLex was chosen because it is the only lexicon specifically developed for the radiology domain, which is the domain under study. It has the disadvantage that no Spanish version has been developed, so it had to be translated from English. The translation is not an easy task, since, particularly, in the medical domain, there exist terms that are used differently in Spanish and in English.

Annotations

Working with languages different than English has, among others, the difficulty of the lack of data and tools. In this case we do not have a Gold Standard for validating the reliability of the new model. Annotating is an expensive task, and domain experts are not always available. The datasets build had to be annotated. The analysis dataset was annotated by two non-experts and the testing dataset

---

[7]http://apps.who.int/classifications/icd10/browse/2016/en
[8]http://www.ihtsdo.org/snomed-ct
[9]http://www.nlm.nih.gov/research/umls/. UMLS is a set of files and software that bring together many health and biomedical vocabularies and standards to enable interoperability between computer systems.
[10]http://rsna.org/RadLex.aspx
[11]http://www.ncbi.nlm.nih.gov/mesh

by an expert of the radiology domain and two non-experts.

All sentences (with previously tagged findings) were annotated as *Affirmed* if it is possible to infer that the finding is present in the patient, *Negated* if the finding is absent, *Probable* if it is not certain that the finding is present, but it probably is, and *Doubt* if the finding corresponds to the past or if it is not clear for the annotator if the finding is present or not. For results evaluation *Probable* annotations were considered as *Affirmed*, since physicians are interested in retrieving them, and sentences categorized as *Doubt* were replaced by other sentences (that were also annotated). In the cases where there was no agreement among annotators, usually the radiology-expert criteria was respected. In case of doubt the annotation criteria was revised by the annotators and the annotation was done according to the results of this process.

In both cases, the annotation process was performed in two stages, so that we could revise the annotation criteria. Some annotated sentences were overlapped, with the objective to calculate the Inter Rater Agreement (IRA) between annotators to measure their level of agreement. As measure for that goal we calculated Cohen's Kappa coefficient (Cohen, 1960).

Figure 2 shows the number of sentences annotated by each annotator individually and by more than one annotator in the testing dataset. Kappa coefficient (κ) was calculated for two sets: 1) 100 sentences annotated by non-expert annotator 1 and radiology domain expert (annotator 3), and 2) 100 sentences annotated by non-expert annotator 2 and annotator 3. Table 1 shows κ measure for the testing dataset. κ measure for the analysis dataset had similar results.

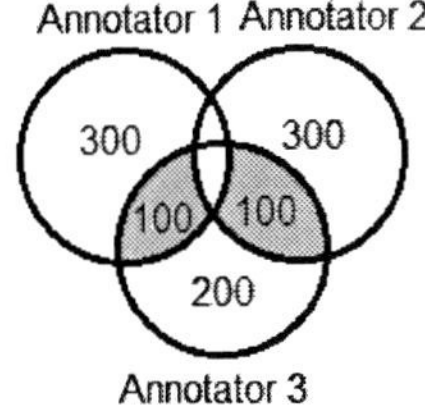

Figure 2: Number of sentences annotated by different annotators in the testing dataset.

annotators	κ
A1 and A3	0.97
A2 and A3	0.96

Table 1: IRA of expert/non-experts annotation in the testing dataset. A1 and A2 are computer science experts (not medical, nor linguistic experts), A3 is a radiology expert.

4   Results

Table 2 shows the performance of our NegEx adaptation and our syntactic methods to Spanish compared to the baseline. We show the best result of NegEx (obtained from the trigger set built from a combination of translated triggers, bi and trigrams and a list of terms suggested by the radiology expert). F1 using NegEx only with translated triggers was similar: 0.91 (81 TP, 76 FP, 144 FN and 699 TN). Results of NegEx with the original triggers (translated) and without the addition of coordinated negations (and tested with another dataset) can be seen in (Stricker et al., 2015).

Precision, Recall and F1 measure are the usual measures in the field and here are based on the interpretation of finding real negations. *F1* measure balances *precision* -how many findings identified as negated, are actually negated- and *recall* -proportion of the negated findings that were retrieved-. *Accuracy* is the rate of correctly classified sentences. True Positive (TP) refers to terms negated by the Gold Standard and correctly predicted by the methods. See Table 4 for the meaning of False Positive (FP), True Negative (TN) and False Negative (FN).

5   Discussion

All algorithms outperform *dictionary lookup*, our baseline algorithm. This makes sense, since the baseline does not take negation scope into account. For example in "*ectasia* pielica izquierda **sin** cambio de diametro postmiccional" what is negated (cambio de diametro postmiccional) is not the finding (ectasia). The baseline algorithm detects the negation (sin) and assumes wrongly that the finding is negated. This scope problem is solved in the rest of the algorithms developed.

Constituent tree patterns and dependency tree patters were tested assuming that they would perform better than PoS tagging patterns and NegEx in the detection of the negation scope, since in

Algorithm	Pattern Matching (baseline)	NegEx (adapted to Spanish)	POS Tagging Patterns	Constituent Tree Patterns	Dependency Tree Patterns
TP	201	220	219	200	194
FP	107	31	31	19	61
FN	24	5	6	25	31
TN	668	744	744	756	714
Accuracy	0.87	0.96	0.96	0.96	0.91
Precision	0.65	0.88	0.88	0.91	0.77
Recall	0.89	0.98	0.97	0.89	0.86
F1	0.75	0.92	0.92	0.90	0.81

Table 2: Performance of different algorithms with testing dataset composed by 1000 sentences.

Algorithm	NegEx (Costumero et al., 2014)	NegEx (Stricker et al., 2015)	NegEx (adapted)
F1	0.74	0.67	0.73

Table 3: Performance of different implementations of NegEx with (Costumero et al., 2014) dataset

	predicted Neg	predicted Aff
actual Neg	TP	FN
actual Aff	FP	TN

Table 4: *actual* stands for Gold Standard annotation, *predicted* for algorithms output.

these two methods we have not to consider fixed windows of words between the negation and the term of interest (as we do consider in NegEx) or each word that forms the sentence (as we do in our PoS tagging method). Nevertheless NegEx and the PoS Tagging based method have better results (not very different from constituent tree patterns). We understand that two factors influence these results: 1) the sentences of the reports are usually in our case relatively short (average: 14 words, longest: 74 words). This explains why having fixed windows of 6 words might be good enough for our data and suggests that we do not need to use more complex methods, that are independent of the length of the sentence and that do not fix word distance. That is, the linear analysis performed by PoS tagging patterns might be enough for these sentences. Dependency and constituent parsing, that perform an analysis based on the sentence structure, might be left for the most complex sentences. 2) MATE, the tool used to do the dependency parsing was trained based on a

general language[12] that includes documents of the medical domain, but that is not restricted to it[13].

Regarding NegEx, another implementation was tried with a very reduced trigger set, in order to try to do it domain independent (see Table 3). F1 is similar when tested with our test set (0.91 instead of 0.92), and it is also similar (0.73) to F1 obtained by (Costumero et al., 2014) (0.74) and better than F1 obtained by (Stricker et al., 2015) (0.67) when tested with Costumero's dataset. This demonstrates that our NegEx implementation with a reduced trigger set could be used for data different that radiology reports.

Further analysis of results shows that: 1) the addition of a line of code to NegEx algorithm allows us to handle complex negations. E.g. in *"no se detectaron finding1 ni finding2"* (*"finding 1 and finding 2 were not detected"*), when *"finding2"* is the term of interest. Those kinds of negations are also handled correctly by the patterns built from our syntactic methods, but in some cases negations are much more complex and are not correctly parsed by the dependency parsing algorithm. 2) Sometimes, negations are not affecting the term of interest, but a modifier of it and the algorithm tags the term of interest as negated. For example, in *"pancreas: no visible por abundante gas"* (*"pancreas: not visible due to abundant gas"*). The

---

[12]https://www.iula.upf.edu/corpus/corpusuk.htm
[13]Besides, the area of documents in the medical domain is broad and the ones used differ from radiology reports.

trigger *"no"* (*"not"*) is applied to *"visible"* (*"visible"*), but the term of interest is *"gas"* (*"gas"*). 3) Constituent tree patterns method has shown to fail where there are no punctuation signs. This shows that the characteristics of the noisy text makes the success of syntactic techniques more complicated.

NegEx shows to perform better than a previous implementation for radiology reports in Spanish (Stricker et al., 2015) and similar than an implementation for general medical texts also in Spanish (Costumero et al., 2014) (see Section 5). Our Pos-Tagging results and the ones reached by (Huang and Lowe, 2007) for radiology reports in English are similar. They obtain 0.90 recall, 0.97 precision and 0.93 F1, while we obtain 0,88 recall, 0,97 precision and 0,92 F1. (Sohn et al., 2012) results for negation detection in clinical texts in English using dependency parsing are also similar to our dependency parser results. They obtain 0.74 recall, 0.97 precision and 0.84 F1, while we obtain 0.77, 0.86 and 0.81 for each of these measures. Nevertheless, it is not easy to compare results with existing papers, since languages and corpora are not the same.

6 Conclusion

Considering the different methods implemented for the detection of negations of terms of interest in radiology reports written in Spanish, NegEx has good results, but only considers partially the negation scope over the target term (since it is calculated based on a fixed-size window of words). Among the pattern methods tested, PoS tags allows us to study the ordering of words in phrases containing negations and to elaborate patterns based on them. But they are dependent on each word of the sentence. Based on a reduced dataset it is not easy to model all type of forms that sentences with negated findings may have. Constituent and dependency tree pattern methods differ from the *PoS tagging method* in that the whole structure of the sentence is used. Constituent tree method segments the sentence in syntactic related groups. These cases do not have to take so many detail into account and are easier to build. Both methods differ in that the second takes into account the dependence among each type of word in the sentence. Dependencies are modeled in a tree and each edge is labeled with the relation that exists among the words.

Detection negation in medical reports is a chal-lenging task as it is characterized by short sentences and informal language often noisy. Furthermore, tools for Spanish in general are less developed than in other languages even more in this specific subdomain. For example, the availability of a large corpus of annotated medical reports (and specifically those in the radiology domain) would enable to have a better behavior of all language related tools (in particular POS tagging as well as constituent/dependency parsers). RadLex, is a comprehensive lexicon of radiology terms that was chosen to detect findings due to its adequacy to our domain of interest. Its translation to Spanish was made locally but unfortunately includes some errors, such as the order of resulting words and issues derived from ambiguity. All these issues made negation detection more difficult.

The high IRA obtained among the annotations performed by the specialist and two non-specialist could imply that this particular type of reports of short sentences could be annotated by non-specialists in the domain. We consider this is an important result, given the scarcity of resources.

We consider that having short sentences (ours have an average of 14 words) may contribute to the fact that NegEx and PoS tagging methods have similar results than the constituent tree method and better results than dependency tree method. An analysis should be performed with more complex sentences in order to test what happens in those cases. The effectiveness of syntactic techniques depends on the compliance of the text to the language grammatical rules. The results obtained support this asseveration.

7 Future Work

We are currently working in analyzing improvements to the dependency parser patterns and we are performing a further analysis of results, evaluating alternative methods (voting method, where the classification (Affirmed/Negated) is based on the tag received by most of the methods) and evaluating the possibility of implementing a hybrid methodology -taking the best of NegEx and syntactic methods- that reduces errors in order to obtain better F1.

We would like to extend our work for dealing with *hedges* and we plan to continue using these methods for other type of medical reports written in Spanish.

References

Zubair Afzal, Ewoud Pons, Ning Kang, Miriam Sturkenboom, Martijn Schuemie, and Jan Kors. 2014. ContextD: an algorithm to identify contextual properties of medical terms in a Dutch clinical corpus.

Blanca Arias, Núria Bel, Mercé Lorente, Montserrat Marimón, Alba Milà, Jorge Vivaldi, Muntsa Padró, Marina Fomicheva, and Imanol Larrea. 2014. Boosting the Creation of a Treebank. *Proceedings of the Ninth International Conference on Language Resources and Evaluation (LREC'14)*, 9(Suppl 11):775–781.

Alan R. Aronson, Thomas C. Rindflesch, and Browne C. Allen. 1994. Exploiting a Large Thesaurus for Information Retrieval. In *Proceedings of RIAO: Recherche d'Information Assistée par Ordinateur. Conférence*, pages 197–217, New York, USA.

Bernd Bohnet, Joakim Nivre, Igor Boguslavsky, Richard Farkas, Filip Ginter, and Jan Hajic. 2013. Joint morphological and syntactic analysis for richly inflected languages. *Transactions of the Association for Computational Linguistics(TACL)*, 1(Suppl 11):415–428.

Xavier Carreras, Isaac Chao, Llus Padr, and Muntsa Padr. 2004. Freeling: An open-source suite of language analyzers. In *Proceedings of the 4th International Conference on Language Resources and Evaluation (LREC'04)*.

Wendy W. Chapman, Will Bridewell, Paul Hanbury, Gregory F. Cooper, and Bruce G. Buchanan. 2001a. A Simple Algorithm for Identifying Negated Findings and Diseases in Discharge Summaries. *Journal of Biomedical Informatics*, 34(5):301–310.

Wendy W. Chapman, Will Bridewell, Paul Hanbury, Gregory F. Cooper, and Bruce G. Buchanan. 2001b. Evaluation of Negation Phrases in Narrative Clinical Reports. In *Proceedings of AMIA, American Medical Informatics Association Annual Symposium*, page 105, Washington, DC, USA.

Wendy W. Chapman, Dieter Hillert, Sumithra Velupillai, Maria Kvist, Maria Skeppstedt, Brian E. Chapman, Mike Conway, Melissa Tharp, Danielle L. Mowery, and Louise Deléger. 2013. Extending the NegEx Lexicon for Multiple Languages. In *Proceedings of the 14th World Congress on Medical and Health Informatics*, pages 677–681, Copenhagen, Denmark.

Jacob Cohen. 1960. A Coefficient of Agreement for Nominal Scales. *Educational and Psychological Measurement*, 20(1):37–46.

Roberto Costumero, Federico López, Consuelo Gonzalo-Martín, Marta Millan, and Ernestina Menasalvas. 2014. An Approach to Detect Negation on Medical Documents in Spanish. In *Brain Informatics and Health*, volume 8609, pages 366–375.

Viviana Cotik, Darío Filippo, and José Castaño. 2015. An Approach for Automatic Classification of Radiology Reports in Spanish. In *Proceedings of 15th MEDINFO*, pages 634–638.

Noa P. Cruz Díaz, Manuel Jesús Maña López, and Jacinto Mata Vázquez. 2010. Aprendizaje Automático Versus Expresiones Regulares en la Detección de la Negación y la Especulación en Biomedicina [Machine Learning versus Regular Expressions in Negation and Speculation Detection in Biomedicine]. *Procesamiento del Lenguaje Natural [Natural Language Processing]*, 45:77–85.

Louise Deléger and Cyril Grouin. 2012. Detecting negation of medical problems in French clinical notes. In *Proceedings of the 2nd ACM SIGHIT International Health Informatics Symposium*, pages 697–702.

Richárd Farkas, Veronika Vincze, György Móra, János Csirik, and György Szarvas. 2010. The CoNLL-2010 Shared Task: Learning to Detect Hedges and Their Scope in Natural Language Text. In *Proceedings of the Fourteenth Conference on Computational Natural Language Learning*, pages 1–12, Uppsala, Sweden.

Henk Harkema, John N. Dowling, Tyler Thornblade, and Wendy W. Chapman. 2009. ConText: An Algorithm for Determining Negation, Experiencer, and Temporal Status from Clinical Reports. *Journal of biomedical informatics*, 42(5):839–851, October.

Yang Huang and Henry J Lowe. 2007. A Novel Hybrid Approach to Automated Negation Detection in Clinical Radiology Reports. *Journal of the American Medical Informatics Association*, 14(3):304–311.

Edward Loper and Steven Bird. 2002. NLTK: The Natural Language Toolkit. In *Proceedings of the ACL-02 Workshop on Effective Tools and Methodologies for Teaching Natural Language Processing and Computational Linguistics*, volume 1, pages 63–70, Philadelphia, Pennsylvania.

Saeed Mehrabi, Anand Krishnan, Sunghwan Sohn, Alexandra M Roch, Heidi Schmidt, Joe Kesterson, Chris Beesley, Paul Dexter, C Max Schmidt, Hongfang Liu, et al. 2015. Deepen: A negation detection system for clinical text incorporating dependency relation into negex. *Journal of biomedical informatics*, 54:213–219.

Roser Morante and Eduardo Blanco. 2012. Sem 2012 shared task: Resolving the scope and focus of negation. In *Proceedings of the First Joint Conference on Lexical and Computational Semantics - Volume 1: Proceedings of the Main Conference and the Shared Task, and Volume 2: Proceedings of the Sixth International Workshop on Semantic Evaluation*, SemEval '12, pages 265–274, Stroudsburg, PA, USA. Association for Computational Linguistics.

Roser Morante and Walter Daelemans. 2009. A Metalearning Approach to Processing the Scope of Negation. In *Proceedings of the Thirteenth Conference on Computational Natural Language Learning*, pages 21–29, Boulder, Colorado.

Christopher Potts. 2011. On The Negativity of Negation. In *Proceedings of Semantics and Linguistic Theory*, volume 20, pages 636–659, New Brunswick, New Jersey.

Thomas C. Rindflesch and Alan R. Aronson. 1994. Ambiguity Resolution While Mapping Free Text to the UMLS Metathesaurus. In *Proceedings of the 18th Annual Symposium on Computer Application in Medical Care*, pages 240–244, Washington, DC, USA.

Lior Rokach, Roni Romano, and Oded Maimon. 2008. Negation Recognition in Medical Narrative Reports. *Journal of Information Retrieval*, 11(6):1–50.

Maria Skeppstedt. 2011. Negation Detection in Swedish Clinical Text: An Adaption of NegEx to Swedish. *Journal of Biomedical Semantics*, 2(Suppl 3):S3, January.

Sunghwan Sohn, Stephen Wu, and Christopher G Chute. 2012. Dependency parser-based negation detection in clinical narratives. *AMIA Summits on Translational Science proceedings AMIA Summit on Translational Science*, 2012:1–8.

Vanesa Stricker, Ignacio Iacobacci, and Viviana Cotik. 2015. Negated Findings Detection in Radiology Reports in Spanish: an Adaptation of NegEx to Spanish. In *IJCAI - Workshop on Replicability and Reproducibility in Natural Language Processing: adaptative methods, resources and software*, Buenos Aires, Argentina.

Anita Sundaram. 1996. Information Retrieval: A Health Care Perspective. *Bulletin of the Medical Library Association*, 84(4):591–593.

Özlem Uzuner, Xiaoran Zhang, and Tawanda Sibanda. 2009. Machine Learning and Rule-based Approaches to Assertion Classification. *Journal of the American Medical Informatics Association : JAMIA*, 16(1):109–115.

Özlem Uzuner, Brett R. South, Shuying Shen, and Scott L. DuVall. 2011. 2010 i2b2/VA Challenge on Concepts, Assertions, and Relations in Clinical Text. *Journal of the American Medical Informatics Association : JAMIA*, 18(5):552–556.

Michael Wiegand, Alexandra Balahur, Benjamin Roth, Dietrich Klakow, and Andrés Montoyo. 2010. A survey on the role of negation in sentiment analysis. In *Proceedings of the Workshop on Negation and Speculation in Natural Language Processing*, NeSp-NLP'10, pages 60–68, Stroudsburg, PA, USA. Association for Computational Linguistics.

2010. In *Proceedings of the Workshop on Negation and Speculation in Natural Language Processing*, NeSp-NLP '10, Stroudsburg, PA, USA. Association for Computational Linguistics.

Andrew S. Wu, Bao H. Do, Jinsuh Kim, and Daniel L. Rubin. 2011. Evaluation of Negation and Uncertainty Detection and Its Impact on Precision and Recall in Search. *Journal of digital imaging*, 24(2):234–242, April.

How to Train Good Word Embeddings for Biomedical NLP

Billy Chiu   **Gamal Crichton**   **Anna Korhonen**   **Sampo Pyysalo**
Language Technology Lab
DTAL, University of Cambridge
{hwc25|gkoc2|alk23}@cam.ac.uk, sampo@pyysalo.net

Abstract

The quality of word embeddings depends on the input corpora, model architectures, and hyper-parameter settings. Using the state-of-the-art neural embedding tool *word2vec* and both intrinsic and extrinsic evaluations, we present a comprehensive study of how the quality of embeddings changes according to these features. Apart from identifying the most influential hyper-parameters, we also observe one that creates contradictory results between intrinsic and extrinsic evaluations. Furthermore, we find that bigger corpora do not necessarily produce better biomedical domain word embeddings. We make our evaluation tools and resources as well as the created state-of-the-art word embeddings available under open licenses from https://github.com/cambridgeltl/BioNLP-2016.

1 Introduction

As one of the main inputs of many NLP methods, word representations have long been a major focus of research. Recently, the embedding of words into a low-dimensional space using neural networks was suggested (Bengio et al., 2003; Collobert and Weston, 2008; Turian et al., 2010; Mikolov et al., 2013b; Pennington et al., 2014). These approaches represent each word as a dense vector of real numbers, where words that are semantically related to one another map to similar vectors. Among neural embedding approaches, the skip-gram model of Mikolov et al. (2013a) has achieved cutting-edge results in many NLP tasks, including sentence completion, analogy and sentiment analysis (Mikolov et al., 2013a; Mikolov et al., 2013b; Fernández et al., 2014).

Although word embeddings have been studied extensively in recent work (e.g. Lapesa and Evert (2014)), most such studies only involve general domain texts and evaluation datasets, and their results do not necessarily apply to biomedical NLP tasks. In the biomedical domain, Stenetorp et al. (2012) studied the effect of corpus size and domain on various word clustering and embedding methods, and Muneeb et al. (2015) compared two state-of-the-art word embedding tools: word2vec and Global Vectors (GloVe) on a word-similarity task. They showed that skip-gram significantly out-performs other models and that its performance can be further improved by using higher dimensional vectors. The word2vec tool was also used to create biomedical domain word representations by Pyysalo et al. (2013) and Kosmopoulos et al. (2015).

Given that word2vec has been shown to achieve state-of-the-art performance that can be further improved with parameter tuning, we focus on its performance on biomedical data with different inputs and hyper-parameters. We use all available biomedical scientific literature for learning word embeddings using models implemented in word2vec. For intrinsic evaluation, we use the standard UMNSRS-Rel and UMNSRS-Sim datasets (Pakhomov et al., 2010), which enable us to measure similarity and relatedness separately. For extrinsic evaluation, we apply a neural network-based named entity recognition (NER) model to two standard benchmark NER tasks, JNLPBA (Kim et al., 2004) and the BioCreative II Gene Mention task (Smith et al., 2008).

Apart from showing that the optimization of hyper-parameters boosts the performance of vectors, we also find that one such parameter leads to contradictory results between intrinsic and extrinsic evaluations. We further observe that a larger corpus does not necessarily guarantee better re-

Proceedings of the 15th Workshop on Biomedical Natural Language Processing, pages 166–174,
Berlin, Germany, August 12, 2016. ©2016 Association for Computational Linguistics

Corpus	Total tokens
PubMed	2,721,808,542
PMC	7,959,548,841
PubMed + PMC	10,681,357,383

Table 1: Corpus statistics

Parameters	Values
neg	1 / 2 / 3 / **5** / 8 /10 / 15
samp	0 / 1e-1 / 1e-2 / **1e-3** / 1e-4 1e-5 / 1e-6 / 1e-7 / 1e-8 / 1e-9
min-count	0 / **5** / 10 / 20 / 50 / 100 / 200 400 / 800 / 1000 / 1200 / 2400
alpha	0.0125 / **0.025** / 0.05 / 0.1
dim	25 / 50 / **100** / 200 / 400 / 500 / 800
win	1 / 2 / 4 / **5** / 8 / 16 / 20 / 25 / 30

Table 2: Hyper-parameters and tested values. Default values shown in bold.

sults in our tasks. We hope that our results can serve as a reference for researchers who use neural word embeddings in biomedical NLP.

2 Materials and Methods

2.1 Corpora and Pre-processing

We use two corpora to create word vectors: the PubMed Central Open Access subset (PMC) and PubMed. PMC is a digital archive of biomedical and life science literature, which contains more than 1 million full-text Open Access articles. The PubMed database has more than 25 million citations that cover the titles and abstracts of biomedical scientific publications. A version of PMC articles is distributed in text format[1] whereas PubMed is distributed in XML. Thus, we use a PubMed text extractor[2] to extract title and abstract texts from the PubMed source XML. Both PubMed and PMC were pre-processed with the Genia Sentence Splitter (GeniaSS) (Sætre et al., 2007), which is optimized for bio-medical text. We further tokenize the sentences with the Tree bank Word Tokenizer provided by the NLTK python library (Bird, 2006). The corpus statistics are shown in Table 1.

2.2 Word vectors

Factors that affect the performance of word representations include the training corpora, the model architectures, and the hyper-parameters. To assess the effect of corpora, we generate three variants of each set of word vectors: one from PubMed, one from PMC, and one from the combination of the two (PMC-PubMed). To study how preprocessing affects word vectors, we create vectors from the original text corpora, lower-cased variants, and variants where sentences are shuffled in random order. We further generate two sets of vectors, one by applying the skip-gram model and one applying the CBOW model, built with the default hyper-parameter values of word2vec. We first evaluate these vectors to determine the better-performing model architecture. Using the better model, we

[1] http://www.ncbi.nlm.nih.gov/pmc/ tools/ftp/#Data_Mining
[2] https://github.com/spyysalo/pubmed

then build vectors by varying values of one hyper-parameter (Table 2) and keeping others as default. We repeat the process for every hyper-parameter under examination. We then report the results of these sets of vectors in our intrinsic and extrinsic evaluations.

2.3 Hyper-parameters

We test the following key hyper-parameters:

Negative sample size (*neg*): the representation of a word is learned by maximizing its predicted probability to co-occur with its context words, while minimizing the probability for others. However, the normalisation of this probability involves a denominator deriving from co-occurrences between words and all their contexts in the corpus, which is time-consuming to compute. To address this issue, negative sampling only calculates the probability with reference to a set number of other randomly chosen negative words (*neg*).

Sub-sampling (*samp*): Sub-sampling refers to the process of reducing occurrences of frequent words. It selects words appearing with a ratio higher than the threshold *samp*, and ignores each occurrence with a given probability. The process is used to minimise the effect of non-informative frequent words in training. Very frequent words (e.g. *in*) are less informative because they co-occur with most words in the corpus. For example, a model can benefit more from seeing an occurrence of *p16* with *CDKN2* than an instance of the frequent co-occurrence of *p16* with *in*.

Minimum-count (*min-count*): The minimum-count defines the minimum number of occurrences required for a word to be included in the word vectors. This parameter allows control the over the size of the vocabulary and, consequently, the resulting word embedding matrix.

Learning Rate (*alpha*): neural networks are trained by gradually updating weight vectors

Vector	Token
PMC-PubMed (Pyysalo et al.)	5,487,486,225 (total)
PMC (Pyysalo et al.)	2,591,137,744 (total)
PubMed (Pyysalo et al.)	2,896,348,481 (total)
PubMed (Kosmopoulos et al.)	1,701,632 (distinct)

Table 3: Baseline word vectors

along a gradient to minimize an objective function. The magnitude of these updates is controlled by the learning rate.

Vector dimension (*dim*): The vector dimension is the size of the learned word vector. While a higher dimension tends to capture better word representations, their training is more computationally costly and produces a larger word embedding matrix.

Context window size (*win*): The size of the context window defines the range of words to be included as the context of a target word. For instance, a window size of 5 takes five words before and after a target word as its context for training.

We refer to Mikolov et al. (2013a) and Levy et al. (2015) for further details regarding these parameters.

2.4 Baseline Vectors

As baselines, we include the biomedical domain vectors created by Pyysalo et al. (2013) and Kosmopoulos et al. (2015). Their corpus statistics are shown in Table 3. All of these vectors are built with the skip-gram model with the default parameter values (see Table 2).

2.5 Intrinsic Evaluation

A standardized intrinsic measure for word representations in the biomedical domain is the UMN-SRS word similarity dataset (Pakhomov et al., 2010). We use its UMNSRS-Sim (Sim) and UMNSRS-Rel (Rel) subsets as our references. They have 566 and 587 word pairs for measuring similarity and relatedness (respectively) whose degree of association was rated by participants from the University of Minnesota Medical School. In UMNSRS, the human evaluation on every word pair is converted to a score to determine its degree of similarity, a higher score implying a more similar pair. The range of the score is on an arbitrary scale. While UMNSRS provides scores to determine the degree of similarity for each word pair, we will measure this by calculating the cosine similarity score for each word pair using the

learned word vectors. Afterwards, we compare the two scores using Spearman's correlation coefficient (ρ), which is a standard metric to compare ranking between variables regardless of scale in word similarity task. We systematically ignore words that appear only in the reference but not in our models.

2.6 Extrinsic Evaluation

Given that the ultimate evaluation for word vectors is their performance in downstream applications, we also assess the quality of the vectors by performing NER using two well-established biomedical reference standards: the BioCreative II Gene Mention task corpus (BC2) (Smith et al., 2008) and the JNLPBA corpus (PBA) (Kim et al., 2004). Both of these corpora consist of approximately 20,000 sentences from PubMed abstracts manually annotated for mentions of biomedical entity names. Following the window approach architecture with word-level likelihood proposed by Collobert and Weston (2008), we apply a tagger built on a simple feed-forward neural network, with a window of five words, one hidden layer of 300 neurons and a hard sigmoid activation, leading to a Softmax output layer. Our word vectors are used as the embedding layer of the network, with the only other input being a low-dimensional binary vector of word surface features.[3] To emphasize the effect of the input word vectors on performance, we avoid fine-tuning the word vectors during training as well as introducing any external resources such as entity name dictionaries. While this causes the performance of the method to fall notably below the state of the art, we believe this minimal approach to be an effective way to focus on the quality of the word vectors as they are created by the tool (word2vec).[4] For parameter selection, we estimate the extrinsic performance of word vectors on the development sets of the two corpora using mention-level F-score. For the final experiment with selected parameters we apply the test sets and evaluation scripts of the two tasks in accordance with their original evaluation protocols.

---

[3]For example, whether a word starts or contains a capital letter or number. For detailed reference, we make our implementation openly available.

[4]It is an interesting question for future work whether the findings from our extrinsic evaluation apply also to state-of-the-art taggers.

Model	PMC-PubMed		PMC		PubMed	
	Sim	Rel	Sim	Rel	Sim	Rel
SG	**0.54**	0.488	0.507	0.453	0.446	**0.497**
CBOW	0.435	0.409	0.348	0.351	0.449	0.446
SG-S	0.555	**0.515**	0.54	0.49	0.551	0.502
SG-L	0.542	0.457	0.502	0.424	0.552	0.47
SG-SL	0.543	0.47	0.52	0.459	**0.56**	0.481
CBOW-S	0.415	0.403	0.434	0.424	0.43	0.414
CBOW-L	0.452	0.404	0.447	0.41	0.461	0.425
CBOW-SL	0.461	0.422	0.45	0.39	0.471	0.426

Table 4: Intrinsic evaluation results for vectors with different pre-processing: Original Text, Sentence-shuffled (S), lowercased (L), and both (SL)

Model	PMC-PubMed		PMC		PubMed	
	BC2	PBA	BC2	PBA	BC2	PBA
SG	60.86	61.89	59.48	62.11	**61.00**	**62.52**
CBOW	55.11	56.97	54.93	58.10	54.25	58.48
SG-S	59.81	62.13	59.23	62.30	60.75	62.11
SG-L	60.52	62.19	59.93	61.64	60.51	**62.64**
SG-SL	**61.33**	62.58	60.23	62.05	61.11	61.65
CBOW-S	51.84	56.78	54.22	58.02	52.82	57.97
CBOW-L	53.72	57.09	54.57	57.51	52.65	57.41
CBOW-SL	52.89	57.15	52.63	56.80	53.21	58.41

Table 5: Extrinsic evaluation results for vectors with different pre-processing: Original text, Sentence-shuffled (S), lowercased (L), and both (SL)

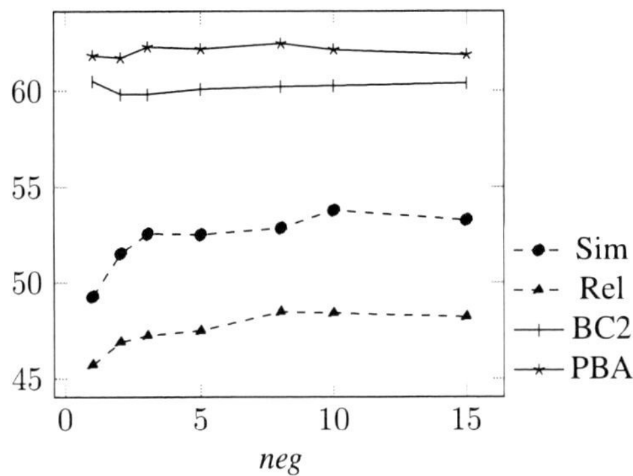

Figure 1: Average intrinsic and extrinsic evaluation results for negative sampling (Unit: ρ: dashed line, F-score: solid line)

neg	PMC-PubMed		PMC		PubMed	
	Sim	Rel	Sim	Rel	Sim	Rel
1	0.52	0.483	0.453	0.405	0.505	0.483
2	0.545	0.493	0.489	0.439	0.511	0.475
3	0.539	0.488	0.506	0.447	0.532	0.482
5	0.538	0.487	0.498	0.444	0.54	0.494
8	0.545	**0.501**	0.497	0.446	0.543	0.507
10	0.543	0.494	0.517	0.459	**0.553**	0.499
15	0.542	0.498	0.514	0.457	0.542	0.491

Table 6: Intrinsic evaluation results for number of negative samples (default = 5)

3 Results

3.1 Skip-grams vs. CBOW

Tables 4 and 5 (first 2 rows) show results comparing the skip-gram and CBOW models with default hyper-parameter values in intrinsic and extrinsic evaluation, respectively. In general, the skip-gram vector shows better results than CBOW in both the word similarity task and in entity mention tagging. In CBOW, the representations of a group of context words are learned through predicting one focus word, with the prediction back-propagated averaged over all context words. By contrast, in skip-gram, the representation of a focus word is learned by predicting every other context word in the window separately, with the prediction error of each context word back-propagated to the target word. This may allow better vectors to be learned as a focus word is trained over more data, but with less smoothing over contexts. Our result is consistent with that of many previous studies, including that of Muneeb et al. (2015), who compared model architectures on different vector dimensions and reported that skip-gram outperforms CBOW in biomedical domain tasks.

From Tables 4 and 5, we see that most vectors benefit from lower-casing and shuffling the corpus sentences. Since in word2vec, the learning rate is decayed as training progresses, text appearing early has a larger effect on the model. Shuffling makes the effect of all text (roughly) equivalent. On the other hand, lower-casing ensures that same word but different cases, such as *protein*, *Protein* and *PROTEIN* are normalised (indexed as one term) for training. Although the shuffled-lower vectors perform better, in the following, we report further results based on the unshuffled-text vector to preserve the comparability of results.

3.2 Hyper-Parameters

We next show that four out of the six hyper-parameters only improve performance notably in the intrinsic task but not the extrinsic one, while one boosts figures in both tasks to a great extent. Lastly, one of them shows opposite effects on intrinsic and extrinsic evaluations.

neg	PMC-PubMed		PMC		PubMed	
	BC2	PBA	BC2	PBA	BC2	PBA
1	60.78	62.29	59.90	61.52	60.80	61.71
2	60.41	62.03	59.44	60.49	59.59	62.63
3	59.37	62.42	59.55	62.02	60.52	62.45
5	60.37	61.90	59.44	62.12	60.44	62.56
8	60.90	62.19	59.49	62.55	60.23	62.68
10	59.65	62.80	59.58	61.61	**61.53**	62.03
15	61.09	61.52	59.92	60.98	60.12	**63.18**

Table 7: Extrinsic evaluation results for number of negative samples (default = 5)

samp	PMC-PubMed		PMC		PubMed	
	Sim	Rel	Sim	Rel	Sim	Rel
None	0.529	0.476	0.465	0.419	0.514	0.451
1e-1	0.542	0.496	0.476	0.42	0.507	0.46
1e-2	0.521	0.464	0.471	0.418	0.513	0.471
1e-3	0.545	0.5	0.497	0.442	0.545	0.494
1e-4	0.56	0.506	0.521	0.459	0.578	0.54
1e-5	0.594	0.542	0.55	0.507	0.589	0.546
1e-6	**0.601**	**0.558**	0.511	0.491	0.546	0.528
1e-7	0.519	0.475	0.401	0.37	0.336	0.306
1e-8	0.09	0.055	0.074	-0.016	-0.061	-0.146
1e-9	-0.074	-0.166	-0.076	-0.183	0.078	0.147

Table 8: Intrinsic evaluation results for sub-sampling (default = 1e-3)

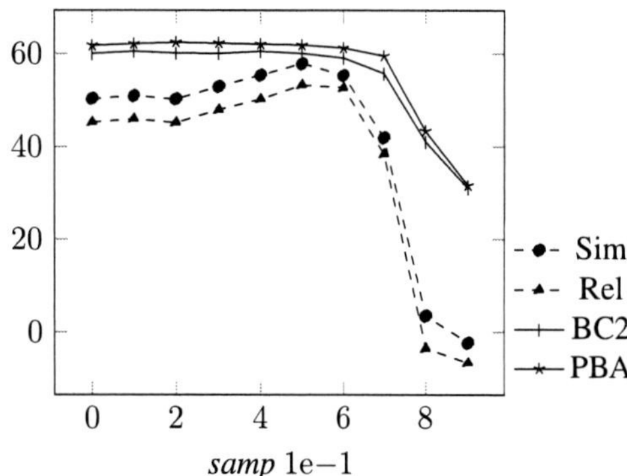

Figure 2: Average intrinsic and extrinsic results for sub-sampling (0 = None)
(Unit: ρ: dashed line, F-score: solid line)

samp	PMC-PubMed		PMC		PubMed	
	BC2	PBA	BC2	PBA	BC2	PBA
None	60.46	61.76	58.83	61.35	60.51	62.00
1e-1	**61.31**	60.99	59.60	62.45	60.47	62.69
1e-2	60.01	62.51	59.86	61.63	60.29	**62.92**
1e-3	60.30	61.99	59.78	61.95	59.87	62.57
1e-4	60.93	62.73	59.87	60.91	60.51	62.22
1e-5	60.58	61.39	60.35	61.26	58.98	62.60
1e-6	60.00	61.67	57.94	60.31	59.02	61.35
1e-7	57.52	61.17	57.04	59.70	52.44	57.34
1e-8	47.35	50.41	44.22	47.23	31.23	32.15
1e-9	33.09	33.13	32.30	32.68	27.40	28.70

Table 9: Extrinsic evaluation results for sub-sampling (default = 1e-3)

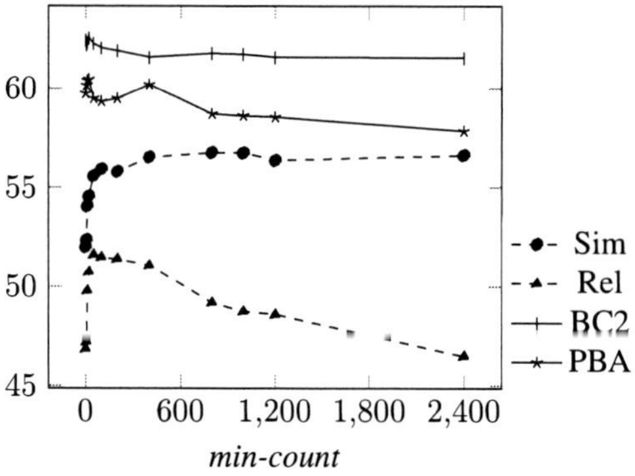

Figure 3: Average intrinsic and extrinsic evaluation results for min-counts
(Unit: ρ: dashed line, F-score: solid line)

3.2.1 Negative Sampling, Sub-sampling, Min-count and Learning Rate

Intuitively, larger values of the *neg* parameter could be expected to benefit the training process by providing more (negative) examples, but we can only see a benefit in the intrinsic result (Figure 1). The performance of word vectors on the intrinsic task generally improves as *neg* increases from 1 to 8 (Table 6), whereas extrinsic task performance remains approximately the same (Table 7). We refer to Levy et al. (2015) for further analysis of the effect of the skip-gram parameter in a general domain context.

Regarding *sub-sampling*, a lower threshold gives more words a probability of being downsampled. From Figure 2, it appears that also sub-sampling has a large effect on the intrinsic task, where most figures increase substantially before *samp* = 1e-6 (Table 8). After *samp* = 1e-7, figures in both measures drop dramatically. While some extremely frequent words (e.g. *the*) are effectively non-informative, other common words may be important for modeling word meaning. Thus, when the sub-sampling threshold decreases continuously, a substantial amount of informative frequent words are downsampled, leading to an ineffective learning of the representation.

Words occurring fewer than *min-count* times will be completely removed from the corpus, resulting in fewer words in the word vectors. From Figure 3, most of the results show limited effect for this parameter, excepting a notable increase for PubMed vectors in the intrinsic task (Table 10).

min-count	PMC-PubMed		PMC		PubMed	
	Sim	Rel	Sim	Rel	Sim	Rel
0	0.543	0.498	0.512	0.444	0.505	0.462
5	0.534	0.485	0.492	0.437	0.544	0.494
10	0.536	0.487	0.528	0.485	0.557	0.521
20	0.531	0.499	0.531	0.492	0.574	0.531
50	0.551	0.523	0.535	0.49	0.581	0.534
100	0.546	0.508	0.553	0.502	0.578	0.534
200	0.547	0.513	0.536	0.49	0.591	**0.538**
400	0.555	0.522	0.543	0.479	0.598	0.531
800	0.55	0.492	0.55	0.467	0.603	0.517
1000	0.551	0.503	0.529	0.443	**0.622**	0.515
1200	0.56	0.506	0.531	0.452	0.601	0.499
2400	0.565	0.485	0.517	0.405	0.616	0.504

Table 10: Intrinsic evaluation results for min-count (default = 5)

min-count	PMC-PubMed		PMC		PubMed	
	BC2	PBA	BC2	PBA	BC2	PBA
0	61.04	62.03	59.73	61.92	59.74	**63.41**
5	60.56	61.83	59.75	61.80	60.52	62.98
10	60.42	62.48	60.22	61.50	60.56	62.98
20	60.64	62.92	60.24	62.17	60.67	62.56
50	**61.32**	62.17	59.58	62.06	59.41	62.59
100	60.59	62.37	58.76	61.47	59.90	62.30
200	59.87	61.39	58.97	61.82	60.00	62.53
400	59.75	62.08	59.95	61.04	60.42	61.62
800	59.35	61.79	59.53	61.75	57.88	61.79
1000	59.98	62.08	58.54	60.98	58.67	62.16
1200	59.26	62.34	58.75	60.74	58.34	61.66
2400	59.49	62.44	58.58	61.54	57.11	60.70

Table 11: Extrinsic evaluation results for min-count (default = 5)

However, our intrinsic evaluations, following the standard protocol, ignore words that are excluded by *min-count*. Hence, for PubMed vectors, when *min-count* $= 400$, only about half of the assessment items are used in intrinsic evaluation. This implies that the result in *min-count* > 400 only reflects the representation of frequent words. By contrast, as the out-of-vocabulary rate in extrinsic tasks is about 2.6%, its influence is less notable.

The learning process will be unstable if the *learning rate* is too large and will be slow if it is too small. From table 12 and table 13, *alpha* $= 0.05$ appears to be an optimal value, for which most of the vectors have their best or second best results in both evaluations.

3.2.2 Vector Dimension (*dim*)

The effect of vector dimension on our vectors is notable in all tasks (Figure 5). In Tables 14 and 15, we see a large improvement in all evaluations when the vector dimension grows. Although the improvement for extrinsic measures stops when $dim > 200$, it is evident that an increase from low

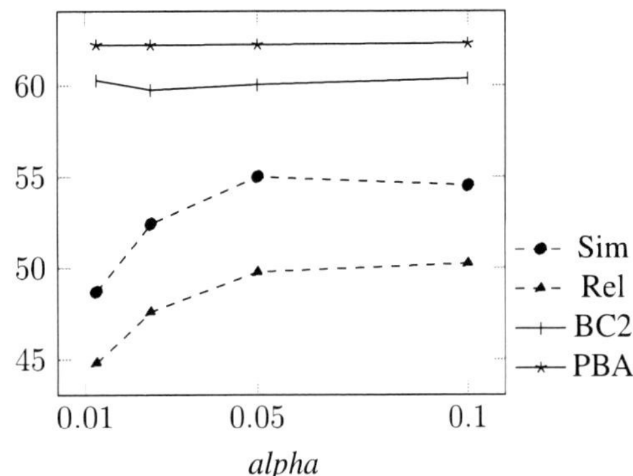

Figure 4: Average intrinsic and extrinsic evaluation results for learning rate (Unit: ρ: dashed line, F-score: solid line)

alpha	PMC-PubMed		PMC		PubMed	
	Sim	Rel	Sim	Rel	Sim	Rel
0.0125	0.511	0.468	0.442	0.401	0.508	0.475
0.025	0.538	0.492	0.492	0.441	0.543	0.493
0.05	0.55	0.501	0.516	0.46	**0.584**	0.532
0.1	0.542	0.504	0.511	0.46	0.583	**0.543**

Table 12: Intrinsic evaluation results for learning rate (default = 0.025)

alpha	PMC-PubMed		PMC		PubMed	
	BC2	PBA	BC2	PBA	BC2	PBA
0.0125	60.03	61.41	60.24	62.04	60.57	**63.29**
0.025	59.57	61.86	59.86	62.16	59.83	62.68
0.05	59.80	62.86	59.54	61.25	**60.77**	62.65
0.1	60.41	62.38	60.40	61.94	60.30	62.64

Table 13: Extrinsic evaluation results for learning rate (default = 0.025)

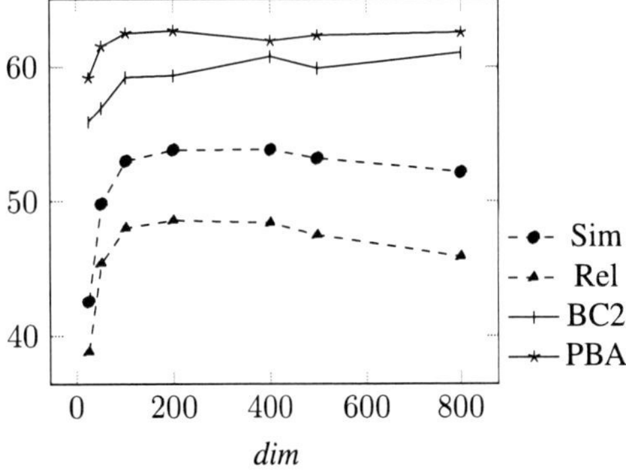

Figure 5: Average intrinsic and extrinsic evaluation results for vector dimension (Unit: ρ: dashed line, F-score: solid line)

dim gives a very substantial improvement.

	PMC-PubMed		PMC		PubMed	
dim	Sim	Rel	Sim	Rel	Sim	Rel
25	0.426	0.38	0.385	0.346	0.466	0.438
50	0.508	0.461	0.452	0.407	0.534	0.494
100	0.537	0.491	0.509	0.459	0.543	0.491
200	0.552	0.504	0.511	0.459	0.551	0.495
400	**0.562**	0.505	0.518	0.469	0.534	0.477
500	0.553	**0.507**	0.511	0.447	0.531	0.47
800	0.544	0.479	0.51	0.448	0.51	0.45

Table 14: Intrinsic evaluation results for vector dimension (default = 100)

	PMC-PubMed		PMC		PubMed	
dim	BC2	PBA	BC2	PBA	BC2	PBA
25	56.33	59.14	55.38	58.06	55.77	60.26
50	59.03	61.38	57.24	61.40	57.57	61.75
100	60.81	62.39	60.84	62.17	60.38	62.88
200	61.22	**63.04**	60.13	62.27	**61.24**	62.68
400	61.17	61.57	60.18	61.61	60.54	62.50
500	60.89	62.21	60.81	62.38	61.03	62.36
800	61.00	62.30	60.43	62.34	60.59	62.92

Table 15: Extrinsic evaluation results for vector dimension (default = 100)

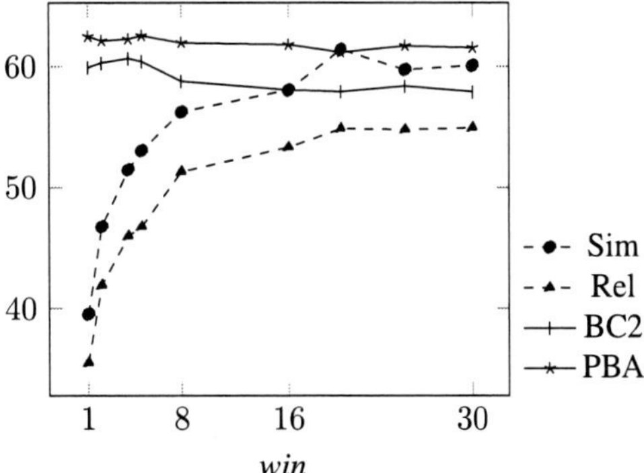

Figure 6: Average intrinsic and extrinsic evaluation results for window size (Unit: ρ: dashed line, F-score: solid line)

	PMC-PubMed		PMC		PubMed	
win	Sim	Rel	Sim	Rel	Sim	Rel
1	0.419	0.377	0.342	0.302	0.425	0.387
2	0.488	0.43	0.422	0.374	0.493	0.454
4	0.528	0.477	0.485	0.425	0.53	0.478
5	0.545	0.494	0.496	0.412	0.55	0.497
8	0.562	0.516	0.544	0.487	0.581	0.536
16	0.589	0.535	0.556	0.506	0.597	0.557
20	**0.66**	0.558	0.562	0.513	0.619	0.574
25	0.6	0.543	0.582	0.531	0.61	0.568
30	0.605	0.541	0.571	0.522	0.627	**0.584**

Table 16: Intrinsic evaluation results for context window size (default = 5)

	PMC-PubMed		PMC		PubMed	
win	BC2	PBA	BC2	PBA	BC2	PBA
1	**61.28**	62.23	60.18	62.44	60.93	62.70
2	60.81	61.74	60.83	61.59	61.11	**63.01**
4	61.29	62.45	60.43	61.43	60.74	62.86
5	59.87	62.25	60.08	62.51	59.47	62.80
8	59.52	61.83	58.78	61.26	60.40	62.74
16	59.82	61.41	59.40	61.30	60.18	62.62
20	59.54	60.80	59.92	60.92	60.02	61.76
25	58.86	60.86	58.91	61.41	58.98	62.79
30	57.83	61.28	57.61	60.53	59.22	62.83

Table 17: Extrinsic evaluation results for context window size (default = 5)

Parameter	Setting
Corpus	PubMed
Architecture	skip-gram
neg	10
dim	200
alpha	0.05
samp	1e-4
win	2, 30
min-count	5

Table 18: Settings selected for comparative evaluation

3.2.3 Context Window Size (*win*)

We find contradictory results from changing the size of the context window parameter (Figure 6). All three sets of vectors show a notable increase in the intrinsic measures when the context window size grows (Table 16). However, the extrinsic evaluation shows the opposite pattern (Table 17): all results in extrinsic tasks have an early perofmance peak with a narrow window (e.g. *win* = 1), followed by a gradual decrease when window size increases. One possible explanation may be that a larger window emphasizes the learning of domain/topic similarity between words, while a narrow context window leads the representa-

tion to primarily capture word function (Turney, 2012). It is possible that for intrinsic evaluation datasets such as UMNSRS it is more important to model topical rather than functional similarity. Conversely, it is intuitively clear that for tasks such as named entity recognition the modeling of functional similarity such as co-hyponymymy is centrally important. For further discussion on the effect of the context window size parameter, we refer to Hill et al. (2015) and Levy et al. (2015).

3.3 Comparative evaluation

Based on the parameter selection experiments covering three corpora (PMC, PubMed and both), various preprocessing options (normal-text, sentence-shuffled text, lower-cased text), two model architectures (skip-gram vs. CBOW) and six hyper-

	Sim	Rel	BC2	PBA
PubMed, win 2 (ours)	0.56	0.507	76.89	**64.13**
PubMed, win 30 (ours)	**0.652**	**0.601**	75.51	63.15
Pyysalo et al. (PMC-PubMed)	0.523	0.48	**77.01**	63.6
Pyysalo et al. (PMC)	0.453	0.396	75.48	63.66
Pyysalo et al. (PubMed)	0.549	0.506	76.47	63.66
Kosmopoulos et al. (BioASQ)	0.589	0.509	75.51	62.85

Table 19: Intrinsic and extrinsic evaluation with comparison to baseline vectors

parameters, we selected the best-performing options for comparative evaluation against the baseline vectors (Table 18). Since the size of the context window (*win*) showed contradictory results between the intrinsic and extrinsic tasks, we created vectors for two different values of this parameter. Note that for this comparative evaluation we use the test sets and test evaluation scripts of the two extrinsic tasks.

Table 19 summarizes the results of the comparative evaluation. For our intrinsic tasks, our vectors with *win* = 30 show the best performance, clearly outperforming the baselines as well as our otherwise identically created vectors with *win* = 2. This further supports the suggestion that a higher context window facilitates the learning of domain similarity for the intrinsic task. For extrinsic tasks, while the difference to the baselines is smaller, our vectors with *win* = 2 show the best results for JNLPBA and the second best in BC2GM, while the vectors with *win* = 30 are clearly less competitive.

The comparative evaluation on test set data thus confirms the indications from parameter selection that the context window size has opposite effects on the intrinsic and extrinsic metrics and indicates that our experiments have succeeded in creating a pair of word embeddings that show state-of-the-art performance when applied to tasks appropriate for each.

3.4 Discussion

In this study, we have created vectors with PubMed, PMC and the combination of the two with a large variety of different model, preprocessing and parameter combinations. While in theory a larger corpus is expected to benefit the learning of word representations, we find that in many cases this does not hold, in particular with the combination of PubMed and PMC showing lower results than PubMed alone. We offer two possible explanations for this surprising find-

ing, which contradicts some prior in-domain results. First, we used PMC texts recently introduced by PubMed Central using an incompletely documented extraction process, and preliminary examination suggests that the proportion of non-prose text in this material may be quite high, potentially affecting learning. An alternative explanation may be that the *word2vec* implementation has a (somewhat hidden) "reduce-vocab" function that triggers rare-word removal when the size of the corpus crosses certain thresholds: the larger the corpus size, the more aggressive the trimming. Preliminary results suggests that this functionality may have affected PMC-PubMed, our largest corpus, to a larger extent than the other corpora. We leave the resolution of this question for future work.

4 Conclusion and future work

In this study, we show how the performance of word vectors changes with different corpora, preprocessing options (normal text, sentence-shuffled text, lower-cased text), model architectures (skip-gram vs. CBOW) and hyper-parameter settings (negative sampling, sub sample rate, min-count, learning rate, vector dimension, context window size). For corpora, sentence-shuffled PubMed texts appear to produce the best performance, exceeding that of the notably larger combination with PMC texts.

For hyper-parameter settings, it is evident that performance can be notably improved over the default parameters, but the effects of the different hyper-parameters on performance are mixed and sometimes counterintuitive. We have previously found a similar result in general domain work (with Wikipedia text) (Chiu et al., 2016).

Several directions remain open for future work. First, in addition to tuning individual parameters in isolation, we can study the effect of tuning two or more parameters simultaneously. In addition, the number of training iterations was not considered in the experiments here, and careful tuning of this parameter both separately and jointly with associated parameters such as *alpha* may offer further opportunities for improvement.

Acknowledgments

This work has been supported by Medical Research Council grant MR/M013049/1

References

Yoshua Bengio, Réjean Ducharme, Pascal Vincent, and Christian Jauvin. 2003. A neural probabilistic language model. *Journal of Machine Learning Research*, pages 1137–1155.

Steven Bird. 2006. Nltk: the natural language toolkit. In *Proceedings of COLING/ACL demos*, pages 69–72.

Billy Chiu, Anna Korhonen, and Sampo Pyysalo. 2016. Intrinsic evaluation of word vectors fails to predict extrinsic performance. *Proceedings of RepEval 2016*.

Ronan Collobert and Jason Weston. 2008. A unified architecture for natural language processing: Deep neural networks with multitask learning. In *Proceedings of ICML*, pages 160–167. ACM.

Javi Fernández, Yoan Gutiérrez, José M Gómez, and Patricio Martınez-Barco. 2014. Gplsi: Supervised sentiment analysis in twitter using skipgrams. In *Proceedings of SemEval*, pages 294–299.

Felix Hill, Roi Reichart, and Anna Korhonen. 2015. Simlex-999: Evaluating semantic models with (genuine) similarity estimation. *Computational Linguistics*.

Jin-Dong Kim, Tomoko Ohta, Yoshimasa Tsuruoka, Yuka Tateisi, and Nigel Collier. 2004. Introduction to the bio-entity recognition task at JNLPBA. In *Proceedings of JNLPBA*, pages 70–75.

Aris Kosmopoulos, Ion Androutsopoulos, and Georgios Paliouras. 2015. Biomedical semantic indexing using dense word vectors in bioasq. *Journal Of Biomedical Semantics*.

Gabriella Lapesa and Stefan Evert. 2014. A large scale evaluation of distributional semantic models: Parameters, interactions and model selection. *Transactions of the Association for Computational Linguistics*, 2:531–545.

Omer Levy, Yoav Goldberg, and Ido Dagan. 2015. Improving distributional similarity with lessons learned from word embeddings. *TACL*, 3:211–225.

Tomas Mikolov, Kai Chen, Greg Corrado, and Jeffrey Dean. 2013a. Efficient estimation of word representations in vector space. In *Proceedings of ICLR*.

Tomas Mikolov, Ilya Sutskever, Kai Chen, Greg S Corrado, and Jeff Dean. 2013b. Distributed representations of words and phrases and their compositionality. In *Proceedings of NIPS*, pages 3111–3119.

TH Muneeb, Sunil Kumar Sahu, and Ashish Anand. 2015. Evaluating distributed word representations for capturing semantics of biomedical concepts. In *Proceedings of ACL-IJCNLP*, page 158.

Serguei Pakhomov, Bridget McInnes, Terrence Adam, Ying Liu, Ted Pedersen, and Genevieve B Melton. 2010. Semantic similarity and relatedness between clinical terms: an experimental study. In *Proceedings of AMIA*, volume 2010, page 572.

Jeffrey Pennington, Richard Socher, and Christopher D Manning. 2014. Glove: Global vectors for word representation. In *Proceedings of EMNLP*, volume 14, pages 1532–1543.

Sampo Pyysalo, Filip Ginter, Hans Moen, Tapio Salakoski, and Sophia Ananiadou. 2013. Distributional semantics resources for biomedical text processing. *Proceedings of LBM*.

Rune Sætre, Kazuhiro Yoshida, Akane Yakushiji, Yusuke Miyao, Yuichiro Matsubayashi, and Tomoko Ohta. 2007. Akane system: protein-protein interaction pairs in biocreative2 challenge, ppi-ips subtask. In *Proceedings of BioCreative II*, pages 209–212.

Larry Smith, Lorraine K Tanabe, Rie Johnson nee Ando, Cheng-Ju Kuo, I-Fang Chung, Chun-Nan Hsu, Yu-Shi Lin, Roman Klinger, Christoph M Friedrich, Kuzman Ganchev, et al. 2008. Overview of biocreative ii gene mention recognition. *Genome biology*, 9(Suppl 2):1–19.

Pontus Stenetorp, Hubert Soyer, Sampo Pyysalo, Sophia Ananiadou, and Takashi Chikayama. 2012. Size (and domain) matters: Evaluating semantic word space representations for biomedical text. In *Proceedings of SMBM*.

Joseph Turian, Lev Ratinov, and Yoshua Bengio. 2010. Word representations: a simple and general method for semi-supervised learning. In *Proceedings of ACL*, pages 384–394.

Peter D Turney. 2012. Domain and function: A dual-space model of semantic relations and compositions. *Journal of Artificial Intelligence Research*, pages 533–585.

An Information Foraging Approach to Determining the Number of Relevant Features

Brian Connolly
Cincinnati Children's
Hospital Medical Center
Burnet Ave
Cincinnati, OH 45229
`brian.connolly`
`@cchmc.org`

Benjamin Glass
Cincinnati Children's
Hospital Medical Center
Burnet Ave
Cincinnati, OH 45229
`benjaminglass`
`@gmail.com`

John P. Pestian
Cincinnati Children's
Hospital Medical Center
3333 Burnet Ave
Cincinnati, OH 45229
`john.pestian`
`@cchmc.org`

Abstract

For many types of high-dimensional data, such as natural language corpora, the vast majority of extracted variables or features are essentially noise. Culling such features can not only reveal important patterns, but also improve the performance of supervised and unsupervised machine algorithms. Most research on feature selection has focused on the statistical measures used to rank features. Meanwhile, little work has been done developing techniques for identifying the optimal subset of features without repeatedly training models. However, developing such techniques is important, as they can significantly decrease computation time while providing a way to determine the features that characterize the classes within a data set, independent of how the data may be classified in the future. Here we introduce a novel method based on information foraging that works in conjunction with existing feature ranking methods to automatically determine a subset of important features. The method is demonstrated on simulated and linguistic data from psychiatric interviews. We show that the method is able to accurately determine the features that characterize the classes within both data sets. The method is fast, simple, and independent of any method of classifying the data, and can be extended to any high-dimensional data set.

1 Background

For many types of high-dimensional data, such as natural language corpora, gene microarrays, and images, the vast majority of extracted vari-ables or features are essentially noise (Yu and Liu, 2004). Culling such features can not only reveal important patterns, but also improve the perfor-mance of supervised and unsupervised machine algorithms (Guyon and Elisseeff, 2003; Saeys et al., 2007). For example, Pestian et al. (Pestian et al., 2016) have recently used natural language processing (NLP) and supervised machine learn-ing methods to automatically distinguish suici-dal from non-suicidal patients using words and phrases from psychiatric interviews (Pestian et al., 2016). In that work, identifying which types of words and phrases were most discriminative not only improved classification performance, but also provided important insights into the language of those at risk of suicide.

Feature selection is usually done in the context of optimizing machine learning models, and so feature selection techniques are divided into three categories by how they relate to the search over such models: filter, wrapper, and embedded meth-ods (Blum and Langley, 1997; Saeys et al., 2007). Filter methods rank features using a statistical measure of relevance (Forman, 2003; Yang and Pedersen, 1997). Typically, lower-ranked features are removed prior to training a machine learning model. By contrast, in wrapper methods, the opti-mal feature subset is identified by repeatedly train-ing a model on multiple feature subsets and eval-uating its performance (Kohavi and John, 1997). The search for an optimal model is "wrapped" in the feature subset search. Finally, in embedded methods the feature search is performed in con-junction with the model search. For example, the number of parameters can be incorporated as a regularization term to be minimized in the objec-tive function (Weston et al., 2003).

Most research on feature selection has focused on the statistical measures used to rank features (Forman, 2003; Yang and Pedersen, 1997). Mean-

175

Proceedings of the 15th Workshop on Biomedical Natural Language Processing, pages 175–180,
Berlin, Germany, August 12, 2016. ©2016 Association for Computational Linguistics

while, little work has been done developing techniques for identifying the optimal subset of features without repeatedly training models (Koller and Sahami, 1996; Ding and Peng, 2005). However, developing such techniques is important, as they can significantly decrease computation time while providing a way to determine the features that characterize classes within a data set, independent of any classification method.

Here we introduce a novel method based on information foraging that works in conjunction with existing feature ranking methods to automatically determine a subset of important features. Information foraging is a behavioral model for maximizing the rate of attaining valuable information (Pirolli and Card, 1999). It assumes that useful information exists in a patchy structure, where the diminishing return of a continued search in a patch must be balanced with the time cost of moving to a new patch.

The utility of our approach is best demonstrated with an example of a typical feature selection approach for text classification. Suppose a large data set of text documents is divided into multiple classes. We want to classify documents into the correct categories using word frequencies. Typically, a text data set may contain many thousands of unique words, most of which have no discriminative power (Scott and Matwin, 1999). Feature selection is used to determine the features that best discriminate between the classes, thereby optimizing classifier performance. A univariate filter method, such as information gain (Fano and Wintringham, 1961) for discrete data, or Analysis of Variance (ANOVA) (Michel et al., 2008) for continuous variables, may be applied to rank the features by their discriminative power. A subset of top-ranked features are then chosen based on some ad-hoc threshold, or by using a wrapper method, where classifiers are built using various sets of top ranked features. The classifier with the best performance then determines the best feature subset. Classifier performance is evaluated using some flavor of bootstrapping, potentially making this method computationally expensive.

In this scenario, the optimal number of features is defined by both the method of ranking features and the classifier; there is no 'objective' determination of which features characterize the classes.

From a computational perspective, no matter how efficient the subset search strategy, a wrapper or embedded method which entails training models will be more costly than a univariate filter subset selection which runs in O(N) time. Other work on filter-only methods for subset selection has been primarily multivariate, identifying correlations between variables and eliminating redundant ones. Hall and Smith (Hall and Smith, 1997) used Pearson's correlation for forward selection filtering, with good results on fairly low-dimensional data. Others (Koller and Sahami, 1996; Yu and Liu, 2004; Ding and Peng, 2005) have used Markov blanket filtering to iteratively remove redundant features via backward elimination. These generally have a complexity of $O(N^2)$.

In this work, we show the proposed foraging-based feature selection leads to performance gains comparable to wrapper methods on a text classification task, while running in linear time. In addition, the algorithm is useful simply for the objective identification of a relevant feature subset, since it is deterministic and entirely independent of the choice of learning algorithm. Further, the method is not tied to a particular feature ranking method, but rather it simply provides a method of determining the optimal number of features *given* a ranking method.

2 Theory

The method of selecting the number of features is based on the Holling's Disk equation (Holling, 1959), which has been used to explain the foraging behavior of both animals (Stephens, 1990; Stephens and Krebs, 1986) and humans (Winterhalder and Smith, 1992). It has also been useful in understanding information foraging (e.g., in web searches (Pirolli, 2007)). The equation is dependent on three variables: the time spent gathering energy from a certain food type i (t_{Wi}), the amount of time it takes to travel to that food type $1/\lambda_i$, and the energy gained from that food type ($g_i(t_{Wi})$). The overall rate of gain for k food sources is then

$$R(k) = \frac{\sum_{i=1}^{k} \lambda_i g_i(t_{Wi})}{1 + \sum_{i=1}^{S} \lambda_i t_{Wi}}. \tag{1}$$

Given S food types, the optimal diet is then found through an algorithm suggested by (Stephens and Krebs, 1986). In this algorithm, the profitability of the food type, given by $g_i(t_{Wi})/t_{Wi}$, is ranked so that $g_1(t_{W1})/t_{W1} > g_2(t_{W2})/t_{W2} > ... > g_S(t_{WS})/t_{WS}$. Food types are added until the rate of gain for a type of top k food types is greater than the $k + 1$ food type; that is, until

$$R(k) > g_{k+1}/t_{Wk+1}. \tag{2}$$

For our purposes, feature subset selection is modeled as a diet optimization task, where features are represented by food types, and a diet is a subset of features. Each feature or food type added to the diet may add gain in terms of the informativeness of the feature, but entails cost in terms of sparseness.

In the present work, the gain is defined by the *informativeness* obtained from feature i, which is broadly defined by any parametrization of the statistical differences between classes. As the class differences for a given feature will be defined in this work as a p-value, we choose two definitions of informativeness which increase with the differences between classes: $1 - p_X$ and $1/p_X$, where p_X is defines as the p-value from either the KS-tests or ANOVA. The time between food types is taken as the mean number of data points between appearances of feature i (Jones, 1987), where each data point equals one time unit. The time spent gathering energy from a food type is arbitrarily set to unity for all i ($t_{Wi} = 1$); λ_i is defined as

$$\lambda_i = \frac{\text{Sum of Non} - \text{ZeroFrequencies for Feature } i}{\text{TotalDataPoints}}. \tag{3}$$

This is the same equation as the reciprocal of the mean time between failures, where "failures" are taken to be non-zero feature frequencies.

3 Experiments

The method is demonstrated on two kinds of data: simulated data sets and a linguistic data set from a clinical trial.

The goal of the simulated experiments is to show that the method is able to accurately identify subsets of features with inter-class statistical differences. In these experiments, the performance of the algorithm is evaluated based on its ability to accurately identify these subsets. The goal of demonstrating the method on clinical trial data is to evaluate the method within a more realistic context of a wrapper method applied to linguistic data. Evaluating the method's performance on such data also illustrates its behavior on data containing redundant and correlated features.

Each simulated data set is comprised of data points from two classes. (The number of data are kept small to reflect the small sample sizes typically found in clinically annotated NLP data sets (Hutton, 2012).) The data from the first class (class A) are generated from a Gaussian distribution with mean 0 and standard deviation σ. The data from the second class (class B) are generated from two Gaussian distributions; $f \times 100\%$ of the features are generated with mean 1 and standard deviation σ, while the rest of the features are generated in the same fashion as those from class A, with mean 0 and standard deviation σ. In this way, $f \times 100\%$ of the features are generated with inter-class differences.

The performance of the algorithm is then evaluated as a function of the definition of gain, sparsity of the data (s), the total number of features (F), number of features with statistical differences (f), and statistical differences between features (parameterized by σ). The gain is define in four ways: as 1-p-value from the Kolmogorov-Smirnov test (Darling, 1957) ($1 - p_{KS}$), 1-p-value from ANOVA (Fisher, 1992) ($1 - p_{ANOVA}$), and the reciprocal of the KS and ANOVA p-values ($1/p_{KS}$ and $1/p_{ANOVA}$, respectively). The influence of λ_i is also studied by setting it to its empirical value and to unity. When they are not being varied, the default values for F, s, σ and f are: $1,000$, 0.5, 0.2 and 0.5, respectively.

The data from the clinical trial are derived from the Suicide Thought Markers study (Pestian et al., 2016). In this study, three hundred seventy-nine adults and adolescents from Cincinnati Childrens Hospital Medical Center (CCHMC), University of Cincinnati (UC), and Princeton Community Hospital (PCH) were enrolled during the course of the study between October 2013 and March 2015. Participants were evenly divided into three subject groups: suicidal, patients with mental illness, and controls. Suicidal subjects consisted of patients who presented in the Emergency Department (ED) with suicidal ideation or behaviors; the mental illness group was not suicidal, but had a mental health diagnosis; and the control group had no mental illness diagnosis and was not suicidal.

Subjects were then asked five open-ended, ubiquitous questions (UQs) (Pestian, 2010; Pestian et al., 2015): Do you have hope?, "Do you have any fear?", "Do you have any secrets?", "Are you angry?", and "Does it hurt emotionally?". These questions were intended to stimulate conversation for language sampling, and would later form the basis of the training sample for the machine learning algorithm. The interviews were transcribed and the subjects words were extracted in a systematic way.

For classification purposes, each subject was characterized by (1) their subject group and (2) a vector of word (1-gram) frequencies. Due to the extreme variability of word frequencies and interview lengths, the frequencies were normalized to smooth the frequency distributions and lessen the classifiers sensitivity to interview length. The word frequencies were therefore logarithmically (log(x+1)) transformed to smooth the frequencies, and further L2-normalized at the subject level as to base the classification on relative word frequencies.

Only suicidal and control patients are used in the present work. To test the method on various sizes and types of data, the data are split three ways: patients from CCHMC (pediatric patients), patients from PCH and UC (adults patients), and patients from all three hospitals. In the end, 2,471, 4,788, and 5,457 unique words were extracted over 84, 169, and 253 suicidal and control subjects from CCHMC, PCH and UC, and all hospitals, respectively.

The number relevant of features are then evaluated using the method presented in this work, and a wrapper method whereby the performance of Support Vector Machine (SVM) classifiers are evaluated using LOO cross-validation. Note the classifications here are simplified versions of the classifications in (Pestian et al., 2016); for instance, the features here are not partitioned based on the questions.

4 Results

Figure 1 show the F_1 scores for selecting features, varying the total number of features (F), the matrix sparsity (s), σ, and the fraction of features with statistical differences. The method is able to determine the features with significant features of a large parameter space when $1 - p_X$ defines the gain. On the other hand, when the reciprocal p-values are used, the method fails spectacularly, indicating that p_X must be bounded or it must possess a more direct statistical interpretation. This aside, performance is, to a degree, invariant to the type of statistic used; the KS test p-value performs better when the matrix is sparse, while the ANOVA p-value works better when the statistical differences are small. This may be less of a reflection on the method, and more to do with the KS test's ability to detect differences in small data samples, and ANOVA's ability to detect statistical differences when the distributions are Gaussian.

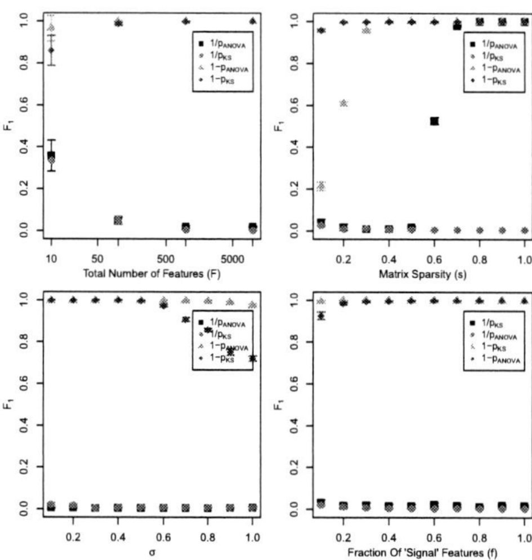

Figure 1: The F_1 score for identifying the features in simulated as a function of F, s, σ, and f. These scores are evaluated using varying definitions of gain: $1/p_{ANOVA}$ (squares), $1/p_{KS}$ (circles), $1 - p_{ANOVA}$ (triangles), and $1 - p_{KS}$ (diamonds).

Figure 2 shows the same plots with the mean time between patches set to unity ($\lambda_i = 1$). The two sets of figures look nearly identical indicating that λ_i does not play a significant role in determining the number of features.

Figure 3 shows the area under the cross-validated receiver operating curve (AROC) of the SVM classifier as a function of the number of top-ranked features. The number of features determined by our method, along with the corresponding AROC, are circled on these plots. In these plots, the relevant number of features are the minimal number of features that optimize classifier performance. When the KS test p-values are used for the gain, the method is unable to predict the optimal features. However, the oscillating performance as the number of features increase indicate the KS test may not be the best choice for feature ranking for this data set. In contrast, the ANOVA p-value is more stable, leading to more monotonic curves, and the method is better able to determine the optimal number of features.

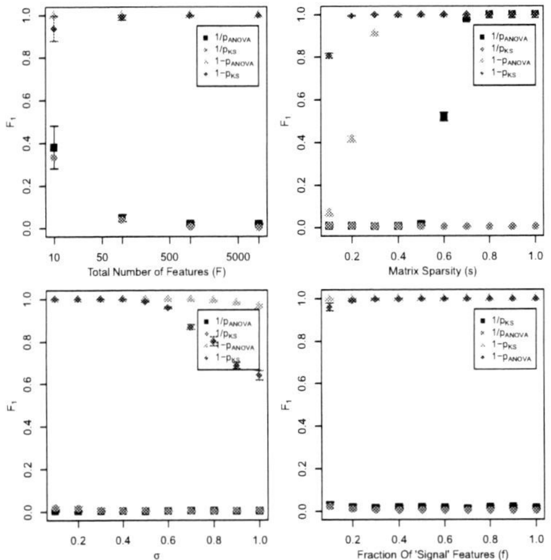

Figure 2: The plots in 1 with the mean time between patches set to unity ($\lambda_i = 1$).

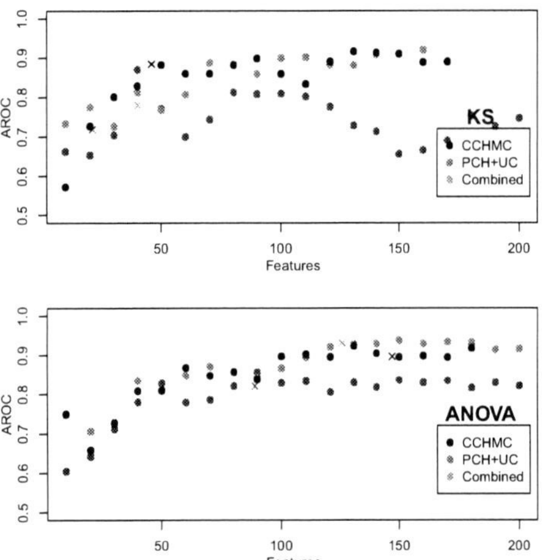

Figure 3: The top plot shows the cross-validated performance of an SVM classifier as a function of the number of top features ranked according to the KS test p-value (top) and ANOVA p-value (bottom) for the CCHMC (diamonds), UC+PCH (circles), and combined data sets (squares). The number of features determined by our method, along with the corresponding AROCs, are circled.

5   Discussion

The results from simulated data indicate there is some flexibility in the definition of informativeness, as long as the statistic gives a proper ranking of features and the statistic is bounded and/or possesses some statistical meaning. The results from real data reflect this conclusion, showing the method performs better when the feature ranking is more accurate. The decrease in classifier performance does not occur until a large number of features are introduced as input to the classifier, which is not shown in the figures. The focus of this study, however, is to determine whether or not the method presented is able to cull superfluous features; the point at which 'gain' in classification performance levels off clearly coincides with the number of features predicted by the method when the ANOVA method is used for feature ranking.

The bad performance of the method when the reciprocal of the p-values are used for the gain, indicates that the gain must be bounded in some way, or that the statistic must have a more direct statistical interpretation. In contrast, the simulated results suggest the method is fairly insensitive to the choice of λ_i, which parametrizes the sparsity of the feature.

Also, although the method is essentially built for univariate data, the performance on real data was good despite the inevitable redundancies and correlations of the features, provided the informativeness measure properly ranked the features.

6   Conclusions

We have presented a simple, fast, and effective method of determining the number of features that characterize classes within a data set where the features are univariate. We have also show it to be useful in determining the features in a linguistic data set, despite the features' inherent redundancies and correlations.

While the method was show to properly identify features that characterize features with interclass statistical differences, its performance is better when the statistic is able to effectively rank the features in terms of statistical relevance. We have also shown that it performs better when p-values are used, as opposed to their reciprocal, showing the definition of informativeness is important. Whether this is because a p-value is a bounded positive number less than 1 or because it has a direct statistical interpretation merits exploration. For instance, the question remains, could any statistic that effectively rank features be inserted into a softmax function and be used to parameterize gain? Also, the method would doubtlessly perform better if correlations and redundancies were somehow accounted for, possibly by grouping correlated features.

References

Avrim L Blum and Pat Langley. 1997. Selection of relevant features and examples in machine learning. *Artificial Intelligence*, 97(1):245–271.

Donald A Darling. 1957. The kolmogorov-smirnov, cramer-von mises tests. *The Annals of Mathematical Statistics*, 28(4):823–838.

Chris Ding and Hanchuan Peng. 2005. Minimum redundancy feature selection from microarray gene expression data. *Journal of Bioinformatics and Computational Biology*, 3(02):185–205.

Robert M Fano and WT Wintringham. 1961. Transmission of information. *Physics Today*, 14:56.

RA Fisher. 1992. Statistical methods for research workers. In *Breakthroughs in Statistics*, pages 66–70. Springer.

George Forman. 2003. An extensive empirical study of feature selection metrics for text classification. *The Journal of Machine Learning Research*, 3:1289–1305.

Isabelle Guyon and André Elisseeff. 2003. An introduction to variable and feature selection. *The Journal of Machine Learning Research*, 3:1157–1182.

Mark A Hall and Lloyd A Smith. 1997. Feature subset selection: a correlation based filter approach. In *International Conference on Neural Information Processing and Intelligent Information Systems*, pages 855–858.

Crawford S Holling. 1959. Some characteristics of simple types of predation and parasitism. *The Canadian Entomologist*, 91(07):385–398.

John J Hutton. 2012. *Pediatric Biomedical Informatics: Computer Applications in Pediatric Research*, volume 2. Springer Science & Business Media.

James V Jones. 1987. *Integrated logistics support handbook*. Tab Books.

Ron Kohavi and George H John. 1997. Wrappers for feature subset selection. *Artificial Intelligence*, 97(1):273–324.

Daphne Koller and Mehran Sahami. 1996. Toward optimal feature selection.

Vincent Michel, Cécilia Damon, and Bertrand Thirion. 2008. Mutual information-based feature selection enhances fmri brain activity classification. In *Biomedical Imaging: From Nano to Macro, 2008. ISBI 2008. 5th IEEE International Symposium on*, pages 592–595. IEEE.

John P Pestian, Jacqueline Grupp-Phelan, Kevin Bretonnel Cohen, Gabriel Meyers, Linda A Richey, Pawel Matykiewicz, and Michael T Sorter. 2015. A controlled trial using natural language processing to examine the language of suicidal adolescents in the emergency department. *Suicide and life-threatening behavior*.

John P. Pestian, Michael Sorter, Brian Connolly, K. Bretonnel Cohen, Cheryl McCullumsmith, Jeffry T. Gee, Louis-Philippe Morency, Stefan Scherer, and the STM Research Group. 2016. A machine learning approach to identifying the thought markers of suicidal subjects: A prospective multicenter *Suicide and Life-Threatening Behavior*.

John Pestian. 2010. A conversation with edwin shneidman. *Suicide and life-threatening behavior*, 40(5):516–523.

Peter Pirolli and Stuart Card. 1999. Information foraging. *Psychological Review*, 106(4):643.

Peter Pirolli. 2007. *Information foraging theory: Adaptive interaction with information*. Oxford University Press.

Yvan Saeys, Iñaki Inza, and Pedro Larrañaga. 2007. A review of feature selection techniques in bioinformatics. *Bioinformatics*, 23(19):2507–2517.

Sam Scott and Stan Matwin. 1999. Feature engineering for text classification. In *ICML*, volume 99, pages 379–388. Citeseer.

David W Stephens and John R Krebs. 1986. *Foraging theory*. Princeton University Press.

DW Stephens. 1990. Foraging theory: up, down, and sideways. *Studies in avian biology*, 13:444–454.

Jason Weston, André Elisseeff, Bernhard Schölkopf, and Mike Tipping. 2003. Use of the zero norm with linear models and kernel methods. *The Journal of Machine Learning Research*, 3:1439–1461.

Bruce Winterhalder and Eric Alden Smith. 1992. Evolutionary ecology and the social sciences. *Evolutionary ecology and human behavior*, pages 3–23.

Yiming Yang and Jan O Pedersen. 1997. A comparative study on feature selection in text categorization. In *Proceedings of the 14th International Conference on Machine Learning (ICML)*, volume 97, pages 412–420.

Lei Yu and Huan Liu. 2004. Efficient feature selection via analysis of relevance and redundancy. *The Journal of Machine Learning Research*, 5:1205–1224.

Assessing the Feasibility of an Automated Suggestion System for Communicating Critical Findings from Chest Radiology Reports to Referring Physicians

Brian E. Chapman[1], Danielle L. Mowery[2], Evan Narasimhan[1], Neel Patel[1],
Wendy W. Chapman[2], Marta E. Heilbrun[1]
[1] University of Utah, Radiology, Salt Lake City, UT
[2] University of Utah, Biomedical Informatics, Salt Lake City, UT
firstname.lastname@utah.edu

Abstract

Time-sensitive communication of critical imaging findings like **pneumothorax** or **pulmonary embolism** to referring physicians is important for patient safety. However, radiology findings are recorded in free-text format, relying on verbal communication that is not always successful. Natural language processing can provide automated suggestions to radiologists that new critical findings be added to a follow-up list. We present a pilot assessment of the feasibility of an automated critical finding suggestion system for radiology reporting by assessing suggestions made by the pyConTextNLP algorithm. Our evaluation focused on the false alarm rate to determine feasibility of deployment without increasing alert fatigue. pyConTextNLP identified 77 critical findings from 1,370 chest exams. Review of the suggested findings demonstrated a 7.8% false alarm rate. We discuss the errors, which would be challenging to address, and compare pyConTextNLP's false alarm rate to false alarm rates of similar systems from the literature.

1 Introduction

The communication of critical imaging findings from the radiologist to the referring physician is a key factor in providing efficacious patient care (Lakhani et al., 2012). Currently, the most common form of communication is a physician-to-physician telephone conversation, initiated by the radiologist at the time of image interpretation. This process is tedious, inefficient, and error prone. A missed communication can result in progressed disease, hospital readmission, and even death. In the United States, the American College of Radiology suggests three hallmarks of effective methods of communication: a) supporting the ordering provider in providing optimal patient care, b) using methods that are tailored to satisfy the need for timeliness, and c) implementing methods to minimize risk of communication errors (American College of Radiology, 2014). Critical findings may result in death or severe morbidity and require urgent or emergent attention (Larson et al., 2014). These critical test results are often documented in free-text imaging notes. Natural language processing (NLP) can automatically extract, track, and report these findings in a timely manner to support patient safety efforts.

2 Related Work

Machine learning and ruled-based NLP techniques have been used to detect critical information from radiology reports to support timely communication.

Yetisgen-Yildiz et al. created a machine-learning based text-processing pipeline that leverages a maximum entropy model to identify and classify sentences conveying clinically important follow-up recommendations concerning unexpected findings (Yetisgen-Yildiz et al., 2013). Pham et al. developed an NLP pipeline to detect and classify mentions of **thromboembolic disease** from angiography and venography reports. They used naive Bayes' feature selection then support vector machines and maximum entropy for classification (Pham et al., 2014). Esuli et al. developed two novel methods for extracting radiological findings from reports: a cascaded, two-stage ensemble of taggers generated by linear-chain conditional random fields (LC-CRFs) and a confidence-weighted ensemble method combining standard LC-CRFs and the two-stage method

Proceedings of the 15th Workshop on Biomedical Natural Language Processing, pages 181–185,
Berlin, Germany, August 12, 2016. ©2016 Association for Computational Linguistics

(Esuli et al., 2013).

Rule-based approaches have also been used to address critical finding detection. Lafourcade et al. created a linguistic-based algorithm to detect semantic relations between radiological findings (Lafourcade and Ramadier, 2016). Lakhani et al. developed an algorithm with finding-specific negation dictionaries to identify nine critical findings in impression sections and demonstrated a mean false alarm rate of 4% (Lakhani et al., 2012). Lacson et al. adapted and compared performance of two NLP systems—A Nearly New Information Extraction system (ANNIE) and Information for Searching Content with an Ontology-Utilizing Toolkit (iSCOUT)—to identify **pulmonary nodules**, **pneumothorax**, and **pulmonary embolus** with overall false alarm rates of 4% (ANNIE) and 10% (iSCOUT) (Lacson et al., 2012).

In this study, we applied a simple NLP system that leverages regular expressions by extending the lexicon for critical findings. We addressed a larger set of critical findings than in prior studies and evaluated not only the accuracy of the NLP system, but the appropriateness of generating a suggestion for critical finding communication.

Our long-term goal is to develop a communication system that identifies a variety of critical findings from radiology exams, facilitates appropriate communication of the findings to referring clinicians, and supports radiologist follow-up regarding the communicated findings. The short-term goal of this study is to build upon prior work by 1) adapting an NLP algorithm to automatically identify critical findings in radiology reports and suggest them to the radiologist for communication to referring physicians, 2) assessing the false alarm rate of the critical findings suggestion system, and 3) characterizing the errors generated by the system to determine feasibility of deploying the suggestion system in a radiology clinic. In this paper, we limit our analysis to imaging of the chest.

3 Methods and Materials

3.1 Data Set

In this IRB-approved study, we obtained all radiology reports from Oct-Dec 2013 generated by a large medical center in the United States. We excluded non-diagnostic exams (e.g., interventional procedures), as well as reports generated from services other than radiology, and reports with empty impressions sections, resulting in 54,459 exams.

Mentions of critical findings in radiology reports are not common. Only 14,815 of the 54,479 reports (27%) contained critical finding expressions from our original knowledge base. Only a small portion of critical finding expressions would be expected to be observations of a new critical finding, because the majority are negated or chronic findings. For instance, from previous studies, we found that approximately 90% of pulmonary embolism mentions in radiology reports are negated.

From the 14,815 reports with critical finding expressions, we selected approximately half of the reports (7,176) for annotation. We built a Flask[1] web application for document-level annotation of the reports. Annotators used the tool to assign the following attributes to each finding mentioned in a report: *Existence* (definite negated existence, probable negated existence, ambivalent existence, probable existence, definite existence) and *Historicity* (new, chronic, historical) (Patel et al., 2016).

We annotated 39 critical findings occurring within abdomen/pelvis, chest, extremity, neuro, and spine exams. Eighteen of these critical findings were relevant to the chest: **aneursym, aortic dissection, cancer, ectasia, epiglottitis, fracture, free air, infarct, inflammation, mediastinal emphysema, pneumonia, pneumothorax, pulmonary embolism, retropharyngeal abscess, ruptured aneurysm, splenic infarct, tension pneumothorax,** and **thrombosis**.

With supervision by an attending radiologist (Author MH), two medical students (Authors NP, FN) independently annotated the impression sections of reports for any of 39 critical findings until acceptable agreement level between annotators was reached (>0.70). Each annotator then annotated reports independently, completing two-thirds of the 7,176 reports for a total 4,786 annotated reports.

From the full set of 7,176 reports, we sampled only reports from chest exams for this study, providing a development and test set of 1,538 chest exam reports. We randomly selected 168 annotated exams as a development set and further extended pyConTextNLP's knowledge base by reviewing pyConTextNLP's disagreements with the annotations. We then tested on the remaining blind set of 1,370 reports.

We split the impression section into sentences using the Python text processing package TextBlob[2] then applied pyConTextNLP [3] to identify acute, positive critical findings.

3.2 Developing an automated critical findings suggestion system with pyConTextNLP

3.2.1 pyConTextNLP

We adapted an existing NLP algorithm, pyConTextNLP, to identify critical findings and their attributes from radiology imaging reports (Chapman et al., 2011). pyConTextNLP is an extension of the NegEx (Chapman et al., 2001) and ConText (Harkema et al., 2009) algorithms and relies on user-defined knowledge bases of targets (e.g., critical finding terms such as "pulmonary embolism"), modifiers (e.g., existence terms such as "may represent"), and lexical terms (e.g., "but") that terminate the scope of the modifiers.

3.2.2 Adapting and refining pyConTextNLP

The pyConTextNLP GitHub repository has a number of database files that have been created for previous projects[4]. We modified existing knowledge bases by comparing our automated classifications using pyConTextNLP against the annotator classifications. First, we reviewed false negative findings in the development set and added new terms to the knowledge base. The number of false negatives in the development set was small. To address potential alert fatigue from false alarms, we then focused our development on evaluating false positives in the development set. An acute, positive critical finding was defined as a mention of a critical finding with the following attributes: *Historicity*-new and *Existence*-probable or definite existence. If there was more than one mention of a given finding in a report, we assigned the report the same value as that of the most positive and most new mention. In reviewing the classifications, we examined the entire pyConTextNLP document for the report so that we could determine if the classification error occurred due to the knowledge base, classification rules, or algorithm implementation. Modifications to the code and knowledge bases were made iteratively to improve positive predictive performance compared to the annotations. Changes to pyConTextNLP primarily

consisted of modifying synonyms and variants for critical findings and corresponding attributes.

3.3 Evaluating pyConTextNLP

We ran pyConTextNLP over the test set and flagged documents with acute, positive critical findings for review. A radiologist (Author MH) was provided the flagged findings and their associated imaging report then asked the question, "Would you include this critical finding in a list of findings to communicate to the referring physician?" This question goes beyond analyzing accuracy of pyConTextNLP's annotations to the more stringent question of whether the finding should be communicated to another physician, which depends not only on accurate identification of the finding, but also on contextual information. We calculated precision (Eq. 1) and false alarm rate (Eq. 2) where FP (false positive) = rejected suggestion and TP (true positive) = accepted suggestion.

$$\text{precision} = \frac{TP}{(TP + FP)} \qquad (1)$$

$$\text{false alarm rate} = 1 - \text{precision} \qquad (2)$$

4 Results

Our primary goal was to assess the false alarm rate of the critical findings suggestion system and to characterize the errors generated by the system.

In total, we detected 77 findings requiring critical communication. These findings came from only five of our 18 categories. The most prevalent flagged findings were **pneumothorax** and **pneumonia** (Table 1).

Table 1: Distribution of flagged critical findings

critical finding	count (%)
pneumothorax	38 (49%)
pneumonia	29 (38%)
fracture	6 (8%)
cancer	3 (4%)
aneursym	1 (1%)
total	**77 (100%)**

Of the 77 observed critical findings, we observed 6 false positives, resulting in a false alarm rate of 7.8% and precision of 92.2%. Of the six false positives three were **cancer**, two were **pneumonia**, and one was an **aneurysm**.

---

5 Discussion and Conclusion

Our false alarm rate (7.8%) resides within the false alarm rates (4%-10%) reported by (Lacson et al., 2012), demonstrating promising results. Our false alarm analysis revealed several challenges for the task of critical finding identification.

For **cancer**, all three cases identified by pyConTextNLP were considered chronic by the radiologist. One report requires coreference resolution to determine that the tumor was not new: "Now with multiple nodular lesions within the bilateral lungs, demonstrating both enlarging of previously seen nodules, and development of new nodules. This is consistent with **metastatic melanoma to the lung parenchyma.**" The other two reports didn't contain explicit linguistic cues indicating chronicity: the radiologist inferred from context that the findings were chronic. One report described two lung lesions consistent with metastases then described another finding, "lytic t11 **lesion**," that should be correlated with a prior MRI. In the other report, a separate finding was described as unchanged: "a small right apical **pneumothorax** persists, and when allowing for differences in angulation, this is either unchanged or slightly increased." The mention of a previous exam, even though not directly in reference to the metastases or pneumothorax, implied that the findings were identified previously. pyConTextNLPs regular-expression-based algorithm cannot address coreference or inference.

For **aneursym**, error resolution would require either mapping different findings to different levels of severity or dropping the general synonym of "dilation" for an aneurysm: "mild **dilatation** of the main pulmonary artery, suggestive of pulmonary arterial hypertension." Identifying new cases of **pneumonia** poses similar challenges: two of 29 reports flagged with a new pneumonia were false positives. One was due to a missed negation, due to an implementation issue related to pruning targets: "*no* focal consolidation to suggest **pneumonia**." In the second, pneumonia was considered present, but as a side effect of cancer and not an infection that should be included in a critical finding follow-up list.

Limitations of this study include review by a single radiologist and only evaluating false alarms. Based on our iterative development, pyConTextNLP also missed valid critical findings, and a follow-up study will evaluate annotated reports to quantify and characterize false negatives.

Successful critical finding identification relies on negation detection, ignoring findings that are mentioned as the reason for exam, accurate differentiation of acute vs chronic findings, and modeling of uncertainty indicated by explicit cues (e.g., "may represent") as well as by linguistic variants used to describe the observation (e.g., "patchy opacity" vs. "pneumonia"). With a false positive rate of 7.8%, we believe pyConTextNLP could feasibly be deployed to suggest critical findings for communication to referring physicians without inducing alert fatigue or irritating radiologists with obvious errors. However, we will formally assess this hypothesis and determine how referring physicians would like to be presented with system recommendations in future user studies. Future work will also include assessment of false negatives, extension and evaluation of all 39 critical findings across all report types, and evaluation of execution speed and work flow integration.

Acknowledgments

We would like to thank the anonymous reviewers for valuable comments. This work was partly funded by the Department of Veteran Affairs (CRE 12-312) and University of Utah Healthcare System Hospital Project funds.

References

American College of Radiology. 2014. ACR practice parameter for communication of diagnostic imaging findings.

Wendy W. Chapman, Will Bridewell, Paul Hanbury, Gregory F. Cooper, and Bruce G. Buchanan. 2001. A simple algorithm for identifying negated findings and diseases in discharge summaries. *Journal of Biomedical Informatics*, 34(5):301–310.

Brian E. Chapman, Sean Lee, Hyunseok Peter Kang, and Wendy Webber Chapman. 2011. Document-level classification of CT pulmonary angiography reports based on an extension of the context algorithm. *Journal of Biomedical Informatics*, 44(5):728–737.

Andrea Esuli, Diego Marcheggiani, and Fabrizio Sebastiani. 2013. An enhanced CRFs-based system for information extraction from radiology reports. *Journal of Biomedical Informatics*, 46(3):425 – 435.

Henk Harkema, John N. Dowling, Tyler Thornblade, and Wendy W. Chapman. 2009. Context: An algorithm for determining negation, experiencer, and temporal status from clinical reports. *Journal of Biomedical Informatics*, 42(5):839–851, Oct.

Ronilda Lacson, Nathanael Sugarbaker, Luciano M Prevedello, Ivan IP, Wendy Mar, Katherine P Andriole, and Ramin Khorasani. 2012. Retrieval of radiology reports citing critical findings with disease-specific customization. *The Open Medical Informatics Journal*, 6:28–35.

Mathieu Lafourcade and Lionel Ramadier. 2016. Semantic relation extraction with semantic patterns experiment on radiology reports. In Nicoletta Calzolari (Conference Chair), Khalid Choukri, Thierry Declerck, Sara Goggi, Marko Grobelnik, Bente Maegaard, Joseph Mariani, Helene Mazo, Asuncion Moreno, Jan Odijk, and Stelios Piperidis, editors, *Proceedings of the Tenth International Conference on Language Resources and Evaluation (LREC 2016)*, Paris, France, may. European Language Resources Association (ELRA).

Paras Lakhani, Woojin Kim, and Curtis P. Langlotz. 2012. Automated detection of critical results in radiology reports. *Journal of Digital Imaging*, 25(1):30–36.

Paul A. Larson, Lincoln L. Berland, Brent Griffith, Charles E. Kahn Jr, and Lawrence A. Liebscher. 2014. Actionable findings and the role of it support: Report of the acr actionable reporting work group. *Journal of the American College of Radiology*, 11(6):552–558, June.

Neel Patel, Evan Narasimhan, Danielle L. Mowery, Wendy W. Chapman, Brian E. Chapman, and Marta E. Heilbrun. 2016. Annotation of critical findings from radiology reports: towards automated communication through the electronic health record. Portland, OR. Society for Imaging Informatics in Medicine.

Anne-Dominique Pham, Aurélie Névéol, Thomas Lavergne, Daisuke Yasunaga, Olivier Clément, Guy Meyer, Rémy Morello, and Anita Burgun. 2014. Natural language processing of radiology reports for the detection of thromboembolic diseases and clinically relevant incidental findings. *BMC Bioinformatics*, 15:266.

Meliha Yetisgen-Yildiz, Martin L. Gunn, Fei Xia, and Thomas H. Payne. 2013. A text processing pipeline to extract recommendations from radiology reports. *Journal of Biomedical Informatics*, 46(2):354–362, April.

Building a dictionary of lexical variants for human phenotype descriptors

Simon Kocbek[1,2]
skocbek@gmail.com

Tudor Groza[1]
t.groza@garvan.org.au

[1]Kinghorn Center for Clinical Genomics, Garvan Institute of Medical Research,
Darlinghurst, Sydney, NSW 2011, Australia
[2]Department of Computing and Information Systems, University of Melbourne,
Melbourne, VIC 3000, Australia

Abstract

Detecting phenotype descriptors in text and linking them to ontology concepts is a challenging task. Current state-of-the art concept recognizers struggle with several issues due the variety of human expressiveness. Here we present initial results of creating a dictionary of lexical variants for the Human Phenotype Ontology. This work is a smaller but important part of a larger project with a goal to improve recall in phenotype concept recognizers.

1 Introduction

Phenotype descriptions (i.e., the composite of one's observable characteristics/traits) are important for our understanding of genetics. These descriptions enable the computation and analysis of a varied range of issues related to the genetic and developmental bases of correlated characters (Mabee et al., 2007). Scientific literature contains large amounts of phenotype descriptions, usually reported as free-text entries.

Concept Recognition (CR) is the identification of entities of interest in free text and their resolution to ontological concepts with the aim of leveraging structured knowledge from unstructured data. Linking from the literature to ontologies such as the Human Phenotype Ontology (HPO) has gained a substantial interest from the text mining community (e.g., Uzuner et al., 2012; Morgan et al., 2008). Although phenotype CR is similar to other tasks such as gene and protein name normalization, it has its specific domain

issues and challenges (Groza et al., 2015). In contrast to gene and protein names, phenotype concepts are characterized by a wide lexical variability. As a result, simple methods like exact matching or standard lexical similarity usually lead to poor results. Additional challenges in performing CR on phenotypes include the use of abbreviations (e.g., *defects in L4-S1*) or of metaphorical expressions (e.g., *hitchhiker thumb*).

Consequently, phenotype CR is an ongoing research area with a demand for improvement. For example, systems such as OBO Annotator (Taboada et al., 2014), NCBO Annotator (Jonquet et al., 2009) and Bio-Lark (Groza et al., 2015) have been evaluated with maximum precision, recall and F-score values of 0.65, 0.49 and 0.56 respectively (Groza et al., 2015).

Here we present initial results of experiments designed to address the lexical variability of phenotype terms. We generate a dictionary of lexical variants for all HPO tokens. When completed, such a dictionary will help improve, in particular, the low recall of phenotype CR systems.

Generating lexical variants for HPO tokens is a fairly challenging task. For example, grouping similar words with classical similarity metrics such as the Levenshtein distance (even when using a high threshold) might group words with different meaning like *zygomatic* (a cheek bone) and *zygomaticus* (cheek muscle) into one lexical cluster. On the other hand, less similar words with same meaning like irregular nouns (e.g. *phalanx*, *phalanges,* or *femur*, *femora*) might be grouped into different clusters. Here, we experiment with the NLM Lexical Variant Generator (LVG) (The Lexical Systems Group, 2016) to generate lexical variants.

186

Proceedings of the 15th Workshop on Biomedical Natural Language Processing, pages 186–190,
Berlin, Germany, August 12, 2016. ©2016 Association for Computational Linguistics

2 Methods

To generate the dictionary of lexical variants, we extracted all concept names and their synonyms from the HPO. The text was then tokenized and a cluster of lexical variants was created for each token. Tokens with overlapping lexical variants were merged into one cluster. We manually analyzed the clusters for their quality and coverage, and performed a preliminary automatic evaluation. In addition, we identified those parts of phenotype terms that display the largest lexical variability. For the latter, we used the following two additional ontologies: Foundational Model of Anatomy (FMA) (Rosse and Mejino, 2003), and the Phenotype and Trait Ontology (PATO) (Gkoutos et al., 2009). Details of data and methods used are described in the following sections.

2.1 The Human Phenotype Ontology

The HPO's primary goal is to offer a tool that allows large-scale computational analysis of the human phenotype (Köhler et al., 2014). The HPO is often used for the annotation of human phenotypes and has repeatedly been adopted in biomedical applications aiming to understand connection between phenotype and genomic variations. Some examples of using the HPO are applications such as linking human diseases to animal models (Washington et al., 2009), describing rare disorders (Firth et al., 2009), or inferring novel drug indications (Gottlieb et al., 2011).

Most terms in the HPO contain descriptions of clinical abnormalities and additional sub-ontologies are provided to describe inheritance patterns, onset/clinical course and modifiers of abnormalities.

Below is an example of part of a term in the OBO format:

```
id: HP:0000260
name: Wide anterior fontanel
def: "Enlargement of the anterior fontanelle with respect to age-dependent norms." [HPO:curators]
synonym: "Large anterior fontanel" EXACT []
...
xref: UMLS:C1866134 "Wide anterior fontanel"
is_a: HP:0000236 ! Abnormality of the anterior fontanelle
property_value: HP:0040005 "Enlargement of the `anterior fontanelle` (FMA:75439) with respect to age-dependent norms."
xsd:string {xref="HPO:curators"}
```

Terms in HPO usually follow the Entity-Quality formalism where they combine anatomical entities with qualities (Mungall et al., 2007) For instance, in the above example, *anterior fontanelle* describes an anatomical entity with the quality *wide*. Entities can usually be grounded in ontologies such as the FMA, while qualities usually belong to the PATO. It is assumed that rich lexical variability comes from the quality part of phenotype terms – due to their wide spread usage in common English.

For this study, we used the OBO versions of the HPO Apr 2016 and the PATO Nov 2015 ontologies, and the FMA OWL version 3.2.1.

2.2 Pre-processing text in ontologies

We extracted labels and synonyms for all HPO, PATO and FMA terms. The OWL API (Horridge and Bechhofer, 2011) was used for parsing.

After a manual inspection of a random subset of names and synonyms, we developed a simple tokenizer that broke each name and synonym into series of lower case tokens. The following characters were removed: . / ( ) ' > < : ; and the space and backslash characters were then used as delimiters. We ignored numbers and short tokens (< 3 characters). The final set contained 8,098 HPO; 1,959 PATO and 8,502 FMA tokens.

2.3 Generating clusters of lexical variants

We use the NLM Lexical Variant Generator (LVG), 2016 release (The Lexical Systems Group, 2016) to create lexical variants for the HPO tokens. LVG is a suite of utilities that can generate, transform, and filter lexical variants from the given input. Its intention is to create robust indexes and to transform user queries into retrievable entries from those indexes. Although LVG focuses on biomedical terms, it is not specialized for phenotype domain.

There are more than 60 functions (flow components) in LVG and each function has a set of parameters. In this work, the following two functions were used with the LVG Java API:
- Generating inflectional variants (IVs), which include the singular and plurals for nouns, the various tenses of verbs, the positive, superlative and the comparative of adjectives and adverbs.
- Generating derivational variants (DVs), which are terms that are related to the original term but do not necessarily, share the same meaning. Often, the derivational variant changes syntactic category from the original term. Only DVs with the same prefix as the original token (i.e., first two characters) were considered.

Both IVs and DVs can be generated with two methods: a) using an internal dictionary, and b) using a set of predefined rules. When generating lexical variants, we experimented with the following three configurations (Cs):

- C1: Generating IVs using the dictionary.
- C2: Generating IVs DVs using the dictionary.
- C3: Generating IVs and DVs using the dictionary first, and using the set of rules for those tokens that did not have any variants in the dictionary.

Generating lexical variants for HPO tokens can be described with the following algorithm:

Generate lexical variants for HPO:

N: number of HPO terms
H_i: single HPO term
S: set of names and synonyms for an HPO term
T: set of unique tokens
M: number of tokens
T_j: single token
ID_j: set of dictionary based IVs for T_j
DD_j: set of dictionary based DVs for T_j
IR_j: set of rule based IVs for T_j
DR_j: set of rule based DVs for T_j
V: sets of lexical variants
V_k: single set of lexical variants, where $0 < k < 4$
C: sets of clusters
C_k: single set of clusters, where $0 < k < 4$

For i = 1 to N do:
   Extract name/synonyms for H_i and save them into S
   Tokenize S and save unique tokens into T
Initialize C_1, C_2 and C_3
For j = 1 to M do:
   Initialize V_1, V_2 and V_3
   Generate dictionary based inflectional variants for T_j and save them into ID_j
   If ID_j is empty then do:
      Generate rule based inflectional variants for T_j and save them into IR_j
   Generate dictionary based derivational variants for T_j and save them into DD_j
   If DD_j is empty then do:
      Generate rule based derivational variants for T_j and save them into DR_j
   $V_1 = ID_j$
   $V_2 = ID_j + DD_j$
   $V_3 = ID_j + DD_j + IR_j + DR_j$
   For k = 1 to 3 do:
      If a cluster in C_k has a variant from V_k then do:
         Put variants from V_k into the existing cluster
      Else do:
         Create a cluster from V_k in C_k

2.4 Inspecting/evaluating lexical clusters

For each configuration we calculated the coverage of extended tokens (i.e., the number of to-

kens for which at least one variant was found), and manually inspected lexical variants for 10 randomly selected tokens. In addition, we inspected clusters for the following two specific tokens of interest that are known to be problematic in phenotype CR: *phalanx* and *shortening*. The former is an irregular noun that changes to *phalanges* in plural form, while the latter represents a participle that is usually not correctly normalized for our need. For example, we would expect *short* and *shortening* in the same cluster (*short finger* vs. *shortening of the finger*). We also inspected variants for *zygomatic* and *zygomaticus* that should not be in the same cluster.

In addition to the manual inspection, we also performed a preliminary automatic evaluation of the clusters. The HPO has been integrated into Unified Medicine Language System (UMLS) Metathesaurus (Humphreys et al., 1998) since the 2015AB version (Dhombres et al., 2015). This potentially gives new synonyms for the HPO terms. The synonyms can contain lexical variants of the HPO term tokens. For example, *acute promyelocytic **leukemia**, does not contain any synonyms in the HPO. However, UMLS contains the synonyms *acute promyelocytic **leukaemia**. Similarly, *ascending **aortic** aneurysm* has no HPO synonyms, while we can find *aneurysm of ascending **aorta*** in UMLS. Therefore, we developed an algorithm for counting those HPO terms that increased the coverage of tokens in UMLS synonyms for these terms (e.g., the above two terms would be counted).

As mentioned in section 2.1, it is assumed that most tokens with rich lexical variability are associated with the quality part of HPO terms. To test this assumption, we finally examined coverage of the HPO tokens in the FMA and the PATO. We then analyzed lexical cluster sizes for these tokens. In case the assumption is true, we expect the cluster sizes of PATO tokens (i.e. quality) larger than tokens found in FMA (i.e. entity).

3   Results and discussion

Table 1 summarizes the number of variants, the number of clusters, the average number of variants in each cluster and the number of tokens with no variants (NV) for different configurations.

Table 1: Results summary for each configuration

	#Variants	#Clusters	Average	#NV
C1	13,471	6,355	2.12	877
C2	18,080	5,620	3.22	877
C3	29,602	6,480	4.57	0

The same tokens with no variants were found using only the dictionary in C1 and C2, which implies that these tokens are not covered with LVG's dictionary. After the manual examination of generated clusters we can identify some examples of tokens without generated variants as follows: *spelling errors* (e.g., accesory, dermititis), *latin words* (e.g., ambiguus), *chemical compounds* (e.g., 23-diphosphoglycerate), *abbreviations* (eg., gnrh, pirc), *roman numbers* (e.g., xii, xiii), and *ordinal numbers* (e.g., 1st, 2nd). Using the rule-based approach in C3 generated variants for these tokens.

Examining the clusters showed that C1 generated several disjoint clusters that should be merged. Some examples are tokens like *abdomen* and *abdominal*, *abnormal* and *abnormality*, *external* and *externally*, and *yellow* and *yellowish*. As for the tokens of our particular interest, *phalanx* contained the following variants in the same cluster: *phalange, phalanges, phalanx,* and *phalanxes*; while *shortening* was clustered with the following variants: *shorten, shortened, shortening,* and *shortenings* and was missing words like *short, shorter* and *shortest*. Variants for *zygomatic* and *zygomaticus* were in separated clusters in all three approaches.

According to Table 1, the C2 approach generated more variants distributed into less clusters when compared to C1. Manual examination revealed that several disjoint clusters from previous paragraph merged into larger clusters (*abdomen* and *abdominal, abnormal* and *abnormality,* and *external* and *externally*). The *phalanx* cluster gained a new variant *phalangeal*, which was previously in a different cluster. There was no change in the *shortening* cluster.

Clusters in C3 extended tokens with no variants in LVG's dictionary with rule generated terms. However, variants for tokens like spelling errors or ordinal numbers were incorrect. For example, *accesory* would be extended with variants like *accesoryed* and *accesoryer*. In addition, participles were not in correct clusters (e.g., shortening). Unfortunately, terms like *brachymesomelia* or *trichromacy* were also extended with wrong variants. This implies that rules defined in LVG might not be appropriate for phenotype terms and we must define our own rules. This investigation is left for future work.

Testing with UMLS, we found that 6,580 (62%) of the HPO terms contained UMLS synonyms. 16% of these terms increased the coverage of synonym tokens with new lexical variants, which indicates that the generated dictionary does include quality variants. We plan to investigate the results in depth in the future.

When testing the coverage of HPO tokens in the PATO and the FMA, we found that 10% and 26% of the HPO tokens can also be found in the PATO and the FMA respectively. Figure 1 shows ratios for different lexical cluster sizes of the overlapping tokens created with the C2 approach (minimum/maximum size of 1 and 11 respectively). One can notice that the PATO tokens tend to form larger clusters, which indicates that these tokens have more lexical variants compared to the FMA tokens. This confirms the assumption from Section 2.1, that the quality part of phenotype term offers more lexical variability than the entity part.

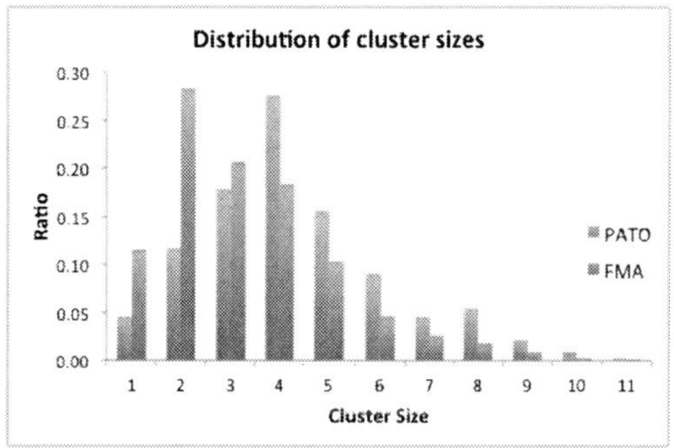

Figure 1: Distribution of lexical cluster sizes for those HPO tokens that were also found in the PATO/FMA.

4   Conclusion

In this paper we presented initial results for creating a dictionary of lexical variants of all tokens in the Human Phenotype Ontology. This task is a part of bigger project with aim to improve phenotype concept recognition. Using the NLM Lexical Variant Generator, we experimented with three configurations where different combinations of inflectional and derivational variants were used to extend original HPO token space. We examined the clusters and performed a preliminary automatic evaluation of these clusters. We also identified parts of phenotype terms that are likely to express more lexical variability.

In the future, we are planning to perform a detailed analysis of the generated clusters and improve the automatic evaluation. As seen in the results section, there are some phenotype tokens that are not covered in external dictionaries such as LVG. We will try to identify patterns of these tokens and see how we can extend them with lexical variants. In addition, we will improve the quality of generated clusters with removing incorrect variants (e.g., results of spelling errors), or tokens that are actually not phenotypes.

Focus of our future work will be the quality part of phenotype terms, since we showed that quality tokens display larger lexical variability than entity tokens. In addition, we have not managed to automatically generate clusters for all participles.

Reference

Ferdinand Dhombres, Rainer Winnenburg, James T. Case, and Olivier Bodenreider. 2015. Extending the coverage of phenotypes in SNOMED CT through post-coordination. In *Studies in Health Technology and Informatics*, volume 216, pages 795–799.

Helen V. Firth, Shola M. Richards, A. Paul Bevan, Stephen Clayton, Manuel Corpas, Diana Rajan, Steven Van Vooren, Yves Moreau, Roger M. Pettett, and Nigel P. Carter. 2009. DECIPHER: Database of Chromosomal Imbalance and Phenotype in Humans Using Ensembl Resources. *American Journal of Human Genetics*, 84(4):524–533.

Georgios V. Gkoutos, Chris Mungall, Sandra Ďolken, Michael Ashburner, Suzanna Lewis, John Hancock, Paul Schofield, Sebastian Ǩohler, and Peter N. Robinson. 2009. Entity/quality-based logical definitions for the human skeletal phenome using PATO. In *Proceedings of the 31st Annual International Conference of the IEEE Engineering in Medicine and Biology Society: Engineering the Future of Biomedicine, EMBC 2009*, pages 7069–7072.

Assaf Gottlieb, Gideon Y Stein, Eytan Ruppin, and Roded Sharan. 2011. PREDICT: a method for inferring novel drug indications with application to personalized medicine. *Molecular systems biology*, 7(496):496.

Tudor Groza, S. Kohler, Sandra Doelken, Nigel Collier, Anika Oellrich, Damian Smedley, Francisco M. Couto, Gareth Baynam, Andreas Zankl, Peter N. Robinson, Sebastian Köhler, Sandra Doelken, Nigel Collier, Anika Oellrich, Damian Smedley, Francisco M. Couto, Gareth Baynam, Andreas Zankl, and Peter N. Robinson. 2015. Automatic concept recognition using the Human Phenotype Ontology reference and test suite corpora. *Database*, 2015(0):bav005–bav005.

Matthew Horridge and Sean Bechhofer. 2011. The OWL API: A Java API for OWL ontologies. *Semantic Web*, 2(1):11–21.

Betsy L. Humphreys, Donald a. B. Lindberg, Harold M. Schoolman, and G. Octo Barnett. 1998. The Unified Medical Language System: An Informatics Research Collaboration. *Journal of the American Medical Informatics Association*, 5(1):1–11.

Clement Jonquet, Nigam H Shah, H Cherie, Mark a Musen, Chris Callendar, and Margaret-Anne Storey. 2009. NCBO Annotator : Semantic Annotation of Biomedical Data. *Iswc*:2–3.

Sebastian Köhler, Sandra C. Doelken, Christopher J. Mungall, Sebastian Bauer, Helen V. Firth, Isabelle Bailleul-Forestier, Graeme C M Black, Danielle L. Brown, Michael Brudno, Jennifer Campbell, David R. Fitzpatrick, Janan T. Eppig, Andrew P. Jackson, Kathleen Freson, Marta Girdea, Ingo Helbig, Jane A. Hurst, Johanna Jähn, Laird G. Jackson, et al. 2014. The Human Phenotype Ontology project: Linking molecular biology and disease through phenotype data. *Nucleic Acids Research*, 42(D1).

Paula M. Mabee, Michael Ashburner, Quentin Cronk, Georgios V. Gkoutos, Melissa Haendel, Erik Segerdell, Chris Mungall, and Monte Westerfield. 2007. Phenotype ontologies: the bridge between genomics and evolution. *Trends in Ecology and Evolution*, 22(7):345–350.

Alexander A Morgan, Zhiyong Lu, Xinglong Wang, Aaron M Cohen, Juliane Fluck, Patrick Ruch, Anna Divoli, Katrin Fundel, Robert Leaman, Jörg Hakenberg, Chengjie Sun, Heng-hui Liu, Rafael Torres, Michael Krauthammer, William W Lau, Hongfang Liu, Chun-Nan Hsu, Martijn Schuemie, K Bretonnel Cohen, et al. 2008. Overview of BioCreative II gene normalization. *Genome biology*, 9 Suppl 2(SUPPL. 2):S3.

Chris Mungall, Georgios Gkoutos, Nicole Washington, and Suzanna Lewis. 2007. Representing phenotypes in OWL. In *CEUR Workshop Proceedings*, volume 258.

Cornelius Rosse and José L V Mejino. 2003. A reference ontology for biomedical informatics: The Foundational Model of Anatomy. *Journal of Biomedical Informatics*, 36(6).478–500.

M. Taboada, H. Rodriguez, D. Martinez, M. Pardo, and M. J. Sobrido. 2014. Automated semantic annotation of rare disease cases: a case study. *Database*, 2014(0):bau045–bau045.

NLM The Lexical Systems Group. 2016. Lexical Tools, 2016, https://lexsrv3.nlm.nih.gov/LexSysGroup/Projects/lvg/2016/web/index.html, accessed June 2016.

Özlem Uzuner, Brett R South, Shuying Shen, and Scott L DuVall. 2012. 2010 i2b2/VA challenge on concepts, assertions, and relations in clinical text. *Journal of the American Medical Informatics Association : JAMIA*, 18(5):552–6.

Nicole L. Washington, Melissa A. Haendel, Christopher J. Mungall, Michael Ashburner, Monte Westerfield, and Suzanna E. Lewis. 2009. Linking human diseases to animal models using ontology-based phenotype annotation. *PLoS Biology*, 7(11).

Applying deep learning on electronic health records in Swedish to predict healthcare-associated infections

Olof Jacobson
Department of Computer
and Systems Sciences, (DSV)
Stockholm University
P.O. Box 7003, 164 07 Kista
olofja@kth.se

Hercules Dalianis
Department of Computer
and Systems Sciences, (DSV)
Stockholm University
P.O. Box 7003, 164 07 Kista
hercules@dsv.su.se

Abstract

Detecting healthcare-associated infections pose a major challenge in healthcare. Using natural language processing and machine learning applied on electronic patient records is one approach that has been shown to work. However the results indicate that there was room for improvement and therefore we have applied deep learning methods. Specifically we implemented a network of stacked sparse auto encoders and a network of stacked restricted Boltzmann machines. Our best results were obtained using the stacked restricted Boltzmann machines with a precision of 0.79 and a recall of 0.88.

1 Introduction

Healthcare-associated infections pose a major problem today within healthcare. In Sweden over ten percent of all in-patients suffer from Healthcare-associated infection (HAI). In Europe, this estimates to three million affected patients per year of which about 50,000 die, (Humphreys and Smyth, 2006).

HAI is defined as "an infection occurring in a patient in a hospital or other healthcare facility in whom the infection was not present or incubating at the time of admission." HAI causes patients to suffer while their healthcare periods are prolonged, it also lays an economic burden on the society. The Swedish National Board of Health and Welfare estimates that HAIs prolong the length of a patients stay at the hospital by an average of 4 days, (Burman, 2006).

A number of tools, using the electronic patient record of the patient to detect and predict HAI, have been developed. Freeman et al. (2013) contain a nice overview of over 44 different approaches, however the authors have not distinguished the technology behind the systems, only the accuracy of each system. Ehrentraut et al. (2014) have focused on describing the different systems in a more technical way, most systems use rules to detect HAI, but some few systems are using machine learning based approaches. Three classification algorithms, SVM, Random Forest and Gradient Tree Boosting algorithms, have been used but no attempt has been made using Deep Learning.

In order to distinguish between different types of documents a classifier may be required to recognize abstract concepts in the text. This is made difficult through the nature of human language.

The concepts themselves may not be directly mentioned in the text or may be abbreviated. For example *Pat was op. two days ago*, meaning *The patient was operated two days ago*, or *Patient has fever* meaning that a patient had fever (a symptom) and might have a infection (disease) that needs to be treated. For example *Penomax was given to pat*, meaning that a patient had an infection or had a suspected infection and therefore were given an antibacterial drug for profylactic reasons.

Deep learning architectures transform input data using several non-linear operations. Theoretically such transformations may enable a system to learn high level abstractions in the data, (Bengio, 2009). Deep learning systems are thereby interesting candidates for text classification tasks.

2 Previous research

A number of different methods have been used to detect healthcare associated infections using electronic patient records (EPRs). These range from manual methods, methods that use the structured fields of the EPRs, the clinical free text, as well as methods using both. Automatic methods have

Proceedings of the 15th Workshop on Biomedical Natural Language Processing, pages 191–195,
Berlin, Germany, August 12, 2016. ©2016 Association for Computational Linguistics

used rules as well as machine learning methods. Proux et al. (Proux et al., 2009; Proux et al., 2011) describe a rule based system for French patient records. Tanushi et al. (2015) describe another rule based system for Swedish patient records.

Tanushi et al. detected urinary tract infections on 1,867 care episodes and they obtained a precision of 0.98, a specificity of 0.99 and negative predictive value 0.99, but a recall (sensitivity) of 0.60. One approach used machine learning based systems as SVM and Random Forest on the Stockholm EPR Detect-HAI Corpus in (Ehrentraut et al., 2014), where the authors obtained the best results using Random Forest with precision 0.83, recall 0.87 and F-score 0.85. In one other approach using GTB Gradient Tree Boosting the authors obtained a precision of 0.80, a recall of 0.94 and an F_1-score of 0.86, (Ehrentraut et al., 2016).

Deep learning has already been successfully used for natural language tasks. Wiriyathammabhum et al. (2012) applied a deep belief network for word sense disambiguation and found that it had better performance than many shallow machine learning architectures including SVMs. They showed that deep learning could be successfull even with few instances and high dimensionality of the data.

3 Materials and Methods

3.1 Materials

We have been using data from the Swedish Health Record Research Bank, (Dalianis et al., 2015) that contain over 2 million patient records from over 800 clinical units covering the year 2006-2014. and specifically a subset called Stockholm EPR Detect-HAI Corpus[1], (Ehrentraut et al., 2014), that contains 213 patient records written in Swedish that were classified by two different domain experts. The domain experts have reviewed and classified each record both separately and jointly and finally decided on the Gold Standard. The two classes into which the records were divided were patients that suffered from HAIs at some point during the time period covered by the record, and patients that did not. In general, the patient records describe patients that are very ill, see Table 1, for details on the corpus. Each record contains both free text but also structured fields as

---

[1]This research has been approved by the Regional Ethical Review Board in Stockholm (Etikprövningsnämnden i Stockholm), permission number 2012/1838-31/3.

body temperatures, microbiological answers and drug prescriptions. The records of HAI positive patients are generally longer than those of patients not suffering from HAIs. This imbalance was not addressed by the methodology of the study.

	HAI	non-HAI
Number of records	128	85
Patient ages [years]	2 - 93	2 - 92
Total number of tokens	1,034,760	230,226
Time in hospital [days]	2 - 144	3 - 93
Total time in hospital [days]	3975	941

Table 1: A table detailing some statistics of the Stockholm EPR Detect-HAI Corpus. Tokens refer to space separated sequences of characters. HAI (Healthcare-associated Infection), non-HAI (no Healthcare-associated Infection)

3.2 Methods

The task of classifying texts through machine learning methods generally involve preprocessing and transforming the text to a numerical representation. Some sort of feature selection may then be used before the actual machine learning system attempts to classify the data.

The performance of the implemented systems were tested using a 10-fold cross validation. In each training fold a grid search, validated using 5-fold cross validation, was used to select the hyper parameters of the classifiers.

The varied parameters for the stacked sparse auto encoder were, the number of hidden layers and the sparsity parameter. The number of nodes in each hidden layer was set to be the mean of the number of nodes in the layers immediately before and after. For the stacked restricted Boltzmann machines the gridsearch was performed over several different network topologies with varying number of hidden layers and nodes in the hidden layers. The gridsearch was also performed over the learning rate in this case. All of the programming in this study was carried out in Python.

3.2.1 Preprocessing

The text of the patient records was preprocessed according to the following steps. First all characters not corresponding to Swedish letters, white space, or digits were removed from the corpus. Since accents were inconsistently used, they were removed from all letters in the corpus. The remaining text was converted to lowercase and all stop words were removed using a stop word list from Python's natural language toolkit (NLTK)[2]. Lowercase conversion and stop word removal are standard preprocessing techniques that simplify the corpus and are usually assumed to not affect the classification (Uysal and Gunal, 2014). Stemming has been shown to work well for many tasks. Specifically for Swedish, that has a rich morphology, it has been shown that stemming improves the results for information retrieval tasks, (Carlberger et al., 2001). Therefore the text in the patient records was stemmed using the Python Snowball stemmer[3] for Swedish.

3.2.2 Conversion to numerical representation

In this study two different models for converting the records into numerical vectors were used. One of these models was a bag of words model where the tf-idf scores for the individual words in the patient records were calculated. Only the 1,000 most common words in the whole corpus were considered. Principal component analysis (PCA) was used to generate acompleted only the encodine tf-idf representation with reduced dimensionality, 99% of the variance was retained.

The other model used for conversion was the word2vec tool in Python's gensim package (Řehůřek and Sojka, 2010) using the skip-gram model, through which each word in a record could be converted into a vector. A record was represented by taking the mean of all such vectors in the record.

3.2.3 Artificial neural networks

Artificial neural networks is a biologically inspired type of machine learning architecture. The basic building blocks of these networks are inspired by the biological neuron and are usually referred to as nodes (Marsland, 2009).

The nodes are connected to other nodes forming a network which can be trained by various methods to perform different tasks (Amaral et al., 2013). In this study two different types of neural networks were used. Stacked sparse auto encoders and stacked restricted Boltzmann machines.

3.2.4 Stacked sparse auto encoders

A stacked sparse auto encoder is a neural network technology which consists of several sparse auto encoders. An auto encoder is a type of neural network that first encodes and then decodes input data. Auto encoders are trained to reproduce the data sent through them (Bengio, 2009). The cross entropy cost function was used for the auto encoders in this study. When the sparse auto encoders were stacked, each encoder was trained to reproduce the encoded data of the previous encoder. This unsupervised training constituted the pretraining phase of the network. After the pretraining was completed only the first half of the auto encoders, responsible for the encoding, was used in the resulting network. A softmax classifier was appended to the last auto encoder and the network was trained in a supervised manner through back propagation.

3.2.5 Stacked restricted Boltzmann machines

A restricted Boltzmann machine (RBM) is a special case of the more general Boltzmann machine. An RBM consists of one layer of visible nodes and one layer of hidden nodes. Nodes in the hidden layer are only allowed to connect to nodes in the visible layer (Bengio, 2009). One notable difference between auto encoders and RBMs is that RBMs have binary activation of the hidden layer nodes during the training phase. In the implemented stacked RBM topology the RBMs were trained to reproduce data input on the visible layer through the means of persistent contrastive divergence (Tieleman, 2008). The stack of RBMs was created in a similar way to the previously described stack of auto encoders. Each RBM was trained on the hidden layer activations of the previous RBM. A softmax classifier was appended at the end of the network and was trained in a supervised manner on the hidden layer representations of the previous RBM. Unlike the stack of auto encoders, the whole network was not fine tuned through supervised training.

4 Results

The results are presented in Table 2 where the acronyms used are the following: SSAE, *Stacked Sparse Auto Encoder classifier*. SRBM, *Stacked Restricted Boltzmann Machine classifier* and tf-idf, the *tf-idf* representation of the data was used. PCA, a version of the tf-idf data that had its dimensionality reduced through *Principal Component Analysis* was used. And finally word2vec, the *word2vec* representation of the data was used.

---

[2]Accessing Text Corpora and Lexical Resources, *http://www.nltk.org/book/ch02.html*

[3]Swedish Snowball stemming algorithm, *http://snowball.tartarus.org/algorithms/swedish/stemmer.html.*

	Precision	Recall	F_1-score
SSAE tf-idf	0.78	0.78	0.78
SSAE PCA	0.71	0.80	0.75
SSAE word2vec	0.78	0.84	0.81
SRBM tf-idf	0.79	0.88	0.83
SRBM word2vec	0.66	0.91	0.77

Table 2: The classification results for the different machine learning methods and preprocessing techniques.

The highest F_1-score was obtained for the stacked restricted Boltzmann machines and that classifier also had the highest precision. The word2vec version of the data gave lower F_1-score than the regular tf-idf data when the SRBM classifier was used but a higher F_1-score for the SSAE classifier. The dimensionality reduced version of the tf-idf data gave a slightly higher recall, and a lower precision as compared to the regular data.

5 Discussion

Deep learning models are computationally expensive to train. This is a disadvantage compared to simpler models. The disadvantage can in many cases be motivated by an increase in classification performance. However that was not observed in this study.

We could see that the SRBM with the tf-idf data gave the best results of our approaches. The results were comparable to those obtained by Ehrentraut et al. (2014), however they were slighly lower than the results presented by Ehrentraut et al. (2016). It was surprising that the classification accuracy for the word2vec representation of the data was higher than for the tf-idf version, for the SSAE classifier. The SRBM classification accuracy was lower for the word2vec representation than for the tf-idf representation. No appealing explanation was found for the difference.

In many cases the gridsearch parameter optimization preferred few layers in the neural networks. This implies that deeper architectures did not help the classifier identify useful abstract patterns in the data. This could be due to the chosen data representations that may not have retained some of the important features of the original data. The lack of training data may also have been contributing to the non efficacity of deeper architectures, since such architectures typically require a lot of training data. In this study the ratio of the number of features, and the number of training examples, may simply have been too large to enable the networks to learn abstract patterns.

6 Conclusion

The classification results were comparable to what has been obtained using more conventional classifiers in previous studies. The SRBM architecture using the bag of words, tf-idf data was the most successfull classifier in this study, see Table 2. SRBM gave a precision of 0.79, a recall of 0.88, and an F_1-score of 0.83. These results were slightly lower than those of the best classifier in previous studies, which had a precision of 0.80, a recall of 0.94 and an F_1-score of 0.86 (Ehrentraut et al., 2016). The word2vec representations scored worse than the bag of words model for the SRBM classifier while giving better scores for the SSAE classifier. The PCA version of the tf-idf data with reduced dimensionality gave worse classification results than the regular data.

In the future we would like to try out a wider range of preprocessing methods, similar the ones tried out by Ehrentraut et al. (2014), to detect negated symptoms or diseases, so called negation detection, for example to perform *negation detection*, or *remove negations*, or to carry out *stop word filtering* and finally *filtering out infection specific terms* to see if that will improve our results.

Acknowledgements

We would like to thank Mia Kvist and Elda Sparrelid both at Karolinska University Hospital. We would also like to thank Claudia Ehrentraut and Hideyuki Tanushi for their ground breaking work to construct the Stockholm EPR Detect-HAI Corpus.

References

Telmo Amaral, Luís M. Silva, Luís A. Alexandre, Chetak Kandaswamy, Jorge M. Santos, and Joaquim Marques de Sá. 2013. Using different cost functions to train stacked auto-encoders. In *MICAI (Special Sessions)*, pages 114–120. IEEE.

Yoshua Bengio. 2009. Learning deep architectures for ai. *Found. Trends Mach. Learn.*, 2(1):1–127, January.

Lars G Burman. 2006. Att förebygga vårdrelaterade infektioner - ett kunskapsunderlag. *Stockholm: Swedish National Board of Health and Welfare. ISBN*, 595547553, (In Swedish).

Johan Carlberger, Hercules Dalianis, Martin Hassel, and Ola Knutsson. 2001. Improving precision in information retrieval for Swedish using stemming. In *Proceedings of NODALIDA '01 - 13th Nordic Conference on Computational Linguistics*.

Hercules Dalianis, Aron Henriksson, Maria Kvist, Sumithra Velupillai, and Rebecka Weegar. 2015. Health bank - a workbench for data science applications in healthcare. In *Proceedings of the CAiSE-2015 Industry Track co-located with the 27th Conference on Advanced Information Systems Engineering - CAiSE 2015*, pages 1–18, Stockholm, Sweden, June. CEUR.

Claudia Ehrentraut, Maria Kvist, Elda Sparrelid, and Hercules Dalianis. 2014. Detecting healthcare-associated infections in electronic health records: Evaluation of machine learning and preprocessing techniques. In *Sixth International Symposium on Semantic Mining in Biomedicine (SMBM 2014), Aveiro, Portugal, October 6-7, 2014*, pages 3–10. University of Aveiro.

Claudia Ehrentraut, Markus Ekholm, Hideyuki Tanushi, Jörg Tiedemann, and Hercules Dalianis. 2016. Detecting hospital acquired infections: A document classification approach using support vector machines and gradient tree boosting. *Health Informatics Journal*, To be published.

Rachel Freeman, Luke S. P. Moore, Laura García Álvarez, Andre Charlett, and Alison Holmes. 2013. Advances in electronic surveillance for healthcare-associated infections in the 21st Century: a systematic review. *Journal of Hospital Infection*.

Hilary Humphreys and Edmund T. M. Smyth. 2006. Prevalence surveys of healthcare-associated infections: what do they tell us, if anything? *Clinical Microbiology and Infection*, 12(1):2–4.

Stephen Marsland. 2009. *Machine Learning: An Algorithmic Perspective*. Chapman & Hall/CRC, 1st edition.

Denys Proux, Pierre Marchal, Frédérique Segond, Ivan Kergourlay, Stéfan Darmoni, Suzanne Pereira, Quentin Gicquel, and Marie-Hélène Metzger. 2009. Natural Language Processing to detect Risk Patterns related to Hospital Acquired Infections. In *Proceedings of the Workshop Biomedical Information Extraction*, pages 35–41. Association for Computational Linguistics.

Denys Proux, Caroline Hagège, Quentin Gicquel, Suzanne Pereira, Stefan Darmoni, Frédérique Segond, and Marie-Hélène Metzger. 2011. Architecture and Systems for Monitoring Hospital Acquired Infections inside a Hospital Information Workflow. In *Proceedings of the Workshop on Biomedical Natural Language Processing. USA: Portland, Oregon*, pages 43–48. Citeseer.

Radim Řehůřek and Petr Sojka. 2010. Software Framework for Topic Modelling with Large Corpora. In *Proceedings of the LREC 2010 Workshop on New Challenges for NLP Frameworks*, pages 45–50, Valletta, Malta, May. ELRA. http://is.muni.cz/publication/884893/en.

Hideyuki Tanushi, Maria Kvist, and Elda Sparrelid. 2015. Detection of Healthcare-Associated Urinary Tract Infection in Swedish Electronic Health Records. *Innovation in Medicine and Healthcare 2014*, 207:330.

Tijmen Tieleman. 2008. Training restricted Boltzmann machines using approximations to the likelihood gradient. In *Proceedings of the 25th international conference on Machine learning*, pages 1064–1071.

Alper Kursat Uysal and Serkan Gunal. 2014. The impact of preprocessing on text classification. *Information Processing and Management*, 50(1):104–112.

Peratham Wiriyathammabhum, Boonserm Kijsirikul, Hiroya Takamura, and Manabu Okumura. 2012. Applying deep belief networks to word sense disambiguation. *CoRR*, abs/1207.0396.

Identifying First Episodes of Psychosis in Psychiatric Patient Records using Machine Learning

Genevieve Gorrell [1], Sherifat Oduola [2], Angus Roberts [1]
Thomas Craig [2], Craig Morgan [2] and Robert Stewart [2]

[1] Department of Computer Science, University of Sheffield, UK
`g.gorrell,angus.roberts@sheffield.ac.uk`

[2] Institute of Psychiatry, Kings College London, UK
`sherifat.oduola,tom.craig,craig.morgan,robert.stewart@kcl.ac.uk`

Abstract

Natural language processing is being pressed into use to facilitate the selection of cases for medical research in electronic health record databases, though study inclusion criteria may be complex, and the linguistic cues indicating eligibility may be subtle. Finding cases of first episode psychosis raised a number of problems for automated approaches, providing an opportunity to explore how machine learning technologies might be used to overcome them. A system was delivered that achieved an AUC of 0.85, enabling 95% of relevant cases to be identified whilst halving the work required in manually reviewing cases. The techniques that made this possible are presented.

1 Introduction

The epidemiology of first episode psychosis (FEP) is the central tenet on which psychiatric research builds an understanding of psychotic disorder, and accurate estimates of incidence rates of psychosis are important to measure the burden of the disease in the population (Baldwin et al., 2005; Hogerzeil et al., 2014). Yet challenges recruiting patients with FEP and variation in incidence rates are widely reported (Patel et al., 2003; Borschmann et al., 2014; Kirkbride et al., 2006). Sampling methods used for estimating incidence of psychosis may contribute to some of the reported challenges. For example, some previous studies have used a first contact sampling frame e.g. first hospital admission or 'first early intervention' (i.e. patients presenting to early-phase psychosis services). However these methods of identifying cases do not take into account individuals who may already be receiving treatment for

non-psychotic disorder but who later manifest psychotic symptoms (Hogerzeil et al., 2014). Electronic health records can help alleviate these problems, whereby clinical information is screened using a diagnostic instrument to identify symptoms of psychosis within a defined period and consequently classify new FEP cases. An example of such an endeavour is work being carried out at the Institute of Psychiatry and South London & Maudsley (SLaM) NHS Trust using the Biomedical Research Centre Clinical Records Interactive Search (CRIS) to identify FEP cases in the CRIS-First Episode Psychosis study (Bourque, 2015). To summarize this work, psychiatric experts manually coded data in the free-text of clinical records between 1st May 2010 and 30th April 2012 for patients presenting to SLaM with compliance for psychotic disorder using a psychiatric diagnostic tool. Whilst the screening of clinical records sampling method comprehensively identifies cases and reduces risk of underestimation, this approach raises resource and efficiency challenges. For example, review of clinical records requires expert level resource (such as a psychiatrist or psychiatric nurse) for annotation, which can be very expensive. On average approximately 80-100 individual clinical records were screened per week by each annotator. It is clear that manual screening of electronic records is resource-intensive and time-consuming.

With these challenges in mind, advances in natural language processing technology have been drawn on in this work to apply techniques to identify and classify FEP cases based on the data generated from the manual screen (Bourque, 2015). An automated screening application has the potential to improve the efficiency of the FEP case identification task, reducing the burden of manual screening as well as saving time and money. Such

Proceedings of the 15th Workshop on Biomedical Natural Language Processing, pages 196–205,
Berlin, Germany, August 12, 2016. ©2016 Association for Computational Linguistics

an approach may also provide a methodological advantage in identifying FEP cohorts who may be followed up longitudinally to answer important questions about outcomes following their experience of psychosis. The use of natural language processing has potential implications for service planning and evaluation for patients with FEP.

CRIS contains both the structured information and the unstructured free text from the SLaM EHR. The free text consists of 20 million text field instances containing correspondence, patient histories and notes describing encounters with the patient. These free text fields contain much information of value to mental health epidemiologists. clinicians often record vital information in the textual portion of the record even when a structured field is designated for this information. For example, a query on the structured fields for Mini Mental State Examination scores (MMSE, a score of cognitive ability) in a recent search returned 5,700 instances, whereas a keyword search over the free text fields returned an additional 48,750 instances. Previous research has noted that free text is convenient, expressive, accurate and understandable (Meystre et al., 2008; Rosenbloom et al., 2011), making it appealing for clinical record data entry despite the greater research value of structured data. Powsner et al (1998) observe that structured data is more restrictive, whereas Greenhalgh (2009) comments that free text is tolerant of ambiguity, which supports the complexity of clinical practice; a particularly relevant factor, perhaps, in psychiatry. Medical language is often hedged with ambiguity and probability, which is difficult to represent as structured data (Scott et al., 2012). For these reasons, the diagnosis structured field in the patient record is of only minor utility in identifying FEP cases. On the other hand, the free text field may often not give a clear initial opinion of the diagnosis. The understanding of this episode as a first episode of psychosis may instead unfold over time; for example, an unclear episode may be more conclusively identified as psychotic in the light of subsequent episodes, or ruled out as psychosis through the finding of organic causes. Furthermore the records we are interested in are those that record the initial psychotic episode, rather than subsequent ones in a patient already diagnosed, though the language surrounding the event may be extremely similar. The task therefore presents challenges for NLP.

1.1 Previous Research

Previous work has attempted to identify relevant cases for research in patient records, and has tended to make use of keyword search and rule-based approaches, though a body of work exists on statistical case classification. Ford et al (2016) note that making use of the free text information consistently improves accuracy compared with structured fields only, but there is little to distinguish the success of rule-based and machine learning approaches to case classification. Of the 67 studies they reviewed, 23 used a data-driven approach to classification, with logistic regression being the most popular choice, but with all of the better known classification algorithms represented. Features on which the classification took place are often bespoke gazetteers, though various established biomedical information extraction systems are used as a preparatory step, most notably cTAKES (Savova et al., 2010), MetaMap (Aronson, 2001) and HITEx (Zeng et al., 2006). Bag-of-words representations and character n-grams are common. Systems often include some form of assertion and/or negation detection, such as NegEx (Chapman et al., 2001). The studies cover a variety of general medical conditions, and results vary, with recalls (sensitivities) and precisions (positive predictive values, or PPVs) typically between around 50% and the high 90s. A further study not included by Ford et al uses word trigrams to achieve a good result in detecting patients with acute lung injury (Yetisgen-Yildiz et al., 2013). Given the varied task conditions, it is difficult to generalize about what constitutes a good result.

Several studies are of more specific relevance to psychiatry. Castro et al (2014) report an AUC of 0.82 classifying patients based on their record according to bipolar status, and an AUC of 0.93 for classifying individual notes (subdocuments) within the patient record, a result they achieved using HITEx for feature generation, along with a bespoke gazetteer, and logistic regression (LASSO) for classification. Among previous work, theirs is perhaps the most comparable to the study presented here, in particular the classification of the entire case, rather than the individual note, since this is a closer parallel to this work, in which a portion of the patient record covering a window of many subdocuments is used to classify the whole case. Bellows et al (2014) focus on terms rather than classifying the whole case to identify binge eating disorder diagnoses. They provide an accuracy figure with no kappa, and a sensitivity (recall) without a specificity or a PPV (precision) so it is hard to compare their outcome with other similar work. Perlis et al (2012) have had some success using bespoke text features and logistic regression to classify patients with major depressive disorder according to their current status, achieving AUCs in the range of 0.85 to 0.88. Huang et al (2014) classify depression patients according to disease severity, and predict 12 month outcomes. Seyfried et al (2009) provide technological support for manual depression case identification, but do not include automated classification.

The approaches used here are in keeping with previous work, whilst applying the techniques to a novel domain with new challenges. Linking to a medical ontology has not been done here since existing vocabularies do not provide a good coverage of terms relevant in this case, but a contextualizer was utilized (discussed in more detail below) to distinguish mentions being experienced by the patient, now, from, for example, those having been experienced by a family member or in the past. Sentence classification has not been used in preference to whole case classification because first episode psychosis diagnoses are so very rarely clearly stated.

2   Data

The manual case identification is described elsewhere (Bourque, 2015). In brief, a three stage screening of clinical records was conducted by three clinically trained researchers (a psychiatrist, a medical doctor and a psychiatric nurse), and a research assistant, overseen by a principal investigator.

Firstly, SQL commands were used to retrieve anonymised information for all persons presenting to all adult mental health services serving the population of interest. Search criteria were weekly search period, service location (i.e. all SLaM services in Lambeth and Southwark), age-range and symptom terms (e.g. psychos*; psychot*, delusion*, voices, hallucinat* paranoia). Once retrieved, individual patient records were screened and reviewed by the aforementioned researchers using a validated diagnostic screening tool, namely, the Item Checklist Group of the Schedule of Clinical Assessment of Neuropsychi-

atry SCAN (WHO, 1994), to identify first episode psychosis cases. Individuals were included as cases if they were: resident in the London boroughs of Lambeth or Southwark; aged 18-64 years (inclusive); experiencing psychotic symptoms of at least one day duration during the study periods and scored at least 2 or more for psychotic symptoms as assessed using the SCAN. This screening process described above enabled the assignment of population at risk into three categories i.e. FEP cases, no psychosis and excluded.

Secondly, two primary researchers (a psychiatric nurse and a psychiatrist) reviewed all the included cases from the first stage screen to ensure cases met all inclusion criteria. An inter-rater reliability test was carried out between the two experts and Cohen's Kappa coefficient of 0.77 (p=<0.01) was achieved. Finally, discrepant or ambiguous cases were resolved by consensus with the principal investigator.

In total, 9109 individual clinical records were screened, of whom 560 screened positive and were FEP cases, 5234 screened negative for psychosis (but remain at risk and allocated to re-evaluation) and 3315 were excluded (because of evidence of any of the following: previous psychosis, organic psychosis, not resident, too young or too old). In the work described below, these 9109 records were split into a tuning set (two thirds of the total) and a test set (the remainder).

3   Experiments

In order to facilitate further identification of relevant cases, a case classification application was created using GATE (Cunningham et al., 2013), since this technology provides a wide variety of different information extraction tools that can be used to create features for machine learning, as well as an integration of LibSVM's (Chang and Lin, 2011) support vector machine and various of the Mallet (McCallum, 2002) and Weka (Hall et al., 2009) algorithms, and has been in use at SLaM for several years.

Due to the challenging nature of the data, systematic exploration of available tools was required to produce a good result. This work focuses on three algorithms; support vector machines (SVMs, in particular LibSVM), and Weka's Random Forest and JRip. In the course of experimentation, many algorithms were tried, but these three have been chosen as the focus here because they formed good practical propositions, both in terms of accuracy of classification and speed, and being diverse, provide insight into the ways that different techniques interact with algorithm choice.

The work is presented here in two parts. Feature selection and parameter tuning is discussed, showing how these can be used to improve the accuracy of the classifiers. Then the problem of bias against the minority class is explored. Since the cases to be identified are by far the minority, and the priority is finding as many of them as possible, optimizing overall accuracy was not sufficient. The second section, therefore, addresses this issue, and concludes with an assessment of the utility of confidence scores for providing fine-grained control over the level of recall achieved.

All experimental software is available to download[1]. A Docker file is provided that builds the experimental environment, with an entrypoint script running the complete experiment set presented in this paper, generating the results shown. The data is however highly confidential and therefore cannot be shared.

3.1 Feature Selection and Parameter Tuning

Early experiments used a feature set including word bigrams and unigrams (trigrams lead to an very high dimensionality of problem, and previous experience has shown that they are likely to overfit all but the largest of corpora, which our 9109 cases, whilst sizeable for an expert-annotated set of complex cases, is not) as well as presence of terms in a comprehensive gazetteer provided by a medical expert. This gazetteer covered symptoms relevant to psychosis, and was further supplemented with a speculative term set relevant to diagnosis and treatment, such as the phrase "first episode [of] psychosis" or phrases relevant to sectioning and hospital admission. ConText (Harkema et al., 2009) was applied to these gazetteer mentions to add information about whether it is the patient that is experiencing the observation or another individual, for example a family member; whether they are stated as experiencing or not experiencing it (e.g. "no evidence of psychosis"); and whether the finding is noted in the present or past (e.g. "had previously experienced auditory hallucinations"). Note that the phrase "first episode psychosis" or "first episode

of psychosis", whilst telling, is extremely rare, occurring only a couple of times in the whole corpus. A typical case record progresses all the way from first presentation through to treatment and management with minimal discussion of the diagnosis.

In addition to these features, there are some structured data fields associated with the cases, including diagnosis, as well as demographic information such as gender, ethnicity and date of birth. A number of quite detailed diagnosis categories correspond to psychosis of the type we are interested in. Diagnosis fields (of which there are several) are utilized to differing extents by clinicians, and may be empty or out of date. Furthermore, diagnosis does not help us to identify that this record describes a first episode of psychosis. However, diagnosis fields are an obvious feature to include.

GATE was used to create feature representations of the tuning instances, which were then exported in ARFF format, in order to experiment with feature extraction techniques available in Weka but not in GATE. Weka's CfsSubsetEval was used with BestFirst feature selection, as this is a pragmatic option. However due to time constraints, this was impractical over the very large dimensionalities necessitated by the inclusion of unigram and bigram features. Instead feature selection was performed without including n-grams. Results are presented for the feature set including unigrams in order to contrast the overall performance, but the feature set across which feature selection was performed was limited to the feature set without n-grams.

Feature selection provides an insight into the data. Note that a feature not being selected does not imply it is of no utility in separating the cases, since it may be redundant in conjunction with a better feature. Note also that the feature selection methods employed may not be congruent with the algorithms we then go on to use, since some algorithms may be able to, for example, combine features differently to produce useful information. Nonetheless it is interesting to note what seems to help to separate the cases. Listed here are the features strongly selected, being found valid over three out of three folds of the data. Below that, the features presented were found valid in two out of three folds. All selected gazetteer features are positive mentions experienced by the patient in the present, as ascertained using ConText.

- Validated in 3/3 folds

- Null or empty values in the following structured fields; borough, ethnicity, gender, postcode, first primary diagnosis
- Age
- First primary diagnosis:
  * bipolar, hypomanic (F31.0)
  * bipolar, unspecified (F31.9)
  * severe depressive w/psychotic symptoms (F32.3)
- Text features, presence of gazetteer terms; "olanzapine", "risperidone", "auditory hallucinations", "voices", "paranoid", "psychotic", "psychosis"

- Validated in 2/3 folds

  - First primary diagnosis
    * bipolar (F31)
    * organic delusional schizophrenia-like disorder (F06.2)
    * organic mood disorder (F06.3)
  - Text features, presence of gazetteer terms; "aripiprazole", "quetiapine", "persecutory", "schizophrenia"

Reflecting on these features, it is interesting that the absence of some structured information, for example an empty value for postcode, enables some separation of the cases. It may be that first episodes of psychosis, perhaps because they often present under troubled circumstances, tend to arrive in the system via a different route that has some systematic differences to more routine cases, resulting in these differences in the case record. It is unsurprising that diagnosis fields are of value, being likely to assist both in finding positive cases and ruling out negative ones (e.g. organic causes). Furthermore, antipsychotic drugs and the more telling of symptoms appear prominently, as do terms such as "psychosis", that suggest a postulated diagnosis. It is also interesting that only a small number of features is selected, the majority being redundant.

Next, the impact feature selection has on accuracy with regards to the three algorithms is investigated. Firstly, the SVM is tuned. Then, feature scaling (normalization) and cost are considered in conjunction with feature selection. The cost parameter of an SVM refers to the importance attached to creating a classifier that correctly classifies the training instances. A high cost results in a better fit to the training data, though may potentially overfit. A low cost may result in a weak classifier that hasn't made the best use of the training data.

Feature normalization describes a process whereby numeric features are brought into a similar statistical distribution with each other, for example by scaling them all to have the same mean and variance. In this case, age is a numeric feature with a very different range than the nominal features that otherwise dominate. Nominal features are expanded out to one dimension per value and assigned counts, which for many fields such as diagnosis fields, amount to ones and zeroes for presence or absence. The greater magnitude of the age feature in no way reflects its greater importance, yet vector space algorithms may attach more importance to larger values. In this work, this is relevant to the SVM. The other two algorithms used here are unaffected by the magnitude of numeric features.

Table 1 shows a sample of the results obtained from evaluating using three-fold cross-validation on the tuning corpus with the large feature set including unigrams, and table 2 shows a sample of the results obtained from evaluating using three-fold cross-validation on the tuning corpus with the reduced feature set. We can see that where the larger feature set is used, including unigrams, cost and feature normalization have an important role to play in getting a competitive result. At lower costs, feature normalization is detrimental, but once cost comes into the right range, it helps. However, on the reduced feature set, obtaining a good result is far easier. Feature normalization does not have much impact any more, and cost, whilst an important parameter to tune, is less critical. This result emphasizes the potential value of selectiveness with features to the SVM, whilst highlighting the role that cost tuning and feature normalization may play in working with a less optimal feature selection.

Having tuned the SVM, this was now compared to the other two algorithms with regards to feature selection. GATE was used to produce a new ARFF file of the tuning instances with the reduced feature set, in addition to the full set with and without unigrams, which were then evaluated in Weka using threefold cross-validation. Adapting

Cost	Feat Norm?	Accuracy	Kappa
1	No	66.3%	0.2496
10	No	70.4%	0.3809
1000	No	74.2%	0.4937
1	Yes	59.86%	0
10	Yes	60.0%	0.0068
1000	Yes	79.6%	0.5629

Table 1: Parameter tuning on the large feature tuning set including unigrams.

Cost	Feat Norm?	Accuracy	Kappa
1	No	81.8%	0.6244
10	No	82.2%	0.6392
1000	No	78.6%	0.5802
1	Yes	76.2%	0.4682
10	Yes	77.8%	0.5105
1000	Yes	81.9%	0.6368

Table 2: Parameter tuning on the reduced feature tuning set.

Algorithm	Feature set	Acc.	Kappa
SVM	Full+uni	79.6%	0.5629
SVM	Full	81.4%	0.6254
SVM	Reduced	81.9%	0.6368
JRip	Full+uni	81.3%	0.6234
JRip	Full	81.5%	0.6296
JRip	Reduced	82.0%	0.6349
Rand. Forest	Full+uni	66.46%	0.2385
Rand. Forest	Full	81.5%	0.6136
Rand. Forest	Reduced	82.2%	0.6274

Table 3: Trying different feature sets with different algorithms.

our GATE application to utilize the features identified as being more useful resulted in an approximation that captures the spirit of what was learned, rather than an exact match, for practical reasons. The GATE Learning Framework machine learning integration [2] makes it easier to simply include the diagnosis field, for example, having shown itself to be of value, rather than picking the diagnoses of interest.

Feature selection wasn't performed on the feature set that included unigrams, so therefore we are interested to see results on this set to get a heuristic feel for whether unigrams are of value, although one can't rule out that had feature selection been performed on the unigrams, some of them would have been found to be of utility. We proceed therefore with three datasets; the full feature set including unigrams (419531 features), the full feature set without unigrams (3256 features) and the reduced set of 2027 features. Note that the reason the reduced feature sets number thousands despite the list being short as above is that a nominal feature is expanded out to a number of numeric (count) features equivalent to one per unique value found in the training set. Table 3 shows the impact of feature set reduction on the results obtained with each algorithm.

In all cases, reducing features results in an improvement, marginal for SVM and JRip but substantial for Random Forest, indeed being required to bring the result obtained up to a competitive standard. The main improvement comes from the removal of unigrams. A further contribution of feature reduction lies in the speed gains obtained at training time. The SVM was trained using a cost of 1000 with feature scaling included. We can see that whilst the algorithms respond differently to feature reduction, using the smaller set there is no very clear winner among them.

3.2 Class Balancing

Having focused evaluation so far around classification accuracy, the question of how effective our classifiers are at obtaining a high sensitivity (recall) on first episode psychosis cases has not yet been considered. The goal of the work is to enable medical researchers to obtain a sample of positive cases with little cost in the way of missing any, whilst reducing the amount of time they spend rejecting negative cases. Finding as near as possible to all of the relevant cases is the main priority. Precision needs to be high enough to justify the exercise, but there is much more flexibility regarding how high is good enough. A classifier that is tuned to produce as high an overall accuracy as possible will tend to favour the dominant class, since in the case of uncertainty, assigning to the dominant class will tend to be right more often than it is wrong. Therefore some innovation must be introduced to counteract this.

Early experimentation focused on the weights parameter on the support vector machine. Figure 1 gives the ROC curve thus obtained, using threefold cross-validation on the tuning corpus. The

[2] https://github.com/GenevieveGorrell/gateplugin-LearningFramework

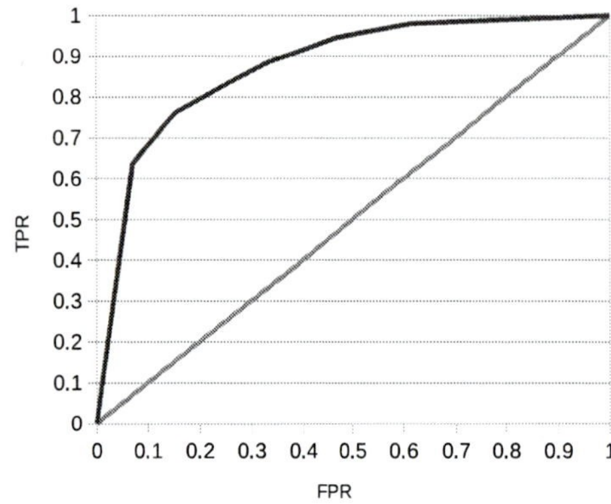

Figure 1: ROC curve for SVM negative class down-weighting

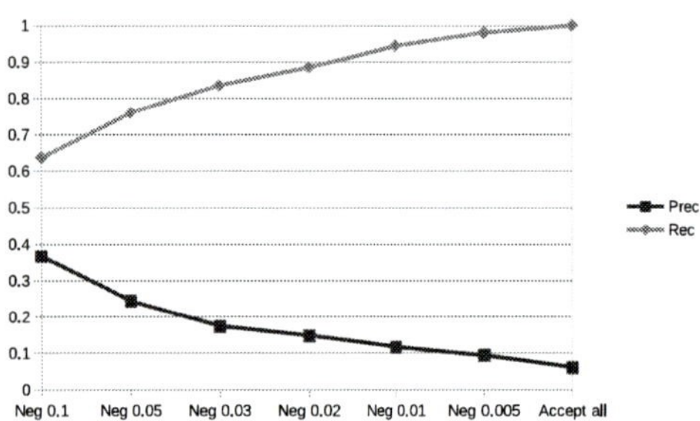

Figure 2: Precision and recall vary with SVM class weighting

AUC (area under the curve) is 0.87. For a recall of 0.944, this gives a specificity of 0.535, which equates roughly to halving the number of cases required to be viewed, at a cost of missing one in 20 cases. It is clear from the graph of precision and recall against weight in figure 2 that the weights parameter provides an effective option for increasing recall of the positive cases to the required level.

Unfortunately this parameter is not available or relevant to the other two algorithms, and also did not transfer easily to the larger training set used to prepare the final application. Further experimentation instead focused on creating a balanced training set that would not penalize the minority class. A balanced training set should lead to a fairer classifier for many algorithms, which aim to minimize the number of misclassified points. Table 4 shows results obtained using Weka to sample the tuning set fairly across classes, having first taken out one third for testing. No replacement of instances was opted for, and the dataset was reduced to 20%,

Algorithm	Conds	Prec	Rec	F1
SVM	0.05	0.244	0.760	0.370
SVM	No	0.544	0.358	0.432
SVM	Yes	0.286	0.675	0.402
JRip	No	0.508	0.258	0.343
JRip	Yes	0.226	0.783	0.351
Rand. Forest	No	0	0	0
Rand. Forest	Yes	0.306	0.725	0.431

Table 4: Class balancing interacts with algorithm choice.

this being large enough to ensure that all positive cases were included, thus thinning the negatives and creating an effect that could be broadly replicated back in GATE by removing some of the negative cases. Separating out a test set is necessary to ensure that the result obtained is indicative of what might be obtained on a naturalistic sample. Had cross-validation been used on an artificially balanced set, the result would have been misleading. The first line in the table gives the most comparable result for weight tuning in SVM ("conds" in this case gives the weight assigned to the negative classes), for comparison. Below that, "conds" indicates whether or not class balancing was used on the training data. We see that for SVM and JRip, class balancing allows recall of the positive class to be improved whilst retaining a broadly similar F1. For Random Forest, class balancing allows us to find the positive cases where previously they were not found at all, and produces a competitive model.

A further option for altering the precision/recall balance lies in making use of the confidence scores provided by the algorithms. However different algorithms are differently able to provide a sensitive and informative confidence score. Confidence scores are made use of in this work to provide the medical researchers with an *ordered* list of cases to review, leaving the power in their hands to progress as far down the list as provides them with the recall they require. This does not negate the need for a classifier tuned to the needs of the task. An appropriately tuned classifier can be expected to give a better F1 for a certain recall than one obtained simply by applying a confidence threshold to a mistuned one. A Random Forest model was trained in GATE using the full tuning set, but with the negative instances thinned to 1 in 13, roughly

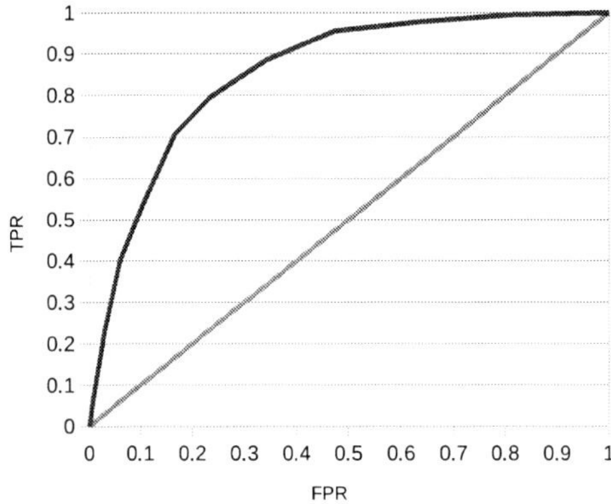

Figure 3: ROC curve based on Random Forest confidence scores in GATE.

balancing the classes. Figure 3 gives a ROC curve based on confidence scores assigned on the test set (AUC 0.85). In keeping with previous results, a recall is obtained of in excess of 95% for a specificity of 0.53, halving the number of cases required to be viewed, by setting the confidence threshold at 0.2. For a recall of almost 0.8, only around a quarter of cases would need to be viewed (specificity 0.77).

4   Conclusion

This paper presents work on a challenging psychiatry domain case classification application. The goal was to facilitate medical researchers' collection of a (further) sample of cases describing a first episode of psychosis, by learning a model from 9109 cases already manually classified. This classification problem, requiring the highest level of domain expertise to accomplish manually, proves challenging for natural language processing techniques. The problem is complicated by the subtlety of distinction between the positive and negative cases; for example, psychotic episodes that are not the first, and those with organic causes are negative instances, although the language surrounding their case is very similar to the positive cases. Furthermore the sample with which we are working is already selected on the basis of psychosis-related keyword search, meaning that the NLP work is required to offer value over and above that. Feature normalization proves essential to making the support vector machine competitive on the task. Feature selection is generally beneficial, in particular making Random Forest competitive, and allowing

a much smaller feature space to be used. Since the task is to identify the minority class with a high recall, an important part of task success focused on tuning the algorithms in favour of the positive class. This was accomplished by thinning the negative instances in the training set. Attempts to use the weights parameter with the SVM were complicated by the apparent sensitivity of this parameter to variations in the task conditions. The final GATE application achieves an AUC of 0.85, a result that compares favourably with previous similar work despite the additional challenges, and allows medical researchers to select their own recall based on the confidence score of the Random Forest algorithm, for example halving the number of cases they are required to examine with a loss of only 5% of positive cases. No one machine learning algorithm notably excelled in this work; success might be attributed to an exceptional training set, both in terms of size and quality, and the freely available machine learning technologies that provided a solution to the problems that arose.

References

Alan R Aronson. 2001. Effective mapping of biomedical text to the umls metathesaurus: the metamap program. In *Proceedings of the AMIA Symposium*, page 17. American Medical Informatics Association.

P. Baldwin, D. Browne, P. J. Scully, J. F. Quinn, M. G. Morgan, A. Kinsella, J. M. Owens, V. Russell, E. O'Callaghan, and J. L. Waddington. 2005. Epidemiology of first-episode psychosis: illustrating the challenges across diagnostic boundaries through the cavan-monaghan study at 8 years. *Schizophr Bull*, 31:624–38.

Brandon K Bellows, Joanne LaFleur, Aaron WC Kamauu, Thomas Ginter, Tyler B Forbush, Stephen Agbor, Dylan Supina, Paul Hodgkins, and Scott L DuVall. 2014. Automated identification of patients with a diagnosis of binge eating disorder from narrative electronic health records. *Journal of the American Medical Informatics Association*, 21(e1):e163–e168.

R. Borschmann, S. Patterson, D. Poovendran, D. Wilson, and T. Weaver. 2014. Influences on recruitment to randomised controlled trials in mental health settings in england: a national cross-sectional survey of researchers working for the mental health research network. *BMC medical research methodology*, 14(1).

F. Bourque. 2015. *A mixed methods study of relation between migration, ethnicity and psychosis*. Ph.D. thesis, Kings College London.

Victor M Castro, Jessica Minnier, Shawn N Murphy, Isaac Kohane, Susanne E Churchill, Vivian Gainer, Tianxi Cai, Alison G Hoffnagle, Yael Dai, Stefanie Block, et al. 2014. Validation of electronic health record phenotyping of bipolar disorder cases and controls. *American Journal of Psychiatry*.

Chih-Chung Chang and Chih-Jen Lin. 2011. Libsvm: a library for support vector machines. *ACM Transactions on Intelligent Systems and Technology (TIST)*, 2(3):27.

Wendy W Chapman, Will Bridewell, Paul Hanbury, Gregory F Cooper, and Bruce G Buchanan. 2001. A simple algorithm for identifying negated findings and diseases in discharge summaries. *Journal of biomedical informatics*, 34(5):301–310.

Hamish Cunningham, Valentin Tablan, Angus Roberts, and Kalina Bontcheva. 2013. Getting more out of biomedical documents with gate's full lifecycle open source text analytics. *PLoS Comput Biol*, 9(2):e1002854.

Elizabeth Ford, John A Carroll, Helen E Smith, Donia Scott, and Jackie A Cassell. 2016. Extracting information from the text of electronic medical records to improve case detection: a systematic review. *Journal of the American Medical Informatics Association*, page ocv180.

T. Greenhalgh, H. W. Potts, G. Wong, P. Bark, and D. Swinglehurst. 2009. Tensions and paradoxes in electronic patient record research: a systematic literature review using the meta-narrative method. *Milbank Quarterly*, 87(4):729–788, Dec.

Mark Hall, Eibe Frank, Geoffrey Holmes, Bernhard Pfahringer, Peter Reutemann, and Ian H Witten. 2009. The weka data mining software: an update. *ACM SIGKDD explorations newsletter*, 11(1):10–18.

Henk Harkema, John N Dowling, Tyler Thornblade, and Wendy W Chapman. 2009. Context: an algorithm for determining negation, experiencer, and temporal status from clinical reports. *Journal of biomedical informatics*, 42(5):839–851.

S. Hogerzeil, A. Van Hemert, F. Rosendaal, E. Susser, and H. Hoek. 2014. Direct comparison of first-contact versus longitudinal register-based case finding in the same population: early evidence that the incidence of schizophrenia may be three times higher than commonly reported. *Psychological medicine*, 44:3481–3490.

Sandy H Huang, Paea LePendu, Srinivasan V Iyer, Ming Tai-Seale, David Carrell, and Nigam H Shah. 2014. Toward personalizing treatment for depression: predicting diagnosis and severity. *Journal of the American Medical Informatics Association*, 21(6):1069–1075.

J. B. Kirkbride, P. Fearon, C. Morgan, P. Dazzan, K. Morgan, J. Tarrant, T. Lloyd, J. Holloway, G. Hutchinson, J. P. Leff, R. M. Mallett, G. L. Harrison, R. M. Murray, and P. B. Jones. 2006. Heterogeneity in incidence rates of schizophrenia and other psychotic syndromes: Findings from the 3-center aesop study. *Archives of General Psychiatry*, 63:250–258.

Andrew Kachites McCallum. 2002. Mallet: A machine learning for language toolkit.

S.M. Meystre, G.K. Savova, K.C. Kipper-Schuler, and J.F. Hurdle. 2008. Extracting information from textual documents in the electronic health record: A review of recent research. *IMIA Yearbook of Medical Informatics*, pages 128–144.

Maxine X Patel, Victor Doku, and Lakshika Tennakoon. 2003. Challenges in recruitment of research participants. *Advances in Psychiatric Treatment*, 9(3):229–238.

RH Perlis, DV Iosifescu, VM Castro, SN Murphy, VS Gainer, Jessica Minnier, T Cai, S Goryachev, Q Zeng, PJ Gallagher, et al. 2012. Using electronic medical records to enable large-scale studies in psychiatry: treatment resistant depression as a model. *Psychological medicine*, 42(01):41–50.

S. M. Powsner, J. C. Wyatt, and P. Wright. 1998. Opportunities for and challenges of computerisation. *Lancet*, 352(9140):1617–1622, Nov.

S. T. Rosenbloom, J. C. Denny, H. Xu, N. Lorenzi, W. W. Stead, and K. B. Johnson. 2011. Data from clinical notes: a perspective on the tension between structure and flexible documentation. *J Am Med Inform Assoc*, 18(2):181–186.

Guergana K Savova, James J Masanz, Philip V Ogren, Jiaping Zheng, Sunghwan Sohn, Karin C Kipper Schuler, and Christopher G Chute. 2010. Mayo clinical text analysis and knowledge extraction system (ctakes): architecture, component evaluation and applications. *Journal of the American Medical Informatics Association*, 17(5):507–513.

Donia Scott, Rossano Barone, and Rob Koeling. 2012. Corpus annotation as a scientific task. In Nicoletta Calzolari (Conference Chair), Khalid Choukri, Thierry Declerck, Mehmet Uur Doan, Bente Maegaard, Joseph Mariani, Jan Odijk, and Stelios Piperidis, editors, *Proceedings of the Eight International Conference on Language Resources and Evaluation (LREC'12)*, Istanbul, Turkey, may. European Language Resources Association (ELRA).

Lisa Seyfried, David A Hanauer, Donald Nease, Rashad Albeiruti, Janet Kavanagh, and Helen C Kales. 2009. Enhanced identification of eligibility for depression research using an electronic medical record search engine. *International journal of medical informatics*, 78(12):e13–e18.

WHO. 1994. Schedules for clinical assessment in neuropsychiatry: version 2.

Meliha Yetisgen-Yildiz, Cosmin Adrian Bejan, and Mark M Wurfel. 2013. Identification of patients with acute lung injury from free-text chest x-ray reports. *ACL 2013*, page 10.

Qing T Zeng, Sergey Goryachev, Scott Weiss, Margarita Sordo, Shawn N Murphy, and Ross Lazarus. 2006. Extracting principal diagnosis, co-morbidity and smoking status for asthma research: evaluation of a natural language processing system. *BMC medical informatics and decision making*, 6(1):1.

Relation extraction from clinical texts using domain invariant convolutional neural network

Sunil Kumar Sahu[Ψ]*, **Ashish Anand**[Ψ], **Krishnadev Oruganty**[♣], **Mahanandeeshwar Gattu**[♣]
[Ψ]Department of Computer Science and Engineering, IIT Guwahati, Assam, India
[♣]Excelra Knowledge Solutions Pvt Ltd, Hyderabad, Telangana, India
{sunil.sahu, anand.ashish}iitg.ernet.in
{krishnadev.oruganty, nandu.gattu}gvkbio.com

Abstract

In recent years extracting relevant information from biomedical and clinical texts such as research articles, discharge summaries, or electronic health records have been a subject of many research efforts and shared challenges. Relation extraction is the process of detecting and classifying the semantic relation among entities in a given piece of texts. Existing models for this task in biomedical domain use either manually engineered features or kernel methods to create feature vector. These features are then fed to classifier for the prediction of the correct class. It turns out that the results of these methods are highly dependent on quality of user designed features and also suffer from curse of dimensionality. In this work we focus on extracting relations from clinical discharge summaries. Our main objective is to exploit the power of convolution neural network (CNN) to learn features automatically and thus reduce the dependency on manual feature engineering. We evaluate performance of the proposed model on i2b2-2010 clinical relation extraction challenge dataset. Our results indicate that convolution neural network can be a good model for relation exaction in clinical text without being dependent on expert's knowledge on defining quality features.

1 Introduction

The increasing amount of biomedical and clinical texts such as research articles, clinical trials, discharge summaries, and other texts created by social network users, represents immeasurable source of information. Automatic extraction of relevant information from these resources can be useful for many applications such as drug repositioning, medical knowledge base creation etc. The performance of concept entity recognition systems for detecting mention of proteins, genes, drugs, diseases, tests and treatments has achieved sufficient level of accuracy, which gives us opportunity for using these data to do next level tasks of natural language processing (NLP). Relation extraction is the process of identifying how given entities are related in considered sentence or text. As given in the example sentence [S1] below, the entities *Lexix* and *congestive heart failure* are related by *treatment administered medical problem* relation. These relations are important for other upper level NLP tasks and also in biomedical and clinical research (Shang et al., 2011).

[S1]: *He was given* **Lexix** *to prevent him from* **congestive heart failure** .

Relation extraction task in unstructured text has been modeled in many different ways. *co-occurrence* based methods due to their simplicity and flexibility are most widely used methods in biomedical and clinical domain. In co-occurrence analysis it is assumed that if two entities are coming together in many sentences, their must be a relation between them (Bunescu et al., 2006; Song et al., 2011). Quite obviously this method can not differentiate types of relations and suffers from low precision and recall. To improve its results, different statistical measures such as point wise mutual information, chi-square or log-likelihood ratio has been used in this approach (Stapley and Benoit, 2000).

Rule based methods are another commonly adapted methods for relation extraction task (Thomas et al., 2000; Park et al., 2001; Leroy et al., 2003). Rules are created by carefully observing the syntactic and semantic patterns in rela-

---

Part of this work was done while *Sunil Kumar Sahu* was doing internship at *Excelra Knowledge Solutions Pvt Ltd*, Hyderabad, Telangana, India.

Proceedings of the 15th Workshop on Biomedical Natural Language Processing, pages 206–215,
Berlin, Germany, August 12, 2016. ©2016 Association for Computational Linguistics

tion instances. *Bootstrapping method* (Xu, 2008) is used to improve the performance of rule based methods. Bootstrapping uses small number of known relation pair of each relation type as a seed and use these seeds to search patterns in huge unannotated text (Xu, 2008) in iterative fashion. Bootstrapping method generates lots of irrelevant patterns too, which can be controlled by *distantly supervised* approach. Distantly supervised method uses large knowledge base such as UMLS or Freebase as an input and extract patterns from huge corpus for all pair of relations present in knowledge base (Mintz et al., 2009; Riedel et al., 2010; Roller and Stevenson, 2014). The advantage of bootstrapping and distantly supervised methods over supervised methods is that they do not require lots of manually labeled training data which is generally very hard to get.

Feature based methods use sentences with predefined entities to construct feature vector through feature extraction (Hong, 2005; Minard et al., 2011b; Rink et al., 2011). Feature extraction is mainly based on linguistic and domain knowledge. Extracted feature vectors are used to decide correct class of relation present between entities in the sentence through any classification techniques. *Kernel methods* are extension of feature based methods which utilize kernel functions to exploit rich syntactic information such as parse trees (Zelenko et al., 2003; Culotta and Sorensen, 2004; Qian and Zhou, 2012; Zeng et al., 2014). State of the art results have been obtained by these class of methods.

However, the performance of feature and kernel based methods are highly dependent on suitable feature set selection, which is not only tedious and time consuming task but also require domain knowledge and is dependent on other NLP systems. Often such dependencies make many existing work less reproducible simply because of absence of the full and finer details of feature extraction. Further often these methods lead to huge number of features and may get affected from curse of dimensionality issues (Bengio et al., 2003; Collobert et al., 2011). Another issue faced by these methods is feature extraction will have to be adjusted according to the data source. As discussed earlier we are having multiple but diverse information resources such as research articles, discharge summaries, clinical trials outcome etc. While in one hand multiple sources bring more information but the other hand it makes it challenging to extract meaningful information automatically simply because of diverse nature of the data source. For example, sentences in research articles are well formed and likely to use only well accepted technical terms. But sentences in clinical discharge summaries may not be well formed sentences instead it could be fragmented sentences with lots of acronyms or terms used only locally. Similarly social media articles may use slang or terms which are not technically used. This makes it difficult for above discussed methods.

Motivated by these issues, this work aims to exploit recent advances in machine learning and NLP domains to reduce such dependencies and utilize convolutional neural network to learn important features with minimal manual dependencies. Convolution neural network has shown to be a powerful model for image processing, computer vision (Krizhevsky et al., 2012; Karpathy and Fei-Fei, 2014) and subsequently in natural language processing it has given state of the art results in different tasks such as sentence classification (Kim, 2014; Kalchbrenner et al., 2014; Hu et al., 2014; Sharma et al., 2016), relation classification (Zeng et al., 2014; dos Santos et al., 2015) and semantic role labeling (Collobert et al., 2011).

In this paper we propose a new framework for extracting relations among *problem, treatment* and *test* in clinical discharge summaries. In particular we use data available under clinical relation extraction task organized by Informatics for Integrating Biology and the Bedside (i2b2) in 2010 as part of i2b2/VA challenge (Uzuner et al., 2011). Extracting relations in clinical texts is more challenging compared to research articles as it contains incomplete or fragmented sentences, and lots of acronyms. Current state of the art methods heavily depend on manual feature engineering and use hundreds of thousands of features (Minard et al., 2011b; Rink et al., 2011). Our result indicates the proposed model can outperform the current state of the art models by using only a small fraction of features. However the main observation is the features used in our model is easy to replicate and adapt as per the data source compared to the feature sets generally used in these tasks.

2 Related Research

i2b2 organized a shared task in 2010 (Uzuner et al., 2011). In this challenge discharge sum-

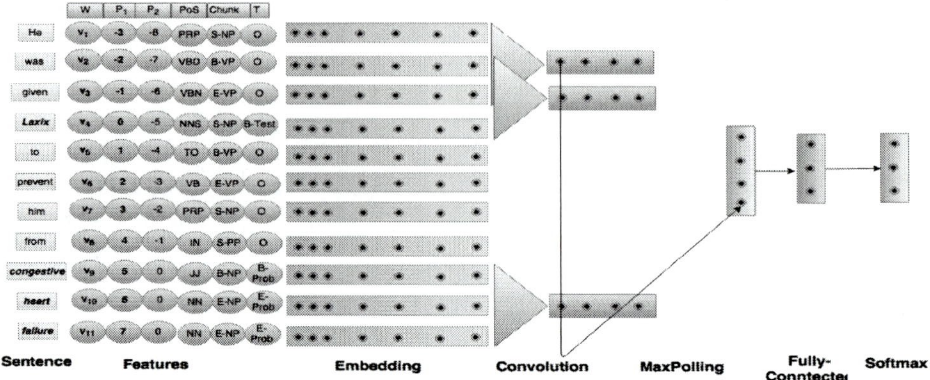

Figure 1: CNN model for relation extraction.

maries from three different sources were annotated for extracting relations among clinical entities such as *problem*, *treatment* and *test*. Most of the participants in this challenge used support vector machine (SVM) with manually designed features (Uzuner et al., 2010). Model proposed by Rink et al. (2011) had first place in this task, which used six classes of features namely, context features, similarity features, nested related relation features, Wikipedia features, single concept features and vicinity features. They formulated the relation extraction task as a multiclass classification problem and SVM with linear kernel were used for classification.

For extracting relation among disease and treatment, Rosario and Hearst (2004) used various graphical and neural network models. They used variety of lexical, semantic and syntactic features for classification and found that semantic features were contributing most among all. The dataset used in this study was relatively smaller and was prepared from biomedical research articles. Li et al. (2008) proposed kernel methods for relation extraction between entities in MEDLINE® articles. They modified the tree kernel function by incorporating trace kernel to capture richer contextual features for classifying the relation. Their results shows that tree kernel outperform other kernel methods such as word and sequence kernels for the considered task.

Conditional random field (CRF) has been used for relation extraction between disease treatment and gene by (Bundschus et al., 2008). In this experiment setting, they did not assume that entities were given, instead their model also predicted entities and its type. They developed two variants of CRF both modeled relation extraction task as sequence labeling task. Recently Bravo et al. (2015) proposed a system for identifying association between drug disease and target in EU-ADR dataset (van Mulligen et al., 2012) and named it BeeFree. BeeFree usese combination of shallow linguistic kernel and dependency kernel for identifying relations.

In contrast to above methods recently there are few work applying convolution neural network based models (Zeng et al., 2014; dos Santos et al., 2015) for *relation classification* in SemEval 2010 relation classification dataset (Hendrickx et al., 2009). Convolution neural network used in this models are using constant length filters, and word embedding and distance embedding as features. Our model leverage on the linguistic features also and we considered *relation extraction* task in clinical notes which is much more informal, rich with acronyms and number of samples for each relations are not stable (Uzuner et al., 2011).

3 CNN for Clinical Relation Extraction

The proposed model based on CNN is first summarized in the next section. Subsequent sections describe it in more detail.

3.1 Model Architecture

The proposed model architecture is shown in the figure 1, which takes a complete sentence with mentioned entities as an input and outputs a probability vector corresponding to all possible relation types. Each feature is having vector representation which is initialized randomly except word embedding feature. For word embedding, we used pre-trained word vector (TH et al., 2015) learned on Pubmed articles using word2vec tool (Mikolov et al., 2013b).

Embedding layer maps every feature value with its corresponding feature vectors and concatenate them. In order to get local features from each part of the sentence we have used multiple filters of different lengths (Kim, 2014) in all possible continuous n-gram of the sentence, where n is the length of filter (We have shown four filters with constant length three in the figure 1). We use max pooling over time to get global features through all filters. Here time indicates filter running over the length of the sentence. Pooled features are then fed to fully connected feed-forward neural network to make inference. In the output layer we use softmax classifier with number of outputs equal to number of possible relations between entities.

3.2 Feature Layer

We represent each word in the sentence with 6 discrete features namely word itself (W), distance from the first entity (P_1), distance from the second entity (P_2), parts of speech tag of the word (PoS), chunk tag of the word (Chunk) and entity type (T). Each feature is briefly described below:

1. W : Exact word appeared in the sentence.

2. P_1: Distance from the first entity in terms of number of words (Collobert and Weston, 2008). For instance in our earlier example [S1] *He* is at -3 distance and *prevent* is at $+2$ distance away from the first entity *Lexis*. This value would be zero for all words which is a part of the first entity.

3. P_2: Similar to P_1 but considers distance from the second entity.

4. PoS: Parts of speech tag of the considered word. We use genia tagger[1] to obtain pos tag of each word.

5. $Chunk$: Chunk tag of considered word. Again we use genia tagger to obtain chunk tag of each word.

6. T: Type of the considered word. For example, it would be entity type such as $B - Prob$, ple, it would be entity type such as $B - Prob$,

$I - Prob$ etc. for entity word and *Other* for rest words following the *BIO* tagging convention.

This way a word $w \in D^1 \times D^2 \times .....D^6$, where D^i is the dictionary for i^{th} local features.

3.3 Embedding Layer

In lookup or embedding layer each feature value is mapped to its vector representation using feature embedding matrix. Lets say $M^i \in \mathbb{R}^{n \times N}$ is the feature embedding matrix for i^{th} local feature (here n represents dimension of feature embedding and N is number of possible values or size of the dictionary for i^{th} local feature). Each column of M^i is vector of corresponding value of ith features. Mapping can be done by taking product of one hot vector of feature value with its embedding matrix (Collobert and Weston, 2008). Suppose $a_j^{(i)}$ is the one hot vector for j^{th} feature value of ith feature then:

$$f_j^{(i)} = M^i \, a_j^{(i)} \tag{1}$$

$$x^i = f_1^{(i)} \oplus f_2^{(i)} .... \oplus f_6^{(i)} \tag{2}$$

Here $\oplus$ is concatenation operation so $x^i \in \mathbb{R}^{(n_1 + ... n_6)}$ is feature vector for i^{th} word in sentence and n_k is dimension of kth feature. For word embedding we used pre-trained word vector obtained after running word2vec tool (Mikolov et al., 2013b; Mikolov et al., 2013a) on huge Pubmed open source articles (TH et al., 2015). Other feature matrix were initialized randomly at the beginning. Since number of elements in all feature dictionary except word dictionary (D^1) are not huge, we assume that while training these vectors will get sufficient updation.

3.4 Convolution Layer

We apply convolution on text to get local features from each part of the sentence (Collobert and Weston, 2008). Consider $x^1 x^2 ..... x^m$ is the sequence of feature vectors of a sentence, where $x^i \in \mathbb{R}^d$ is a vector obtained by concatenating all feature vector of ith word. Let $x^{i:i+j}$ represents concatenation of $x^i ..... x^{i+j}$ feature vectors. Suppose there is a *filter* parameterized by weight vector $w \in \mathbb{R}^{cd}$ where c is the length of filter (in figure 1 filter length is three). Then output sequence of convolution layer would be

$$h^i = f(w \cdot x^{i:i+c-1} + b) \tag{3}$$

---

[1] http://www.nactem.ac.uk/GENIA/tagger/

Where $i = 1, 2, \ldots m - c + 1$, . is dot product, f is rectify linear unit (ReLu) function and $b \in \mathbb{R}$ is biased term. w and b are the learning parameters and will remain same for all $i = 1, 2, \ldots m - c + 1$.

3.5 Max Pooling Layer

Output of convolution layer length $(m - c + 1)$ will vary based on number of words m in the sentence. We applied max pooling (Collobert and Weston, 2008) over time to get fixed length global features for whole sentence. The intuition behind using max pooling is to consider only most useful feature from entire sentence.

$$z = \max_{1 \leq i \leq (m-c+1)} [h^i] \qquad (4)$$

We have just explained the process of extracting one feature from a whole sentence using one filter. In figure 1 we extracted four features using four filters of the same length three. In our experiment we use multiple such filters of variable length (Kim, 2014; Yin and Schtze, 2015). The objective of using different length filter is to accommodate context in varying window size around words.

3.6 Fully Connected Layer

The output of max pooling layer is sequence z came with different filters. We call this global feature because it came by taking max over entire sentence. To make classifier over extracted global feature, we used fully connected feed forward layer. Suppose $z^i \in \mathbb{R}^l$ is output of max pooling layer for entire filters then output of fully-connected layer would be

$$o^{(i)} = W^\upsilon z^i + b^\upsilon \qquad (5)$$

Here $W^o \in \mathbb{R}^{[r] \times l}$ and $b^o \in \mathbb{R}^{[r]}$ are parameters of neural network and $[r]$ denotes number of classes.

3.7 Softmax Layer

In output layer we used softmax classifier for which objective function would be minimization of

$$L_i = -\log \left(\frac{e^{o_{y_i}^{(i)}}}{\sum_{\forall j} e_j^{o^{(i)}}} \right) \qquad (6)$$

for i^{th} sentence. Here y_i is correct class of relation for i^{th} instance.

3.8 Implementation

We experiment with filter lengths in two different experiment settings. In first, we use 100 different filters of a fixed length in the convolutional layer, while in another set of experiments we use varying length filters, but used 100 different filters for each varying length. So, in the first setting, we obtain 100 features after max pooling, while in the second, we obtain 100 times number of different length filter features. For regularization (Srivastava et al., 2014), we follow (Kim, 2014) and use *dropout* technique in output of max pooling layer. Dropout prevents co-adaptation of hidden units by randomly dropping few nodes. We set this value to 0.5 during training and 1 while testing. We use Adam technique (Kingma and Ba, 2014) to optimize our loss function. Entire neural network parameters and feature vectors are updated while training. We have implemented the proposed model in Python language using tensorflow package (Abadi et al., 2015) and will make it available on request. Results of each filter length were explained in results section. Dimension of word vector is set to 50 and rest all feature embedding size is kept to 5.

4 Dataset and Experimental Settings

In recent years several challenges have been organized to automatically extract information from clinical texts (Uzuner et al., 2007; Uzuner et al., 2008; Uzuner et al., 2011; Uzuner et al., 2010; Sun et al., 2013). i2b2 has released dataset for clinical concept extraction, assertion classification and relation extraction as a part of i2b2-2010 shared task challenge. This dataset was collected from three different hospitals and was manually annotated by medical practitioners for identifying problems, treatments and test entities, and eight relation types among them. These relations were: *treatment caused medical problems* (**TrCP**), *treatment administered medical problem* (**TrAP**), *treatment worsen medical problem* (**TrWP**), *treatment improve or cure medical problem* (**TrIP**), *treatment was not administered because of medical problem* (**TrNAP**), *test reveal medical problem* (**TeRP**), *Test conducted to investigate medical problem* (**TeCP**), *Medical problem indicates medical problems* (**PIP**). (Uzuner et al., 2011) has given the exact definition of each relation type.

While during the challenge original dataset had 394 documents for training and 477 documents for testing but when we downloaded this dataset from i2b2 website we got only 170 documents for training and 256 documents for testing. After preliminary experiment we found that we did not have

Name	Number instances
TeCP	503
TrCP	525
PIP	2202
TrAP	2616
TeRP	3052
No Relation	55600

Table 1: Relation types and number of instances of i2b2 dataset (partial)

enough training samples for all relation classes present in the dataset, therefore we decided to remove 3 relation classes along with their instances (*TrWP* (132 instances), *TrIP* (202 instances) and *TrNAP* (173 instances)). Statistics of the dataset is shown in the Table 1.

For extracting relations among entities we considered all sentences having more than one entities in each discharge summary to check whether any relation exists between them or not. In our experiment we assume that entities and their types are already known like other existing works (Rink et al., 2011; Minard et al., 2011a; Minard et al., 2011b). We created data sample for every pair of entities present in the sentence and labeled it with the existing relation type. For example in sentence [S2] (all continuous bold phrases are entities) entity pairs (*"her white count", "elevated"*) label would be *"TeRP"*, for entity pair (*"her g-csf", "elevated"*) label would be *"TrNAP"* and for (*"her white count", "her G-CSF"*) label would be "None".

[S2]: **Her white count** *remained* **elevated** *despite discontinuing* **her G-CSF** .

5 Results and Discussion

5.1 Influence of filter lengths

We combined the training and testing data and performed five-fold cross-validation on the available limited i2b2 dataset for all our evaluations. First we evaluate the influence of filter lengths. We experiment with selection of filter length using all features. Results as average of five-fold experiment are shown in the Table 2.

In case of single filter, the results indicate increasing the size of filter length generally tends to improve the performance. Using only single filter the best performance with F1 score as 70.43% was obtained by using filter length of 6. However further increasing the filter length did not improve the

Filter length	Precision	Recall	F Score
[3]	74.54	64.29	68.44
[4]	74.90	65.50	69.19
[5]	76.17	64.68	69.61
[6]	76.05	66.56	70.43
[7]	76.76	64.49	69.23
[3,4]	74.96	64.65	68.91
[3,5]	74.66	66.81	70.10
[4,5]	74.90	68.20	70.91
[4,6]	**76.34**	**67.35**	**71.16**
[5,6]	76.08	65.31	69.77
[3,4,5]	75.83	65.10	69.30
[4,5,6]	76.12	65.68	70.15
[2,3,4,5]	74.99	65.19	69.34
[3,4,5,6]	75.88	65.98	70.13

Table 2: Comparative performance of the proposed model using filters of different lengths separately and together. Each of the models used all features (WV+P_1+P_2+PoS+Chunk+Type) and 100 filters for each filter length.

result. Intuitively it also seems that selection of either of too small or too large filter length may not be a good option. Filter length gives the window size to capture context features. One can expect that too small filter length (window size) may not capture enough good context feature and too big filter length may include noise or irrelevant contexts.

Further, we used multiple filters to see whether it improves the result. Results indicate that combination of small and mid-length filter size is perhaps the better choice. For example, combination of filter lengths 3 and 4 together did not improve the performance compared to the single filter length of 3 or 4. On the other hand combination of filter lengths 3 and 5, and 4 and 5 improved the performance compared to use of single filters of either length. It can be seen, the best result with F1 score as 71.16% is obtained by using filter lengths of 4 and 6 together. But adding more than two filters did not lead to performance improvement.

5.2 Classwise Performance

We took the best combination of filter lengths and looked at the classwise performance. Results are described in the Table 3.

We see from the results that as number of training examples (see Table 1) increases, performance of the model also improves. The relation class

Name	Precision	Recall	F Score
TeCP	63.48	43.67	50.56
TrCP	63.60	43.67	56.44
PIP	67.32	63.30	64.92
TrAP	73.49	65.83	69.23
TeRP	82.74	79.88	81.25

Table 3: Class wise performance with all features (filter size : [4,6] each with 100 filters)

TeRP has the maximum number of training examples and the model obtained quite a good F1 score. On the other hand, the model could not perhaps able to learn well for the relation classes *TeCP* and *TrCP* having relatively lesser number of training examples.

5.3 Contribution of Each Features

In order to investigate the contribution of each feature in final result we gradually include one feature in our model and compared the performance. Table 4 shows the obtained results. First we use only random vector (RV) representation along with entity types (T) (first row in the table) as a baseline for our comparison. Adding position features (2nd row) lead to approximately 15% increase in recall, 7% in precision and 11.7% in F1 score. However including PoS and Chunk features although improved recall and F1 score by 4.3% and 1.3% but precision was decreased by 3.6%. In the second set of experiments, we first use pre-trained word vectors along with entity types (4th row) and later repeated the similar experiments as previously. Here again, inclusion of position features improved the recall by more than 14% and F1 score by around 11%. This clearly indicates word position relative to the entities of interest plays important role in deciding their influence in the context. Further including PoS and Chunk features also led to performance improvement.

Name	P	R	F
RV + T	67.21	52.97	57.87
+(P₁+P₂)	71.86	60.69	64.66
+(PoS+Chunk)	69.25	63.34	65.52
WV + T	70.75	59.17	63.82
+(P₁+P₂)	75.54	67.69	70.97
+(PoS+Chunk)	**76.34**	**67.35**	**71.16**

Table 4: Contribution of each features (filter size : [4,6] each with 100 filters)

5.4 Comparison with Feature Based Method

We could not compare our results directly with the state of the art results obtained on the i2b2 dataset as we did not have the complete dataset. We build a linear SVM classifier using similar features as defined in earlier studies (Rink et al., 2011) as a baseline for comparison. The following features are used for each entity pair instance:

- Any word between relation arguments

- Any PoS tags between relation arguments. We used genia tagger for PoS

- Any bigram between relation arguments

- Word preceding first and second argument

- Any three words succeeding the first and second arguments

- Sequence of $chunk$ tags between relation arguments. We used genia tagger for $chunk$ tag

- String of words between relation arguments

- First and second argument type (problem, treatment and test)

- Order of argument type appeared in sentence

- Distance between two arguments in terms of number of words

- Presence of only punctuation sign between arguments.

This way we prepared attribute-value and numerical features for each instances. Table 5 shows the comparison of best results obtained by the proposed model and SVM based model. Linear SVM classifier with different cost parameter C was implemented using scikit learn (Pedregosa et al., 2011). Here again results shown are average over the 5-folds.

Name	P	R	F
CNN (FL=[4,6])	**76.34**	**67.35**	**71.16**
SVM (Linear, C=0.01)	72.23	57.75	58.96
SVM (Linear, C=0.1)	73.75	64.18	67.35
SVM (Linear, C=1)	73.17	64.18	67.32

Table 5: Comparative performance of SVM and CNN with filter length [4,6] each with 100 filters

Based on the results, We can make following observations:

- Instead of SVM, other classifier could have been also used. We decided to use SVM as SVM based model with similar features obtained the best performance in the 2010 challenge.

- In any case we still would have to define huge number of features and only few of them would have non-zero values in any given sample or instance.

- The proposed model with limited number of features (75 * number of words in the sentence; 5 dimensional vector for 5 features other than word embedding, which is 50 dimensional vector) still gave the better performance.

- Consistent with our observations in the section 5.1, too many features trying to capture more contexts adversely affect the performance of classifier. If we look at the features defined above it includes features which try to capture context of all possible window size between the mentioned entities.

6 Conclusion

In this work we present a new framework based on CNN for extracting relations among clinical entities in clinical texts. The proposed model has shown better performance by using only a small fraction of features compared to the SVM based baseline model. Our results indicate that CNN is able to learn global features which can capture contextual features quite well and thus helps in improving the performance.

Acknowledgments

We would like to thank i2b2 National Center for Biomedical Computing funded by U54LM008748, for providing the clinical records originally prepared for the Shared Tasks for Challenges in NLP for Clinical Data organized by Dr. Ozlem Uzuner, i2b2 and SUNY.

References

Martın Abadi, Ashish Agarwal, Paul Barham, Eugene Brevdo, Zhifeng Chen, Craig Citro, Greg S Corrado, Andy Davis, Jeffrey Dean, Matthieu Devin, et al. 2015. Tensorflow: Large-scale machine learning on heterogeneous systems, 2015. *Software available from tensorflow.org.*

Yoshua Bengio, Réjean Ducharme, Pascal Vincent, and Christian Janvin. 2003. A neural probabilistic language model. *J. Mach. Learn. Res.*, 3:1137–1155, March.

Àlex Bravo, Janet Piñero, Núria Queralt-Rosinach, Michael Rautschka, and Laura I Furlong. 2015. Extraction of relations between genes and diseases from text and large-scale data analysis: implications for translational research. *BMC bioinformatics*, 16(1):1.

Markus Bundschus, Mathaeus Dejori, Martin Stetter, Volker Tresp, and Hans-Peter Kriegel. 2008. Extraction of semantic biomedical relations from text using conditional random fields. *BMC bioinformatics*, 9(1):1.

Razvan Bunescu, Raymond Mooney, Arun Ramani, and Edward Marcotte. 2006. Integrating co-occurrence statistics with information extraction for robust retrieval of protein interactions from medline. In *Proceedings of the Workshop on Linking Natural Language Processing and Biology: Towards Deeper Biological Literature Analysis*, pages 49–56. Association for Computational Linguistics.

Ronan Collobert and Jason Weston. 2008. A unified architecture for natural language processing: Deep neural networks with multitask learning. In *Proceedings of the 25th international conference on Machine learning*, pages 160–167. ACM.

Ronan Collobert, Jason Weston, Léon Bottou, Michael Karlen, Koray Kavukcuoglu, and Pavel Kuksa. 2011. Natural language processing (almost) from scratch. *J. Mach. Learn. Res.*, 12:2493–2537, November.

Aron Culotta and Jeffrey Sorensen. 2004. Dependency tree kernels for relation extraction. In *Proceedings of the 42nd Annual Meeting on Association for Computational Linguistics*, page 423. Association for Computational Linguistics.

Cıcero Nogueira dos Santos, Bing Xiang, and Bowen Zhou. 2015. Classifying relations by ranking with convolutional neural networks. In *Proceedings of the 53rd Annual Meeting of the Association for Computational Linguistics and the 7th International Joint Conference on Natural Language Processing*, volume 1, pages 626–634.

Iris Hendrickx, Su Nam Kim, Zornitsa Kozareva, Preslav Nakov, Diarmuid Ó Séaghdha, Sebastian Padó, Marco Pennacchiotti, Lorenza Romano, and Stan Szpakowicz. 2009. Semeval-2010 task 8: Multi-way classification of semantic relations between pairs of nominals. In *Proceedings of the Workshop on Semantic Evaluations: Recent Achievements and Future Directions*, pages 94–99. Association for Computational Linguistics.

Gumwon Hong. 2005. Relation extraction using support vector machine. In *Natural Language Processing–IJCNLP 2005*, pages 366–377. Springer.

Baotian Hu, Zhengdong Lu, Hang Li, and Qingcai Chen. 2014. Convolutional neural network architectures for matching natural language sentences. In *Advances in Neural Information Processing Systems*, pages 2042–2050.

Nal Kalchbrenner, Edward Grefenstette, and Phil Blunsom. 2014. A convolutional neural network for modelling sentences. *arXiv preprint arXiv:1404.2188*.

Andrej Karpathy and Li Fei-Fei. 2014. Deep visual-semantic alignments for generating image descriptions. *arXiv preprint arXiv:1412.2306*.

Yoon Kim. 2014. Convolutional neural networks for sentence classification. In *Proceedings of the 2014 Conference on Empirical Methods in Natural Language Processing (EMNLP)*, pages 1746–1751, Doha, Qatar, October. Association for Computational Linguistics.

Diederik Kingma and Jimmy Ba. 2014. Adam: A method for stochastic optimization. *arXiv preprint arXiv:1412.6980*.

Alex Krizhevsky, Ilya Sutskever, and Geoffrey E Hinton. 2012. Imagenet classification with deep convolutional neural networks. pages 1097–1105.

Gondy Leroy, Hsinchun Chen, and Jesse D Martinez. 2003. A shallow parser based on closed-class words to capture relations in biomedical text. *Journal of biomedical Informatics*, 36(3):145–158.

Jiexun Li, Zhu Zhang, Xin Li, and Hsinchun Chen. 2008. Kernel-based learning for biomedical relation extraction. *Journal of the American Society for Information Science and Technology*, 59(5):756–769.

Tomas Mikolov, Kai Chen, Greg Corrado, and Jeffrey Dean. 2013a. Efficient estimation of word representations in vector space. *arXiv preprint arXiv:1301.3781*.

Tomas Mikolov, Ilya Sutskever, Kai Chen, Greg S Corrado, and Jeff Dean. 2013b. Distributed representations of words and phrases and their compositionality. In *Advances in Neural Information Processing Systems*, pages 3111–3119.

Anne-Lyse Minard, Anne-Laure Ligozat, Asma Ben Abacha, Delphine Bernhard, Bruno Cartoni, Louise Deléger, Brigitte Grau, Sophie Rosset, Pierre Zweigenbaum, and Cyril Grouin. 2011a. Hybrid methods for improving information access in clinical documents: concept, assertion, and relation identification. *Journal of the American Medical Informatics Association*, 18(5):588–593.

Anne-Lyse Minard, Anne-Laure Ligozat, and Brigitte Grau. 2011b. Multi-class svm for relation extraction from clinical reports. In *RANLP*, pages 604–609.

Mike Mintz, Steven Bills, Rion Snow, and Dan Jurafsky. 2009. Distant supervision for relation extraction without labeled data. In *Proceedings of the Joint Conference of the 47th Annual Meeting of the ACL and the 4th International Joint Conference on Natural Language Processing of the AFNLP: Volume 2 - Volume 2*, ACL '09, pages 1003–1011, Stroudsburg, PA, USA. Association for Computational Linguistics.

Jong C Park, Hyun Sook Kim, and Jung-Jae Kim. 2001. Bidirectional incremental parsing for automatic pathway identification with combinatory categorial grammar. In *Pacific Symposium on Biocomputing*, volume 6, pages 396–407.

F. Pedregosa, G. Varoquaux, A. Gramfort, V. Michel, B. Thirion, O. Grisel, M. Blondel, P. Prettenhofer, R. Weiss, V. Dubourg, J. Vanderplas, A. Passos, D. Cournapeau, M. Brucher, M. Perrot, and E. Duchesnay. 2011. Scikit-learn: Machine learning in Python. *Journal of Machine Learning Research*, 12:2825–2830.

Longhua Qian and Guodong Zhou. 2012. Tree kernel-based proteinprotein interaction extraction from biomedical literature. *Journal of Biomedical Informatics*, 45(3):535 – 543.

Sebastian Riedel, Limin Yao, and Andrew McCallum. 2010. Modeling relations and their mentions without labeled text. In *Machine Learning and Knowledge Discovery in Databases*, pages 148–163. Springer.

Bryan Rink, Sanda Harabagiu, and Kirk Roberts. 2011. Automatic extraction of relations between medical concepts in clinical texts. *Journal of the American Medical Informatics Association*, 18(5):594–600.

Roland Roller and Mark Stevenson. 2014. Applying umls for distantly supervised relation detection. In *Proceedings of the 5th International Workshop on Health Text Mining and Information Analysis (Louhi)*, pages 80–84, Gothenburg, Sweden, April. Association for Computational Linguistics.

Barbara Rosario and Marti A. Hearst. 2004. Classifying semantic relations in bioscience texts. In *Proceedings of the 42Nd Annual Meeting on Association for Computational Linguistics*, ACL '04, Stroudsburg, PA, USA. Association for Computational Linguistics.

Yue Shang, Yanpeng Li, Hongfei Lin, and Zhihao Yang. 2011. Enhancing biomedical text summarization using semantic relation extraction. *PLoS ONE*, 6(8):1–10, 08.

Ranti D Sharma, Samarth Tripathi, Sunil K Sahu, Sudhanshu Mittal, and Ashish Anand. 2016. Predicting online doctor ratings from user reviews using convolutional neural networks. *International Journal of Machine Learning and Computing*, 6(2):149.

Qiang Song, Yousuke Watanabe, and Haruo Yokota. 2011. Relationship extraction methods based on co-occurrence in web pages and files. In *Proceedings of the 13th International Conference on Information Integration and Web-based Applications and Services*, iiWAS '11, pages 82–89, New York, NY, USA. ACM.

Nitish Srivastava, Geoffrey Hinton, Alex Krizhevsky, Ilya Sutskever, and Ruslan Salakhutdinov. 2014. Dropout: A simple way to prevent neural networks from overfitting. *The Journal of Machine Learning Research*, 15(1):1929–1958.

Benjamin J Stapley and Gerry Benoit. 2000. Biobibliometrics: information retrieval and visualization from co-occurrences of gene names in medline abstracts. In *Pac Symp Biocomput*, volume 5, pages 529–540.

Weiyi Sun, Anna Rumshisky, and Ozlem Uzuner. 2013. Evaluating temporal relations in clinical text: 2012 i2b2 challenge. *Journal of the American Medical Informatics Association*, 20(5).

MUNEEB TH, Sunil Sahu, and Ashish Anand. 2015. Evaluating distributed word representations for capturing semantics of biomedical concepts. In *Proceedings of BioNLP 15*, pages 158–163, Beijing, China, July. Association for Computational Linguistics.

James Thomas, David Milward, Christos Ouzounis, Stephen Pulman, and Mark Carroll. 2000. Automatic extraction of protein interactions from scientific. In *Pacific symposium on biocomputing*, volume 5, pages 538–549.

Özlem Uzuner, Yuan Luo, and Peter Szolovits. 2007. Evaluating the state-of-the-art in automatic deidentification. *Journal of the American Medical Informatics Association*, 14(5):550–563.

Özlem Uzuner, Ira Goldstein, Yuan Luo, and Isaac Kohane. 2008. Identifying patient smoking status from medical discharge records. *Journal of the American Medical Informatics Association*, 15(1):14–24.

Özlem Uzuner, Imre Solti, and Eithon Cadag. 2010. Extracting medication information from clinical text. *Journal of the American Medical Informatics Association*, 17(5):514–518.

Özlem Uzuner, Brett R South, Shuying Shen, and Scott L DuVall. 2011. 2010 i2b2/va challenge on concepts, assertions, and relations in clinical text. *Journal of the American Medical Informatics Association*, 18(5):552–556.

Erik M. van Mulligen, Annie Fourrier-Reglat, David Gurwitz, Mariam Molokhia, Ainhoa Nieto, Gianluca Trifiro, Jan Kors, and Laura I Furlong. 2012. The EU-ADR corpus: Annotated drugs, diseases, targets, and their relationships. *Journal of biomedical informatics*, 45(5):879–884.

Fei-Yu Xu. 2008. *Bootstrapping Relation Extraction from Semantic Seeds*. Ph.D. thesis, Saarland University.

Wenpeng Yin and Hinrich Schtze. 2015. Multichannel variable-size convolution for sentence classification. In Afra Alishahi and Alessandro Moschitti, editors, *CoNLL*, pages 204–214. ACL.

Dmitry Zelenko, Chinatsu Aone, and Anthony Richardella. 2003. Kernel methods for relation extraction. *The Journal of Machine Learning Research*, 3:1083–1106.

Daojian Zeng, Kang Liu, Siwei Lai, Guangyou Zhou, and Jun Zhao. 2014. Relation classification via convolutional deep neural network. In *COLING 2014, 25th International Conference on Computational Linguistics, Proceedings of the Conference: Technical Papers, August 23-29, 2014, Dublin, Ireland*, pages 2335–2344.